Handbook of the Eurolaser Academy

Handbook of the Eurolaser Academy

Volume 1

Edited by D. Schuocker
Director of the Eurolaser Academy
Vienna
Austria

SPRINGER-SCIENCE+BUSINESS MEDIA, B.V. 

First edition 1998

© 1998 Springer Science+Business Media Dordrecht
Originally published by Chapman & Hall in 1998
Softcover reprint of the hardcover 1st edition 1998

ISBN 978-1-4613-7416-9   ISBN 978-1-4615-5295-6 (eBook)
DOI 10.1007/978-1-4615-5295-6

A catalogue record for this book is available from the British Library

∞ Printed on permanent acid-free text paper, manufactured in accordance with ANSI/NISO Z39.48-1992 and ANSI/NISO Z39.48-1984 (Permanence of Paper).

Contents

List of contributors

Volume Editor

Professor D. Schuöcker
Department of High Power Beam Technology
Vienna University of Technology
Austria

Principal authors in alphabetical order

Dr. R.C. Crafer
Abington Consultants, Cambridge
England

Professor C. Fotakis
Laser and Applications Division
Foundation for Research and Technology Hellas (FORTH)
Greece

Dr. J.M. Green
Pro Laser, Abingdon
England

Professor D. Schuöcker
Department of High Power Beam Technology
Vienna University of Technology
Austria

Professor H. Weber
Optisches Institut
Technische Universität Berlin
Germany

Professor E. Wintner
Institut für Allgemeine Elektrotechnik
Technische Universität Wien
Austria

Preface

The European Community regards training as a priority area and has therefore developed a series of programmes in the field of vocational training.

This book is the result of a pilot project selected under two of these Community Action Programmes. It was initially selected under the COMETT programme, concerned with the development of continuing vocational training in the European Community. Moreover, it was one of the few selected projects to receive further funding under a second selection in the context of the LEONARDO DA VINCI Action Programme for the implementation of a European Community Vocational Training policy.

It is with great pleasure that I present the outcome of this project which embodies one of the fundamental objectives of the LEONARDO DA VINCI Programme - training for new technologies in SMEs, which make a significant contribution to economic development in Europe.

K DRAXLER
Director
Directorate General XXII
European Commission

Acknowledgements

The Volume Editor gratefully acknowledges funding by the LEONARDO DA VINCI Programme of the Commission of the European Community and by the Austrian Federal Ministry of Science and Transport whose financial support has made the EuroLaser Academy a reality and has led directly to the generation of this handbook.

He is also indebted to Director Dr. Klaus Draxler, Head of the LEONARDO DA VINCI Programme, DG XXII of the Commission of the European Community, moreover to Director General Raul Kneucker, Minister's Advisor Helmut Schacher and Mrs. Friederike Pranckl-Kloepfer from the Austrian Federal Ministry of Science and Transport.

Grateful thanks are due to authors and their organisations for their contributions, and for their efforts in updating their manuscripts from lecture form to chapter form.

A number of people behind the scenes have been instrumental in reading and correcting the manuscripts, suggesting improvements and compiling the index. Specific thanks are due to Ms. Tanja Altreiter, Gerhard Liedl, Andreas Penz, Kurt Schröder and Wendelin Weingartner of the Department of Laser Technology, Vienna University of Technology.

Alexander Kaplan of the Department of Laser Technology and Ms. Gabriele Schmid of ARGELAS, the Austrian Laser Association, have provided constant organisational support throughout this project, while Roger Crafer from Abington Consultants has provided substantial support in editing the book.

Finally the Volume Editor extends grateful thanks and acknowledgements to all those other co-workers who are not mentioned here.

Introduction

High power lasers can be used to perform various manufacturing processes with significant advantages such as flexibility in terms of material, geometry and processing tasks. Furthermore, laser processing is fast, yields high processing quality that often permits post process machining to be eliminated, and is environmentally friendly. High power laser sources used for conventional processes as cutting, welding and surface treatment have reached a high degree of industrial maturity. New processes that extend the range of applications of high power lasers in material processing, such as 3-D processing, rapid prototyping, forming and micromachining are rapidly developing. For these reasons, laser processing is replacing more and more mechanical or thermal production processes, and leading in general to higher manufacturing competitiveness.

Although laser processing has important advantages, it is nevertheless a difficult task compared to conventional processing:

- The energy source, a high power laser, is a complex system with powerful high frequency electronics, ultra high temperature plasmas, fast gas flows, ultra precise optics and sophisticated sensor and control systems.
- The production tool, the laser beam itself, is an invisible electromagnetic light wave with very specific properties, and entirely different from natural light. For example power densities up to 100 MW/cm² determine the quality and performance of the production process, in turn depending strongly on the appropriate operation of the laser source.
- The results of the manufacturing process depend not only on the quality of the laser beam, but also on the properties of the workpiece material and on the performance of the manipulation system that moves the workpiece along the desired contour.

The production process itself results from the combined action of various phenomena, e.g. absorption of light by the workpiece, heating of the workpiece, subsequent melting and evaporation, plasma formation, action of external gas flows and fast flow and resolidification of the molten material. The optimum use of these phenomena in production depends on the appropriate choice of process parameters such as laser power, focus diameter, external gas flow conditions and manipulation speed. The application of high power lasers is also accompanied by potential hazards such as high voltage electricity, invisible and harmful radiation and toxic emissions. Nevertheless, appropriate safety precautions can totally avoid these dangers.

Laser sources, laser material processing systems and the processes themselves are highly complex and have their own inherent dangers. Application, optimisation and operation in industry demands very specific knowledge and experience which is essentially interdisciplinary covering topics such as high-frequency electronics, electronic control techniques, thermodynamic machinery, precision optics, electromagnetic waves, materials sciences, thermodynamics and fluid flow mechanics. Engineers who have graduated in Electrical, Mechanical or Production Engineering, in Physics or Materials Sciences, therefore need additional knowledge to meet the above requirements.

In these two volumes, all major theoretical aspects of high power laser technology are treated in a manner that is consistent with the curriculum of the EuroLaser Academy, a postgraduate education system founded and funded by the European Union to promote the application of lasers in Europe. Practical experience is also essential to complement the theoretical aspects and assist engineers in establishing the use of lasers in industry. This practical experience can be acquired in the laboratory training provided by the EuroLaser Academy.

D SCHUÖCKER
Director
EuroLaser Academy

1

Basic laser mechanisms

D. Schuöcker

1.1 BASIC WAVE PHENOMENA

1.1.1 The wave function

It is a common experience that a disturbance in a solid or a fluid propagates outwards in all directions with a certain speed. An example is the everyday experience of a stone thrown into water.

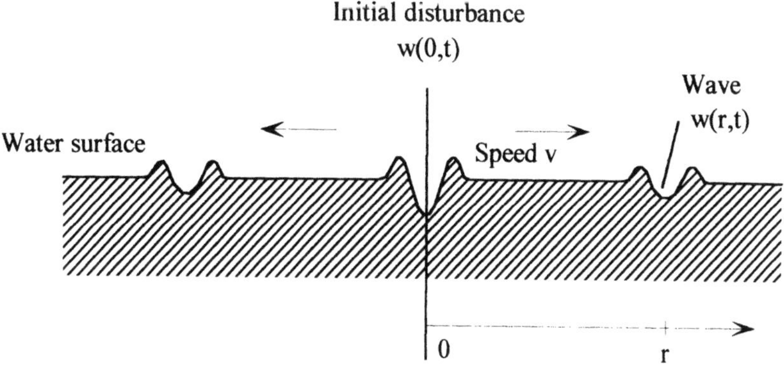

Fig. 1.1 Wave excited by a short time disturbance on a water surface.

If the disturbance is represented by

$$w = g(t) \tag{1.1}$$

and a rectangular coordinate system is assumed where

$$r = \sqrt{x^2 + y^2 + z^2} \tag{1.2}$$

and the initial disturbance takes place at the origin, then the disturbance at a distant point $P(x,y,z)$ is

$$w(r,t) = A(r)g(t - r/v) \tag{1.3}$$

The resulting disturbance is constant over a sphere of radius r. The function $A(r)$ describes the decay of the wave amplitude with distance from the origin. In many cases the square of the amplitude is proportional to the so-called energy density which is the energy per unit area transported by the wave (e.g. the electric or magnetic field). Assuming the absence of loss or gain, the energy of the initial disturbance is conserved and distributed evenly over the spheres. For this reason the energy density must decay with $1/r^2$ meaning that the amplitude decays as $1/r$, yielding $A(r) = A_0/r$.

The expression $w(r,t)$ is called a **wavefunction**. The expression $t - r/v$ is called the **phase** of the wave, since it determines the amplitude at any point and time. The spheres over which the phase is constant are called **wavefronts**. They are always perpendicular to the direction of propagation. In the case treated above, the wave is called a **spherical wave**.

1.1.2 Harmonic waves

If the disturbance is periodic with frequency f or angular frequency ω as in the case of an electromagnetic wave, where the electric field strength oscillates, then equation (1.3) can also be written

$$w(r,t) = A_0 \frac{1}{r} \cos 2\pi f \left(t - \frac{r}{c} \right)$$

$$= A_0 \frac{1}{r} \cos 2\pi \left(ft - \frac{r}{\lambda} \right) \tag{1.4}$$

$$= A_0 \frac{1}{r} \cos(\omega t - kr)$$

where c is the speed of propagation. $\lambda = c/f$ is the **wavelength** since it determines the distance between subsequent maxima or minima. $\omega = 2\pi f$ is the **angular frequency**, since it gives the angle traced out per unit time. $k = 2\pi/\lambda$ is the **wavenumber**, since it gives the number of full waves that fit on a circle with unit radius.

1.1.3 Dispersion relationship

The last expression in equation. (1.4) is the usual description of a wave propagating in r direction with frequency f and wavelength λ. These latter two quantities are combined by the so called **dispersion relationship**

$$f = \frac{c(\lambda)}{\lambda} \tag{1.5}$$

This describes how the actual speed of light c in a transparent material depends on the speed of light c_0 in a vacuum and the **refractive index** n, through the relationship $c = n(\lambda)c_0$. As c depends on the wavelength, a white light beam is thus dispersed into separate beams with all colours during refraction at an interface e.g. between air and glass.

1.1.4 Spherical and plane waves

So far, spherical waves have been discussed where a point source generates wave fronts that are spheres which propagate such that they become larger and larger the further they are from the origin (Fig. 1.2a).

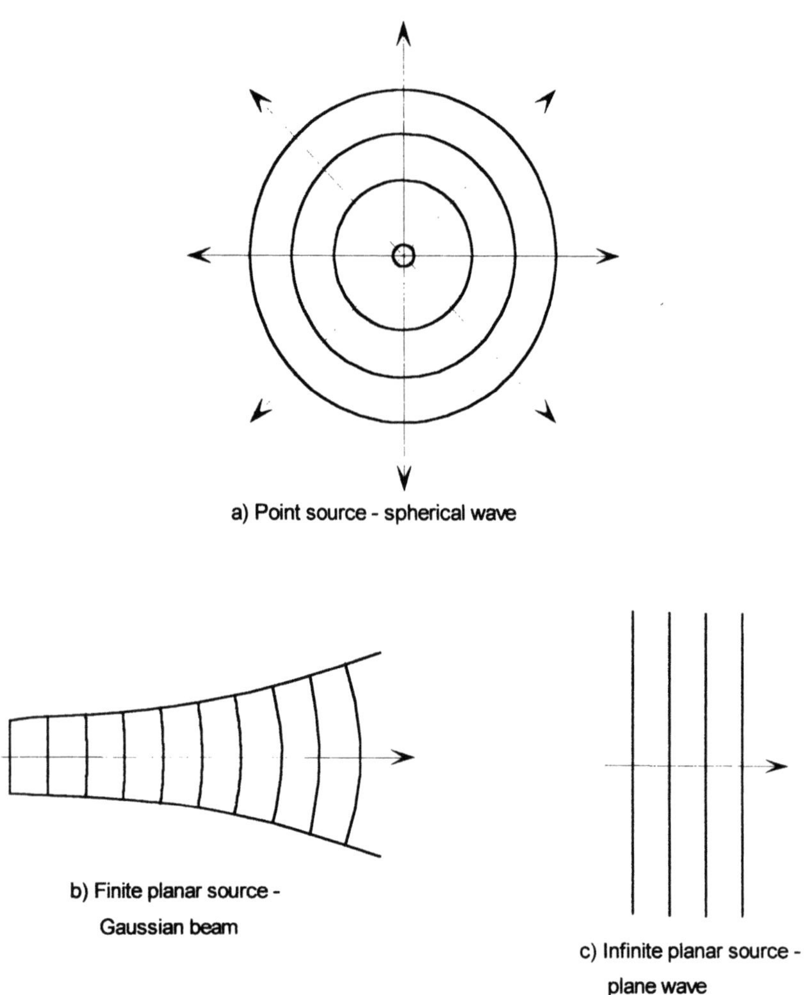

Fig. 1.2 Wave fronts of a spherical wave, a planar wave and a Gaussian beam.

In the case of a wave source with infinite extension, for instance in the yz plane, the wavefront must be plane, therefore a so called **plane wave** is generated (Fig. 1.2c). If the beam source has a finite dimension, a situation that applies to lasers, then a wave type is produced that resembles a plane wave very near to the source and also if the distance from the source is small compared to the lateral extension of the source. At a distance that is large compared to the lateral extension of the source, it resembles a spherical wave. This kind of wave (Fig. 1.2b) is usually called a **Gaussian beam**.

1.1.5 Complex waves

Finally, it should be mentioned that waves generated by harmonic oscillation are usually described mathematically by complex functions such as

$$w(r,t) = A\frac{1}{r}\exp i(\omega t - kr) \qquad (1.6)$$

Of course, only the real part of this wavefunction has a physical meaning and yields the actual wavefunction as given by Eq. (1.4), but the addition of the imaginary part allows an enormous mathematical simplification in the mathematical treatment of the waves.

1.1.6 Phase velocity

The velocity v in equations (1.3) and (1.4) is called the **phase velocity** since it determines the propagation of a certain value of the phase. It depends on the properties of the medium where the wave propagates, such as fluid mechanical properties in the above example, and on the electric and magnetic properties of vacuum or matter in the case of electromagnetic waves. In the latter case the velocity depends on the dielectric constant ε and on the permeability μ as will be explained in the next section. It should be mentioned that the frequency of the wave is determined by the initial disturbance that produces the wave and is thus imposed from outside and must be constant everywhere the wave propagates. The wavelength is then given by the dispersion relation and depends on the properties of the medium and is thus subject to changes if the wave leaves one kind of medium and enters another. It thus depends on the material properties at every point that the wave reaches (Klein and Furtak, 1986).

1.1.7 Reflection and transmission

When a wave leaves a medium with phase velocity v_1 and enters a second medium with a different velocity v_2 due to a change of material property such as refractive index, then the energy flow, which for a simple harmonic wave is also determined by the phase velocity, changes from the first to the second medium. To obtain a continuous energy flow at the intersection and to ensure conservation of energy, a third wave must appear to satisfy the boundary conditions. In general, both a change of phase velocity and amplitude of the incident, transmitted and reflected waves must be considered.

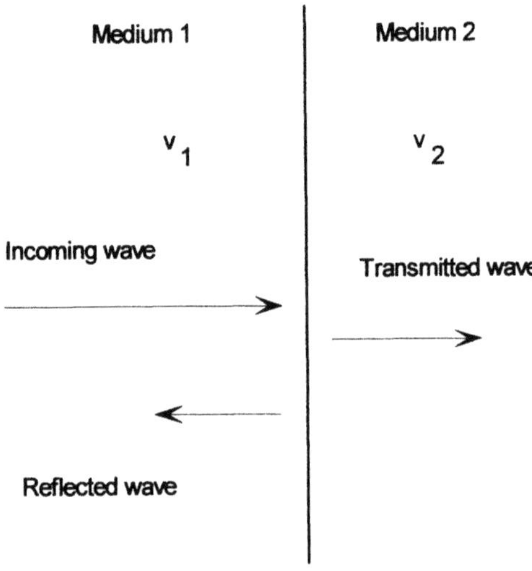

Fig. 1.3 Reflection and transmission at the interface between two media.

Each change of phase velocity resulting from changes of material property leads to reflection of the initial wave. In the case of non perpendicular incidence onto a reflecting interface, the requirement of continuity of energy flow remains valid, both perpendicular and parallel to the interface. This leads to the well known laws of reflection and refraction.

It should be mentioned that if a wave arrives at an interface where its further propagation is impossible due to the specific properties, then 100% reflection of the initial wave must take place. This is for instance the case in the example of waves on the surface of a liquid where the wave arrives at a solid surface, or if an electromagnetic wave arrives at a perfect electric conductor.

1.1.8 Deformation of wavefronts during reflection

Figure 1.4 illustrates what happens if a wave with a curved wavefront is reflected by a spherical mirror:

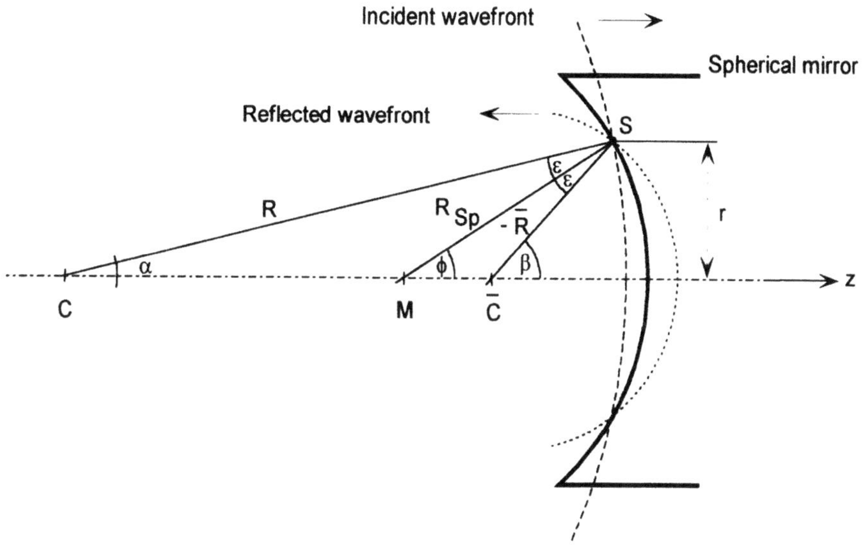

Fig. 1.4 Spherical wave reflected at a spherical mirror with different radius of curvature.

The incident wavefront shown here with radius R and angle of beam spread α intersects the mirror along a circle and is reflected according to the law of reflection. The ray reaching the mirror and inclined at angle α to the axis, is reflected at angle β where the average of both these angles equals the inclination ϕ of the line drawn from the point of intersection to the centre of curvature M of the mirror. For **paraxial beams** with low inclination and low distance from the axis, the angles α, ϕ and β are small and therefore the sines can be replaced by the angles themselves, yielding equations (1.7) to (1.9). The angular sums of the triangles CMS and $\overline{\text{MCS}}$ are

$$\alpha + \left(\pi - \phi\right) + \varepsilon = \pi$$

$$\phi + \left(\pi - \beta\right) + \varepsilon = \pi$$

By eliminating the angle ε from these two expressions, equation (1.10) for the angles α, ϕ and β is derived.

$$\alpha = r/R \tag{1.7}$$

$$\phi = r/R_{Sp} \tag{1.8}$$

$$\beta = r/\left(-\overline{R}\right) \tag{1.9}$$

$$\alpha - 2\phi + \beta = 0 \tag{1.10}$$

In addition it must be considered that a spherical wave which propagates from the left to the right with its origin on the left of the wavefront usually gets a positive sign. With the origin to the right, a negative sign is agreed upon. For a wave propagating in the counter direction this rule must be applied accordingly (Klein and Furtak, 1986). As a consequence the radius R of the incident wave possesses a positive sign whereas the radius $\overline{R}$ of the reflected wave is negative. Therefore it is denoted as ($-\overline{R}$) in Fig. 1.4.

By substituting equations (1.7) – (1.9) into (1.10) the transformation of a spherical wave of radius of curvature R by a spherical mirror can be calculated

$$\overline{R} = \frac{R}{-2\dfrac{R}{R_{Sp}} + 1} \tag{1.11}$$

So the wavefront of the reflected wave shows a smaller radius of curvature than the incident wave. From these considerations it follows that only a wavefront that fits precisely the surface of the mirror remains unchanged and a wavefront with stronger curvature as the mirror becomes less curved after reflection.

1.1.9 Interference and standing waves

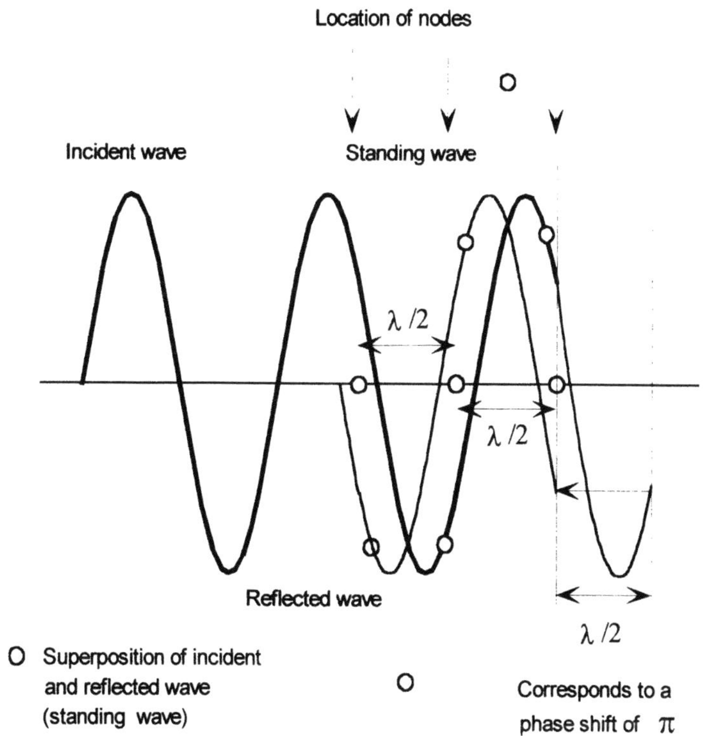

Fig. 1.5 Standing wave.

When a wave arrives at an obstacle formed by changes of the properties of the medium, then a reflected wave appears in the opposite direction to the initial wave. At the reflecting interface the wavefunction changes its polarity or suffers from a sudden phase change by π thus ensuring that the wavefunction remains always zero at the interface. This prevents penetration into the second medium where propagation is inhibited due to the material properties. If a point is considered at distance $\lambda/4$ from the interface, the phase difference between this point and the interface is always $\pi/2$ for the incoming wave. At the interface the wave suffers a sudden change of phase by π and the reflected wave shows also a phase difference between the interface and the point under consideration of $\pi/2$. Therefore the total phase difference between the incoming wave and the reflected wave at the point mentioned above is 2π, which means that at this point the incoming wave and the reflected wave add in a way such that they amplify each other. This is usually referred to as **constructive interference**.

According to Fig. 1.5, at point P a distance of $\lambda/4$ from the reflecting interface, the wave function simply oscillates with twice the amplitude of the incoming wave, whereas at the interface the wave function remains always zero. It can easily be verified according to the above considerations that there are always nodes at distances that are multiples of $\lambda/2$ from the interface and that there are always wave functions with maximum amplitude at distances that are odd multiples of $\lambda/4$. It is thus said that the incoming wave and the reflected wave show constructive interference and build up a so called **standing wave**. Standing waves can easily be observed at the surface of a liquid and are of major importance for lasers where standing electromagnetic waves appear.

When two waves that have their origin in the same source experience a phase difference of π they extinguish themselves totally, an effect usually referred to as **destructive interference**. This can occur for instance when one part of the wave propagates along a short path, and a second part of the wave is split off, and forced to propagate along a longer path, extending its way by $\lambda/2$, after which they are recombined. It should be mentioned that interference, either constructive or destructive is solely associated with the wave phenomenon. Its experimental observation is thus the proof of the wave nature of matter as for instance in the case of light or also in the case of matter waves.

1.1.10 Diffraction

Each point that is reached by a wave in a medium, also acts as a point source of spherical waves that propagate in all directions of the medium under consideration. In the case of electromagnetic light waves this is known as Huygen's principle.

According to this principle, a plane wave that arrives at a solid diaphragm with a narrow slot of width d (see Fig. 1.6) gives rise to the emission of spherical waves originating from every point of the aperture of the diaphragm. When the resulting intensity pattern at a great distance is to be derived, the beams reaching an arbitrary point may be approximated by parallel beams only characterised by the angle α and not by the three-dimensional coordinates of the point. The regime where this approximation can be made is called **far field**.

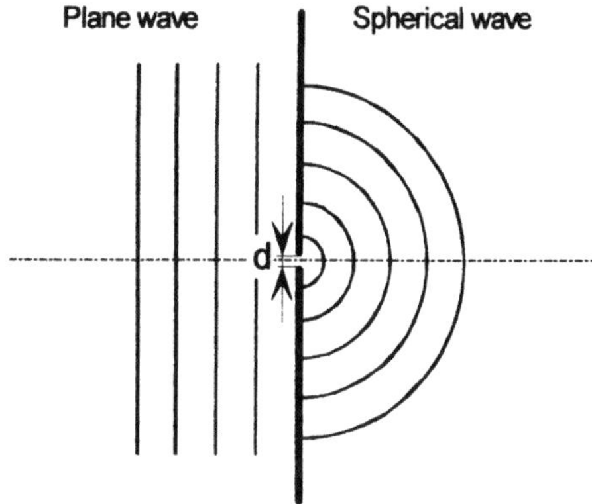

Fig. 1.6 Diffraction of a plane wave at a pointlike hole.

To obtain the intensity in the far field these parallel beams must be superimposed, considering both amplitude and phase. Due to this approximation the phase differences between the beams depend only on the angle α which can be seen in Fig. 1.7.

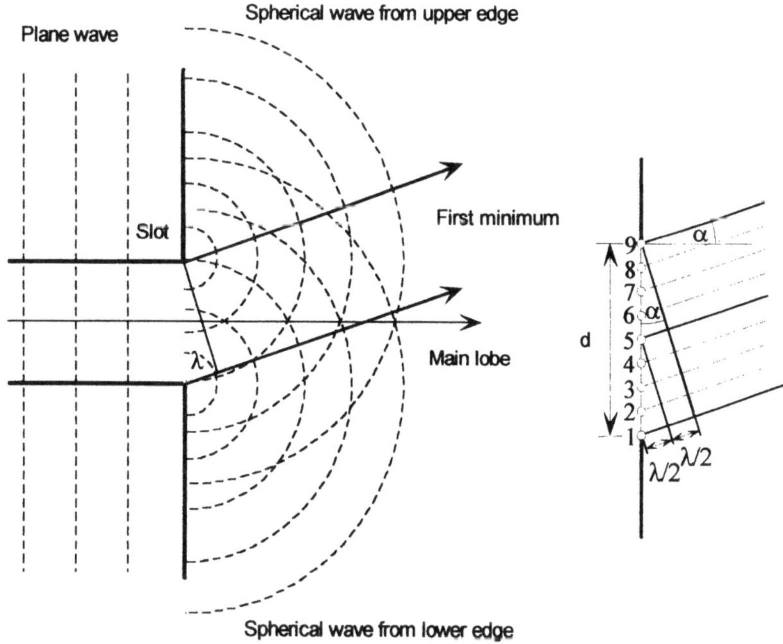

Fig. 1.7 Diffraction at a narrow slot.
Left: Total view, Right: Two pencils of rays with a path difference of $\lambda/2$.

To give a qualitative description of the intensity pattern in the far field it is straightforward to split the beams originating from the aperture into pencils of rays of phase difference $\lambda/2$. According to the path difference, rays 1 and 5, 2 and 6, 3 and 7, ... are interfering destructively, meaning that in sum both pencils of rays are extinguishing each other. Obviously this intersection into discrete rays can be made infinitesimally small in order to approximate the real situation.

In the case of Fig. 1.7 the path difference between the upper and the lower rays amounts to λ which yields destructive interference, according to the first minimum. A path difference of a whole multiple of the wavelength yields minima of higher order, since an even number of pencils of rays with a path difference of $\lambda/2$ exist which extinguish themselves in pairs. The angle λ_n for the minimum of the order n can be calculated when the gap width d is given

$$\sin \alpha_n = \frac{n\lambda}{d} \tag{1.12}$$

where n is a positive integer.

When on the contrary, the path difference between upper and lower ray is set to an odd number of half wavelengths, one pencil of rays is remaining which is not extinguished. The intensities of the respective rays are summed at a distant point. These angles α_n are related to the side lobes which can be calculated in a similar manner to the angles in equation (1.12)

$$\sin \alpha_n = (2n+1)\frac{\lambda}{2d} \tag{1.13}$$

where n is a positive integer.

To sum up, the introduction of a diaphragm with a small slot changed the initially plane wave to a regular pattern of waves propagating in all directions. In some distinct directions amplification takes place and in some directions extinction appears. So considering light for example, a picture of the slot is obtained that is associated with narrow stripes of light, the so called **interference fringes**, which lead to a smearing of the sharp edges of the slot in the optical picture. This phenomenon, usually referred to as **diffraction**, is very important for laser sources, where it leads to energy losses, and for laser beams where it gives rise to a certain widening of the initially narrow laser beams. Of course, all kinds of waves are subject to diffraction, not only electromagnetic waves and light but also in a fluid, where waves can also be observed behind an obstacle, e.g. a pier that protects a harbour, or in acoustics where sound can be heard behind an obstacle of finite extension such as a wall. Usually, diffraction experiments give clear evidence for the wave nature of a certain phenomenon as in the case of light, or equally significantly for matter waves.

1.1.11 Waves caused by non periodic disturbances

It has hitherto been assumed that waves are mainly caused by periodically oscillating disturbances. Nevertheless, waves can also be caused by single events, such as a pulse of any shape, (see Fig. 1.1), a situation that corresponds to the example of a stone thrown to the surface of water.

This more complicated situation can be reduced to the much simpler case of periodic waves by describing the pulse as a superposition of periodic oscillations using a Fourier transformation (Klein and Furtak, 1986). This yields amplitude and phase, again in a complex representation of all frequencies, contained in the wave function at the origin of the wave.

$$S(f) = \int_{-\infty}^{+\infty} s(t) \exp(-2\pi i f t)\, dt \tag{1.14}$$

It is a general property of Fourier transformation that the bandwidth of the frequencies that are contained in the wave function, the so called **spectrum** increases as the pulse becomes shorter, since shorter pulses lead to faster changes that must be described by higher frequency oscillations. If finally the single event wave function is only a sharp pulse, a so called **Dirac function**, the spectrum obtained by Fourier transformation contains all frequencies with evenly distributed amplitude. This property of Fourier transformation is of crucial importance for the appearance of distinct energy levels in atoms and their broadening due to finite lifetimes, and determines therefore the bandwidth of laser radiation.

Oscillations with various frequencies contained in the spectrum of the single event wavefunction give rise to the propagation of waves with various frequencies that add to each other at all points thus yielding the resulting wavefunction. The latter process is carried out by an integral called the **Inverse Fourier transformation**.

$$s(t) = \int_{-\infty}^{+\infty} S(f)\,\exp(2\pi i f t)\,df \tag{1.15}$$

Evaluation of this integral, for instance for a rectangular pulse at the origin of the waves, shows that due to wave propagation the pulse is smeared and no longer shows sharp rectangular borders. This smearing is caused by the fact that the many frequencies contained in the pulse propagate with different wavelengths and are thus subject to somewhat differing phase changes that disturb the initial spectrum of the pulse, leading to distortions of the initial time dependence. It should be added that the only function that remains unchanged by Fourier transformation is a Gaussian distribution (Fig. 1.8), although its width in the time domain and in the frequency domain are inversely proportional.

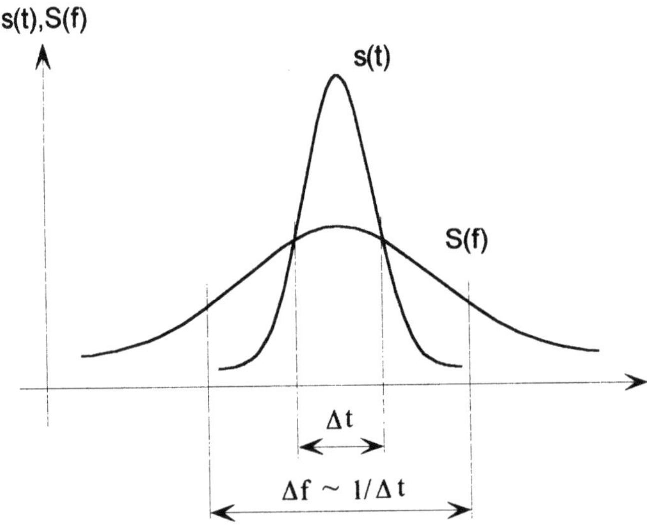

Fig. 1.8 Time function and spectrum obtained by Fourier transformation.

1.1.12 Author's note

The author admits that all points discussed in this lecture are covered in the basic physics education of engineers. Nevertheless he has experienced some confusion from students from all over the world, even at graduate level about the main phenomena associated with and caused by wave propagation. He has tried therefore to set out a moderately simple but quite general description of waves, that demonstrates that these phenomena are independent of the nature of the wave, e.g. electromagnetic waves, acoustic waves, fluid waves or matter waves, and can be applied to all kinds of waves, particularly those that must be treated later in these lectures.

1.2 ELECTROMAGNETIC WAVES AND LIGHT

1.2.1 Generation of electromagnetic waves

One of the most important phenomena of electromagnetism is **electromagnetic induction** (Duffin, 1990), where a temporal change of magnetic field built up by an alternating electric current generates an electric field with field lines surrounding the initial magnetic field. If this initial field oscillates with a certain frequency f the induced electric field oscillates with the same frequency but is delayed by one quarter of the periodic time $T=1/f$ due to its dependence on the temporal change of the magnetic field. In a similar manner, temporal changes in the induced electric field generate in turn a magnetic field, whose field lines surround the electric field and which is also delayed by one quarter of the periodic time. So, an initial oscillation of a magnetic field at the origin generates a system of perpendicular electric and magnetic field lines that extends in all directions with a certain speed v. During the propagation of this **electromagnetic wave,** the building up of the electric and magnetic fields is associated with electric and magnetic polarisation of the medium in which the wave propagates. These polarisations are determined by the dielectric constant ε and the magnetic permeability μ of the medium.

The latter polarisation, either electric or magnetic, is associated with the movement and displacement of electric charges by an amount given by the dielectric constant or the permeability of the medium. The movement of these charges takes some time and therefore delays the building up of the electric and magnetic fields. The phase velocity c of the electromagnetic wave depends on the inverse of the dielectric constant and the permeability.

$$c = \frac{1}{\sqrt{\varepsilon\mu}} \tag{1.16}$$

Since the building up of the electric and magnetic fields during wave propagation takes time even in the absence of a polarisable medium, this implies that the dielectric constant ε and the permeability of the vacuum μ are not zero, but ε_0 and μ_0 respectively. It follows that the vacuum speed of light is not infinite, but shows a finite value

$$c_0 = \frac{1}{\sqrt{\varepsilon_0\mu_0}} \tag{1.17}$$

Usually, in those materials, where electromagnetic waves propagate, the permeability is that of the vacuum μ_0, whereas the dielectric constant ε is larger than in vacuum as for instance in glasses or transparent crystals. Since the change of dielectric constant at an interface such as between vacuum and glass leads to refraction due to the changing phase speed according to section 1, the quantity

$$n = \sqrt{\frac{\varepsilon}{\varepsilon_0}} \tag{1.18}$$

is called the **refractive index**. As mentioned above for an electromagnetic wave, the oscillating electric field strength is associated with a perpendicular magnetic field strength. The relation between the amplitudes of the electric and the magnetic field is also determined by the polarisation mechanism, and depends thus on the dielectric constant and permeability. The amplitude of the electric field is given by the so called **characteristic wave impedance** multiplied by the amplitude of the magnetic field. The characteristic wave impedance for a vacuum can be written

$$Z = \sqrt{\frac{\mu_0}{\varepsilon_0}} \tag{1.19}$$

From the dispersion equation (1.5) and equation (1.18) the following expression for the wavelength results

$$\lambda = \frac{c_0}{nf} \tag{1.20}$$

So far only spherical electromagnetic waves originating from a point source have been treated. An infinitely extended planar source yields a plane wave. In the latter wave different orientations of the electric field, that is perpendicular both to the magnetic field and the direction of propagation, are possible. An important case concerns a situation where the electric field always oscillates in one and the same direction, and which is referred to as **linear polarisation**. A second possibility is an electric field that describes a full circle during one period of oscillation, and which is referred to as **circular polarisation**. Polarisation has a strong influence on the reflection of an electromagnetic wave and is thus important for laser techniques.

In many applications of electromagnetic waves, such as energy transmission, the energy transport per unit time through a unit area, the **intensity** of the wave, is most important. It can be calculated from the electric and the magnetic energy σ contained in a unit volume due to the presence of electric and magnetic fields. A cylinder with unit cross sectional area, filled with energy density σ, transports energy $c\sigma$ through a perpendicular plane in unit time. The intensity of the wave is thus

$$I = c\sigma \tag{1.21}$$

The energy density of the electric field can now be determined by a straightforward calculation from the energy stored in a simple capacitor consisting of two parallel plates with a field strength F between them (Duffin, 1990)

$$\sigma_{el} = \frac{\varepsilon F^2}{2} \tag{1.22}$$

The energy density of the magnetic field can now be determined similarly from the energy stored in a simple coil, that is subject to a current flow and generates thus a magnetic field (Klein and Furtak, 1986)

$$\sigma_{mag} = \frac{\varepsilon H^2}{2} \tag{1.23}$$

With equation (1.19), the energy density given by (1.22) becomes equal to that of (1.23) and the total energy density of an electromagnetic wave is

$$\sigma = \varepsilon F^2 \tag{1.24}$$

The intensity of the wave is then (Klein and Furtak, 1986)

$$I = c\varepsilon F^2 \tag{1.25}$$

An electromagnetic wave can thus be characterised by the:

- mode of propagation, for instance spherical or planar;
- frequency determined by the source of the wave;
- wavelength depending on the medium;
- amplitude of the electric field that determines the intensity;
- energy transported;
- mode of polarisation, either linear or circular;

1.2.2 Light as a visible electromagnetic wave

Electromagnetic waves are subject to interference phenomena typical for all kinds of waves. In consequence they show diffraction which can be used for instance as direct proof of the wave nature of light, which is also an electromagnetic wave, since it shows refraction determined by equation (1.18).

The only difference between light and other electromagnetic waves is that usually the human eye shows no sensitivity for electromagnetic waves with the exception of a narrow band between roughly 0.4 µm and 0.8 µm which is registered as coloured light, the lower wavelength associated with blue, and the upper associated with red light. Radiation with a wavelength somewhat above or below the visible range, behaves similarly to visible light and are called **infrared** and **ultraviolet** light respectively.

An electromagnetic wave, originating from an artificial source, for instance a periodically moving charge that generates an oscillating electric field, consists of a continuous train of maxima and minima, which goes on for a long time, and under idealised conditions, even for infinity. According to the properties of Fourier transformation, a wave of that kind is associated with one distinct frequency and is called **monochromatic**. Nevertheless, under natural conditions this is by no means the case. Natural light is emitted by short time atomic transitions, which will be discussed later, and therefore consists only of fully independent wave trains with rather short length and no fixed phase. Due to the short length of these wave trains or **wave packages** according to the properties of Fourier transformation, a wide range of frequencies is contained giving the human eye the impression of white light. Due to the absence of a phase correlation of these wave packages there is no continuous train of waves and therefore natural light is referred to as **incoherent** (Born and Wolf, 1989).

1.2.3 Quantisation of light energy and photons

The property of natural light to be a stream of wave packages, each associated with a certain energy, leads directly to the idea that a light beam consists of a stream of entities, each transporting a certain amount of energy and moving like particles, which will later on be called **light quanta** or **photons**. There are in fact certain experiments that give evidence for this picture, always associated with the interaction of light with matter. One example is the photoelectric effect (Wichmann, 1971), where light impinging on a metal surface leads to the liberation of electrons, if the light frequency exceeds a lower limit. This experiment can be explained if the light interacting with metal is regarded as a stream of energy quanta, with a quantum energy E_q proportional to the light frequency f.

$$E_q = hf \tag{1.26}$$

In this case only if the frequency is high enough, is the quantum energy, with Planck's constant $h = 6.626 \times 10^{-34}$ Js, sufficient for the electrons to overcome the attraction of the positive atoms and leave metal surface.

If a beam of light really consists of a stream of particles, it must also be possible to associate a momentum with these particles. There are in fact some experiments that prove the mechanical momentum of light quanta.

For instance, the Compton experiment (Wichmann, 1971) where monochromatic light hits a thin sheet of monocrystalline germanium. After transmission through the material the light is no longer monochromatic but also contains some parts with longer wavelength. This can be explained if it is assumed that photons carry a certain momentum inversely proportional to the wavelength, which is partly transferred to the metal atoms during collisions, thus leaving the transmitted light quanta with lower momentum, and therefore longer wavelength. A numerical evaluation of the actual experiment shows that the constant of proportionality between the momentum of the photons and the reciprocal wavelength is again Planck's constant h. The momentum of light quanta can also be calculated using the relativistic expression for the energy of a photon $E_q = mc^2$

$$E_q = (mc)c = p_q c = hf \quad p_q = \frac{hf}{c} = \frac{h}{\lambda} \tag{1.27}$$

So finally, it appears that light can consist of a stream of particles, the so called photons, that carry a certain energy E_q and momentum p_q, if light interacts with matter. If light interacts with light it appears as a wave characterised by frequency f and wavelength λ. Both pictures are related to each other by equations (1.26) and (1.27), the so called **duality relations**.

1.3 ELECTRONS, ATOMS AND MOLECULES

1.3.1 Matter waves

Electrons behave in most experiments as particles with elementary charge e and momentum p. Nevertheless, there are some peculiar cases, that show a wavelike behaviour. For instance, in the famous experiment of Davisson and Germer (Feynman, 1965) a beam of electrons is directed towards a monocrystalline copper layer, see Fig. 1.9.

Although it was initially expected that the reflected electrons would be evenly distributed over all possible angles with a clear maximum at the angle of perpendicular incidence of the initial electron beam, a distribution was found with side maxima at certain angles and associated minima similar to diffraction patterns obtained with X-rays transmitted through a crystal.

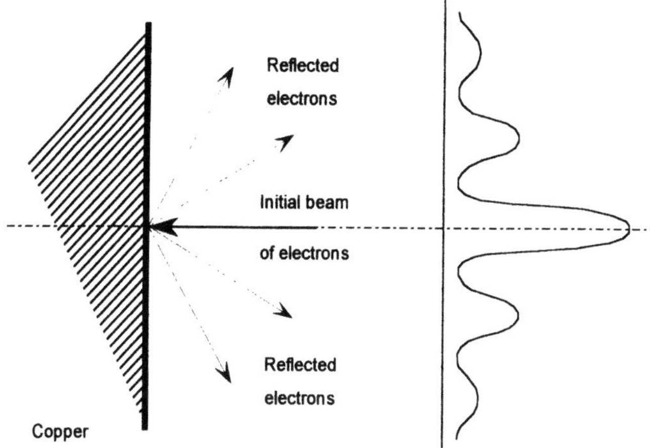

Fig. 1.9 Reflection of an electron beam on a copper monocrystal.

Therefore it must be considered that in this experiment the electrons interact with matter as a wave. It was therefore assumed by De Broglie that the duality relations found for light and given in equation (1.27) can also be applied to electrons. In particular the wavelength λ_{el} can be associated with electrons that have been accelerated by a voltage u

$$\lambda_{el} = \frac{h}{p} = \frac{h}{\sqrt{2meu}} \tag{1.28}$$

A numerical evaluation of equation (1.28) yields an electron wavelength of the order of the wavelength of X-rays for a medium electron energy of 50 keV. This fits excellently to the above experiment. So it must be assumed that electrons that interact with matter behave like waves, whereas if they interact for instance with electric fields they behave like particles. Thus the same duality applies for electrons as for light, due to the fact that particles and light are solely different kinds of energy.

Due to the duality relations mentioned above, the angular frequency associated with electrons can be calculated from the energy, and the wave number can be calculated from the momentum, leading to the following dispersion relation for electrons

$$\omega_{el} = \frac{h}{2m} k^2 \tag{1.29}$$

It is clear that the angular frequency of an electron depends on the square of the wave number, whereas in the case of electromagnetic waves, the angular frequency depends linearly on the wave number, see equation (1.20), thus excluding that matter waves could be electromagnetic waves. It should be mentioned here, that Schrödinger's equation (Feynman, 1965) allows the calculation of a wave function for matter waves that has unfortunately no physical meaning, since it is usually complex, due to the complex nature of the latter equation. Nevertheless, the square of the absolute value of the wave function can easily be interpreted as the probability of finding the electron in a certain volume.

1.3.2 Electron energies in an atom

If an initial plasma is cooled down, electrons are attracted by positive ions and finally caught. They rotate around the positively charged nucleus, thus establishing an equilibrium between electric attraction and mechanical centrifugal forces (Feynman, 1965).

$$\frac{e^2}{4\pi\varepsilon_0 r^2} = mr\omega_{el}^2 \tag{1.30}$$

If the wave nature of electrons is considered, it can be argued that a multiple of a full electron wavelength λ_{el} must fit on the circumference of the electron orbit, see Fig. 1.10, thereby determining the radius of the electron orbit

$$2\pi r = n\frac{h}{p_{el}} = n\frac{h}{mr\omega_{el}} \tag{1.31}$$

If the circumference of the orbit equals only one wavelength, it is the smallest possible orbit with lowest distance from the nucleus and has thus the lowest possible energy and is therefore called the **ground state** . If two or more wavelengths fit on the circumference of the electron orbit, the distance from the nucleus becomes larger, therefore higher energy values are obtained. The kinetic energy of the electrons in the atomic orbits is determined by the electron momentum, and the potential energy is given by the orbital radius.

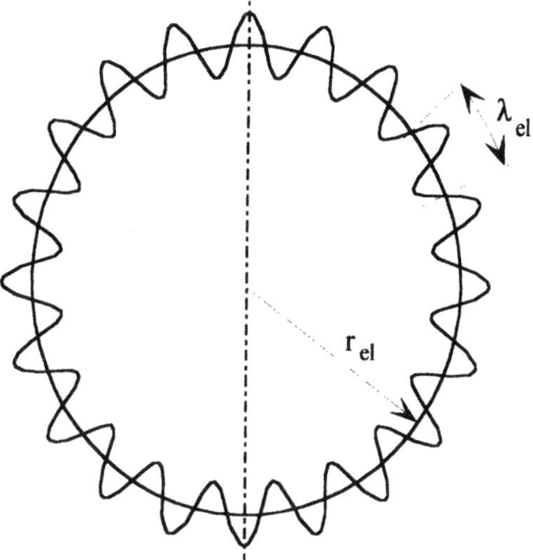

Fig. 1.10 Electron waves on a circular orbit.

The total electron energy is thus

$$E = -\frac{e^2}{4\pi\varepsilon_0 r} + \frac{p_{el}^2}{2m} \tag{1.32}$$

The electron that rotates around the positive nucleus shows always a negative potential energy which is compensated to some part by the positive kinetic energy of the rotation. Nevertheless, the total energy remains always negative as long as the electron is bound to the nucleus, since otherwise it could easily leave the atom. The electron energy composed from potential and kinetic energy can be written in terms of the number n, which gives the number of wavelengths that fit the circumference of the orbit and determines the radius of the orbit as follows

$$E_n = -\frac{1}{2}\frac{e^4 m}{\left(4\pi\varepsilon_0\dfrac{h}{2\pi}\right)^2}\frac{1}{n^2} \tag{1.33}$$

The energy levels of the electron in an atom show a distinct lowest possible value and additionally higher distinct values that are called **excited states**, since they can only be reached if energy is delivered to the atom. Usually the atom occupies the ground state, which is the only stable one due to its lowest energy. The excited states are in principle unstable and can only be occupied for a very short time, the so called **lifetime**, that is in the order of magnitude of 10^{-9} s with the exception of so called **metastable state**s, that exhibit lifetimes up to several minutes or more. n is called the principal quantum number. Experimental investigations by absorption measurement with continuously tuned light frequencies show that these energy levels, which have been obtained by a simple plausibility treatment, agree with experimental results. Moreover, the exact treatment with the help of Schrödinger's equation yields precisely the same energy levels for the hydrogen atom, for the special case of wave functions with a spherical symmetry, which means that the electron does not rotate, so the angular momentum is zero. At this point of course the plausibility treatment fails, because it assumes a rotation of the electron around the nucleus. The precise treatment of the atom shows that it can also exhibit a certain angular momentum that is also quantised as is the electron energy and is described by the second quantum number l, that gives its absolute value. There is also a third quantum number m, which determines the direction of the angular momentum. Finally, electrons rotate around their own axis, leading to an additional angular momentum, that is a universal constant, but can be oriented in two directions, characterised by a fourth quantum number which can assume only two values, according to the two orientations. The angular momentum of the atom and of the electrons lead to the generation of magnetic fields since they are associated with a rotation of electric charges. Angular momenta are therefore of major importance for the magnetic properties of matter and also for the interaction with electromagnetic radiation.

If atoms with more than one electron are considered, then the possible energy values are filled with electrons, as long as the electrons do not coincide in all four quantum numbers as mentioned above (Pauli's Exclusion Principle). According to the results of the precise treatment of the atom with Schrödinger's equation, the ground state E_I shows spherical symmetry and can therefore have no rotation and no angular momentum.

Electrons with zero angular momentum are called s electrons. The second-lowest energy level, the first excited one, can then be thought of as having the energy of the ground state plus some additional energy caused by a certain angular momentum of the atom. So rotations in the atom with $l = 1,2,..$ appear only at higher levels than the ground state which is also shown by the precise treatment. Electrons with higher angular momentum are called for instance p and d electrons, showing the lowest possible energy E_2 and E_3 respectively since stronger rotations increase the energy of the atom.

It must again be mentioned that the very simple plausibility treatment that yields a rotation of the electron, even in the ground state associated with non-zero angular momentum is not valid according to the precise treatment by Schrödinger's equation, which allows rotations that are associated with angular momenta for higher energy levels. Nevertheless, the plausibility treatment was able to give the precise energy levels and to explain why they are distinct. So, the plausibility treatment used here proves to be quite useful to understand those properties of the atoms which are most important for lasers, namely the discrete nature of the atom.

Therefore plausibility treatments will be used throughout Chapters 1 – 3 to explain those phenomena that are of crucial importance for the understanding of laser sources and their radiation.

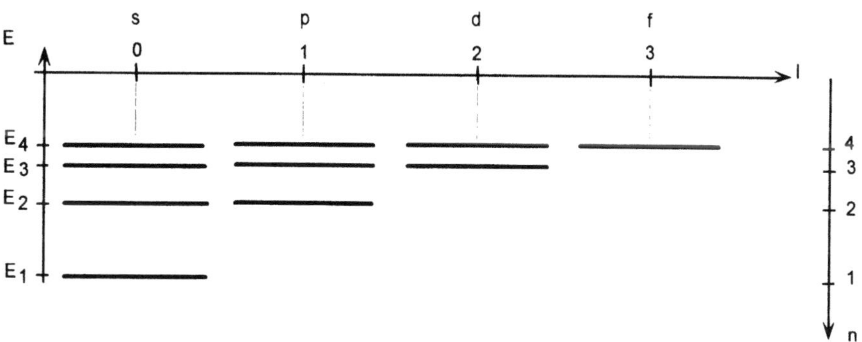

Fig. 1.11 Electron energy levels in an atom.

1.3.3 Molecular energy levels

Usually atoms assemble in molecules, whereas a most simple example is the nitrogen molecule that contains two nitrogen atoms. These atoms are bound together by elastic forces and therefore allow a vibration of the atoms around their equilibrium positions. These vibrations are associated with a certain vibrational energy, that is also quantised in a similar manner to the energy levels of the electron shell of the atoms, where the typical order of magnitude of the energy differences is 0.1 eV. Since molecules show always one or more axes of symmetry, rotations of the molecule can take place, whose energy is also quantised with a typical increment of 0.01 eV. So the total energy of a molecule is composed from the energy of the electrons in the atomic shell, the molecular vibrations and the molecular rotations, whereas the energy of the nucleus that is usually constant in the case of laser physics, but plays an important role of course in nuclear technology, has been excluded (Fig. 1.12).

It should be mentioned that the quantisation of atomic and molecular energies is related to the fact that light can exchange energy with matter only by one, or much less probably, more energy quanta, given by the photon energy and dependent on the light frequency.

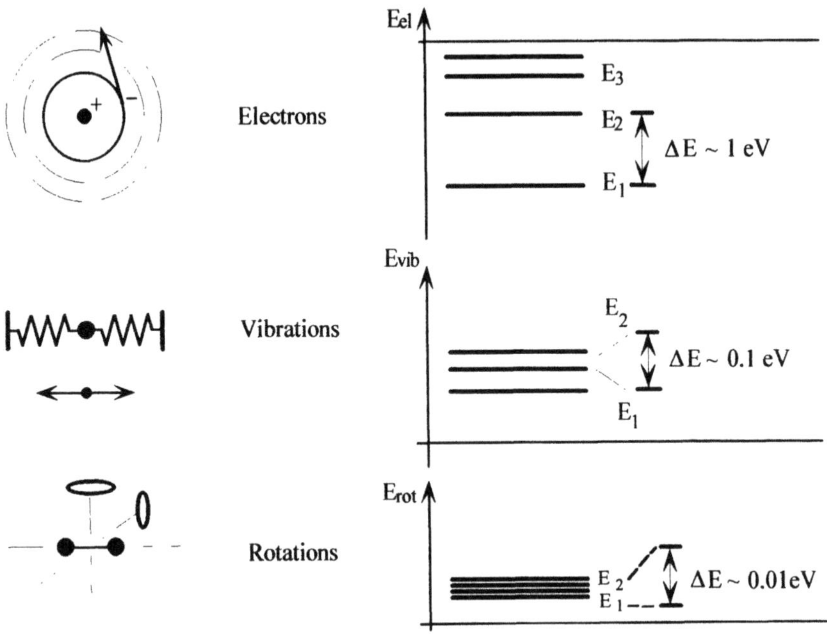

Fig. 1.12 Energy levels of the atomic electrons and the molecular vibrations and rotations.

1.4 INTERACTION BETWEEN LIGHT AND MATTER

1.4.1 Transitions between atomic energy levels.

An atom usually occupies the lowest possible energy, the ground state. If energy is delivered to the atom either mechanically by collisions with a particle, or electromagnetically by the absorption of light, the atom is excited to a higher energy level. The energy transferred during the collision or by a photon corresponds to the difference between the higher energy level and the ground state to fulfil the energy balance of the collision process.

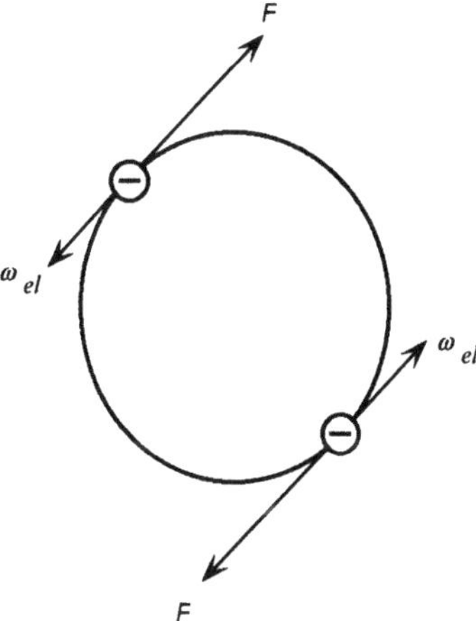

Fig. 1.13 Acceleration of an electron in its orbit by an electric field with $\omega = \omega_{el}$.

The atom remains in the excited state for an average lifetime, which is usually rather short as mentioned above, and falls then back spontaneously without any external influence while the excess energy is carried away by a collision of the second kind, i.e. with a change of inner energy or by a photon. The latter process is called **spontaneous emission**. If the collision that leads to an excitation of the atom is strong enough then the energy supplied to the electron can be sufficient to overcome the attraction of the nucleus, which means that the electron is liberated and can move away freely. In this case a **collisional ionisation** or a **photoionisation** takes place, and a positively charged atom remains. For laser physics those processes, that are due to the interaction with light, such as absorption and spontaneous emission are most important (Siegman, 1986).

Absorption of light by an atom can be understood by a simple plausibility explanation, that takes into account the interaction of the electromagnetic field F of the light wave with the electron, for instance of a one electron atom (Fig. 1.13). The electron is accelerated in its orbit by the electric field of the wave at a certain time, if the electric field has the opposite direction than the speed of the electron due to the negative charge of the electron. If the electron arrives after a time π/ω_{el} where ω_{el} is the angular speed of the electron at the opposite side of its orbit, it is still accelerated if during that time the electric field of the wave has changed its polarity, i.e. if the following equation is true for the angular frequency of the wave ω

$$\frac{\pi}{\omega_{el}} = \frac{\pi}{\omega} \tag{1.34}$$

So if equation (1.34) is fulfilled, that means if the frequency of the wave is in resonance with the rotating electron, then the electron is always accelerated by the wave, and light energy is absorbed by the atom. The above condition of resonance corresponds to the quantum mechanical description, where the quantum energy of light must fit to the difference of the energies of the atomic levels between which the transition takes place. If it is now assumed that the electric field of the wave has initially the same direction as the velocity of the electron, then the electron is decelerated and if equation (1.34) is also true in this case, then the electron is always decelerated by the electric field of the wave, what means that the atom loses energy that must be transferred to the wave, thus amplifying the initial wave. This process, the reverse phenomenon of absorption is called **stimulated emission** since an incoming wave whose frequency is in resonance with the atom stimulates the emission of additional light energy with the same frequency and fixed phase correlation, thus amplifying the initial wave as before. In terms of quantum mechanics an atom occupies a certain higher energy level and carries out the transition to a lower energy level under the stimulating effect of an incoming wave with a quantum energy that is equal to the difference of the two atomic levels under consideration. This phenomenon is totally different from spontaneous emission since the latter takes place without any external influence on the atom from outside. Usually in lasers both emission mechanisms take place at the same time.

According to the properties of Fourier transformation, a wave with limited lifetime as from $-\tau/2$ to $+\tau/2$ shows a frequency bandwidth that is inversely proportional to the lifetime mentioned before

$$\Delta f = \frac{2\pi}{\tau} \tag{1.35}$$

If now the wave particle duality of matter waves is considered, it appears that for a finite lifetime there is no distinct energy, but an energy band ΔE given by the electron energy related to the electron frequency and the lifetime τ

$$\tau \Delta E = 2\pi h \tag{1.36}$$

This relation is well known as the **uncertainty relation** for time and energy. It is a direct consequence of this relation that stationary states where the time is entirely undefined, show discrete energy levels (Wichmann, 1971).

Since the electrons in the excited states of an atom have a finite lifetime, the matter waves associated to the electrons on their orbit exist solely during the lifetime of the excited states, which in turn leads to a widening of the energy levels to bands with an average width given by the lifetime according to equation (1.36)

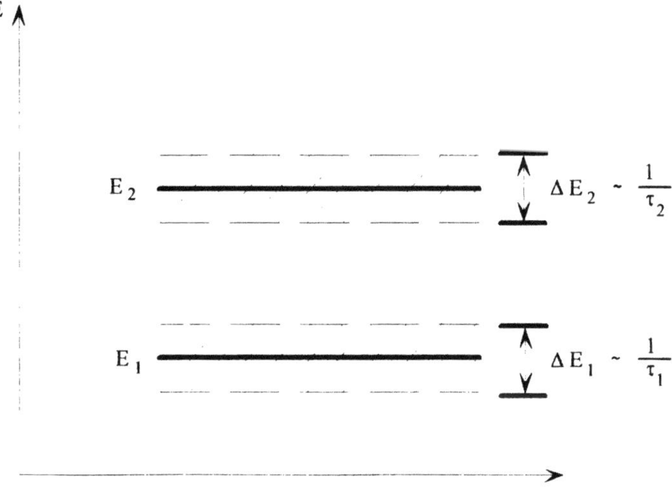

Fig. 1.14 Spread of the atomic energy levels due to finite lifetimes.

Due to this unsharpness of the energy levels caused by the finite lifetime of excited states, absorption and stimulated emission take place not only for one single resonance frequency, but for a band of frequencies centred at the resonance frequency (Fig. 1.14). A detailed analytical treatment of the transition processes (Feynman, 1965) shows that the dependence of absorption and stimulated emission on the frequency is described by a resonance curve $g_0(f)$ that depends solely on the lifetimes of the relevant energy level and the frequency of the light wave (Fig. 1.15).

$$g_0(f) = \frac{\tau_{12}^2}{\dfrac{1}{\tau_{12}^2} + 4\pi^2(f - f_{12})^2}$$

$$\int_{-\infty}^{+\infty} g_0(f)\,df = 1 \tag{1.37}$$

$$f_{12} = \frac{E_2 - E_1}{h}$$

τ_{12} is the lifetime for transitions from energy level 2 to 1. There are also lifetimes for transitions to all other energy levels, whereas τ_1 and τ_2 are the resulting total lifetimes. To obtain the total lifetime for instance of E_2, the reciprocal values of all lifetimes must be added to yield the inverse of the total lifetime, since the shorter the lifetime, the more atoms leave the energy level per unit time (Siegman, 1986). It can easily be seen that the above resonance function yields a frequency band width as explained above.

1.4.2 Rate equations

For an analytical description of absorption and emission between the wave and matter, the interaction of a large number of atoms contained in a certain volume must be considered instead of the treatment of a single atom. It is then sufficient to consider only the two energy levels involved in the interaction process with their energies E_1 and E_2. The number of atoms that occupy the lower energy level E_1 per unit volume at a certain time is N_1 and the number of atoms that occupy the energy level E_2 again per unit volume is N_2.

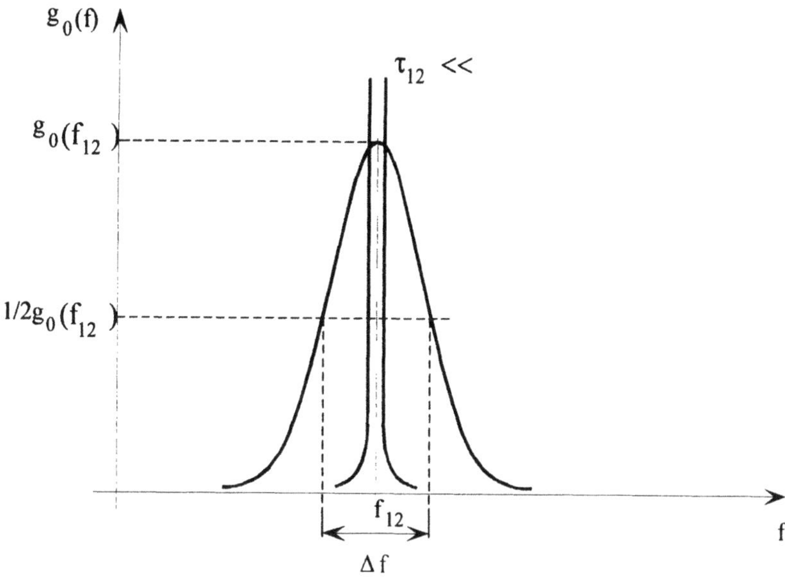

Fig. 1.15 Line shape $g_0(f)$ for two different lifetimes τ_{12}.

The number of absorption events per unit time and volume leads to a decrease of the population density N_1 and must be proportional to the number of atoms occupying the lower energy and to the radiation energy density $\sigma(f)$, whereas the proportionality constant B_{21} describes the properties of the atom and the energy level under consideration and is called the **Einstein coefficient of absorption** (Siegman, 1986). The resonant nature of absorption is described by the function $g_0(f)$ as mentioned above and therefore integration over the frequency is necessary

$$\left(\frac{dN_1}{dt}\right)_{abs} = -\int_0^\infty B_{21} N_1 g_0(f)\sigma(f)df \tag{1.38}$$

Stimulated emission can be described in a similar way with the **Einstein coefficient for a stimulated emission.**

$$\left(\frac{dN_2}{dt}\right)_{stim} = -\int_0^\infty B_{12} N_2 g_0(f)\sigma(f)df \tag{1.39}$$

Due to reciprocity of absorption and stimulated emission the two Einstein coefficients B_{21} and B_{12} are equal.

Spontaneous emission from the upper level E_2 to the lower level E_1 is given by the number of atoms that occupy the upper level E_2 and must be inversely proportional to the lifetime τ_{12} of this level. According to the unsharpness of the energy level given by its lifetime as mentioned above, there is also a frequency distribution of the light emitted and therefore the resonance function $g_0(f)$ also enters the equation for spontaneous emission.

$$\left(\frac{dN_2}{dt}\right)_{spont} = -\int_0^{\infty} \frac{N_2}{\tau_{12}} g_0(f) df \tag{1.40}$$

All three kinds of interaction between light and matter take place at the same time, whereas their relative importance depends on the densities N_1 and N_2 of atoms occupying the energy levels E_1 and E_2. The net absorption of radiation energy per unit time $d\sigma/dt$ is given by absorption reduced by stimulated emission as described above:

$$-\frac{1}{hf}\frac{d\sigma}{dt} = \int_0^{\infty} B_{12}(N_2 - N_1) g_0(f)\sigma(f) df \tag{1.41}$$

If no energy is delivered to the volume of matter considered here, the situation is adiabatic and an equilibrium distribution of the occupation of the energy levels is achieved called a **Boltzmann equilibrium** (Siegman, 1986), where the higher level shows a lower occupation than the lower level, meaning $N_1 > N_2$. In this situation absorption, which depends on the occupation of the lower level N_1, prevails over stimulated emission that depends on the occupation of the upper level N_2 and therefore a net absorption takes place.

1.5 BASIC LASER MECHANISMS

1.5.1 Inversion and light amplification

If now in contrast energy is supplied to the volume of matter considered above, either of optical, electrical or any other nature, all energy levels of the atom, especially the two levels E_1 and E_2, are excited and the distribution of atoms over the energy levels is changed. If the two energy levels under consideration show only a moderate energy difference, it can be assumed that an equal number of atoms is excited per unit time and volume to each of these two energy levels. If now the lifetime of the upper level τ_2 is relatively high, more atoms are excited to energy E_2 from outside than are lost per unit time and volume and therefore a relatively high number of atoms will occupy this energy level. If on the contrary the lifetime of the energy level E_1 is comparatively small, $\tau_1 < \tau_2$, a low population number of the lower level will result and finally the population of the higher level will be larger than that of the lower level, a situation that is called a **population inversion**. In this case, where energy is pumped into the volume and causes inversion of the population numbers, the net absorption given by equation (1.41) turns to a net amplification of radiation, according to the acronym **LASER**, meaning light amplification by stimulated emission of radiation.

For the analysis of amplification the light wave can be described by an intensity I, giving the energy crossing unit area per unit time as treated in section 1.2. If an energy balance of a volume with unit cross section and unit length inside the medium with population inversion is set up according to Fig. 1.16, the difference of the beam energy entering the volume at x and leaving the volume of $(x + dx)$ per unit time must be equal to the light energy generated in the volume per unit time $d\sigma/dt$ as given by equation (1.41)

$$\frac{dI}{dx} = \frac{d\sigma}{dt} \tag{1.42}$$

With Eq. (1.41) the following differential equation for the gradient of the intensity is obtained

$$\frac{dI}{dx} = \frac{hf}{c} \int_0^{\infty} B_{21}(N_2 - N_1) I(f) g_0(f) df \tag{1.43}$$

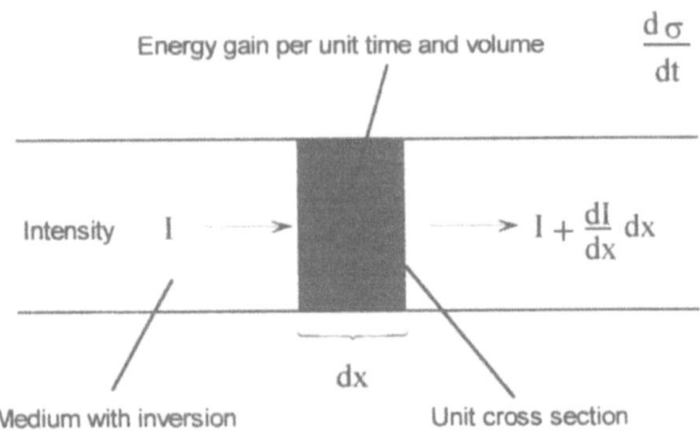

Fig. 1.16 One dimensional model for amplification of intensity I.

In practice, the bandwidth of the laser mode is much smaller than the amplification bandwidth given by $g_0(f)$ due to the resonance properties of optical resonators and therefore the integral in equation (1.42) can be replaced by the product

$$\frac{dI}{dx} = \frac{hf}{c} B_{21}(N_2 - N_1)g_0(f)I \tag{1.44}$$

If the length of the amplifying medium is L and the intensity at the input is I_0, the intensity at the output can easily be calculated from equation (1.44), where N_1 and N_2 are assumed to be constant

$$I = I_0 \exp(Gx) \tag{1.45}$$

From this equation the **gain of the laser** G is defined as follows

$$G = \frac{hf}{c}(N_2 - N_1)g_0(f) \tag{1.46}$$

The population densities N_1 and N_2 that enter the gain in equation (1.46) are given by the strength of the pumping process and remain constant as long as the intensity of the beam that is amplified remains quite low, but they change their magnitude if the beam intensity becomes considerably high, since in this case a depletion of the upper laser level takes place due to enhanced stimulated emission and the opposite appears with the population of the lower laser level thus leading to a reduced inversion N_2 - N_1 and in consequence to reduced gain of the laser. For this the gain given by equation (1.46) is called the **small signal gain**. The latter depends on the line shape and in consequence on the frequency. To describe the reduction of the gain due to increasing beam intensity, a balance of the upper laser level must be considered, that contains atoms brought to the upper laser level per unit volume and time by the pumping source and on the loss side atoms lost per unit volume and time due to stimulated emission, that is proportional to the intensity of the beam and also due to spontaneous emission. This balance yields the population density of the upper laser level, dependent on the strength of pumping $p = dN_2/dt$ and the intensity of the amplified beam

$$N_2 = \frac{p + B_{21} N_1 g_0(f)\dfrac{I}{c}}{\dfrac{1}{\tau_2} + \dfrac{I}{c} B_{21} g_0(f)} \tag{1.47}$$

A similar treatment for the lower laser level that is also excited by the pumping source yields for N_1

$$N_1 = \frac{p + B_{21} N_2 g_0(f)\dfrac{I}{c}}{\dfrac{1}{\tau_1} + \dfrac{I}{c} B_{21} g_0(f)} \tag{1.48}$$

A small intensity yields the population densities $N_{10} = p\tau_1$ and $N_{20} = p\tau_2$ and from equation (1.46) the small signal gain G_0 is obtained

$$G_0 = \frac{hf}{c} B_{21} p(\tau_2 - \tau_1) g_0(f) \tag{1.49}$$

If the intensity becomes larger, the following **large signal gain** G (Siegman, 1986) is obtained from equations (1.46) to (1.48), where $k = 1$

$$G = \frac{G_0}{\left(1 + \dfrac{I}{I_S}\right)^k} \tag{1.50}$$

For broadening other than homogeneous k is different from 1. This equation is illustrated by Fig. 1.17 that shows that the initially relatively high small signal gain G_0 decreases with rising beam intensity, reaches one half of the small signal gain for a characteristic intensity, the so called **saturation intensity** I_S and approaches zero for infinitely rising intensity. The saturation intensity depends also on the strength of the pumping source and the properties of the amplifying medium and is given by

$$I_S = \frac{1}{2} \frac{c}{B_{21} \tau_2 g_0(f)} \tag{1.51}$$

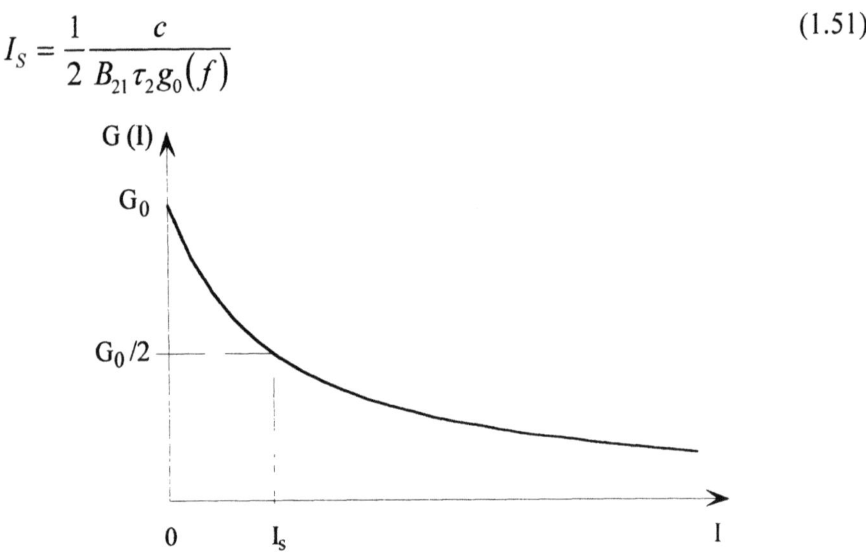

Fig. 1.17 Dependence of laser gain *G(I)* on intensity *I*.

It should be mentioned that the above equation for the large signal gain has been derived under the assumption that all atoms suffer from the same reduction of atomic lifetimes, for instance due to collisions with other particles, that increase the bandwidth of the energy levels of each atom to the same amount. This is called **homogeneous line broadening,** and is predominant in high power gas lasers, employing a relatively high gas pressure leading to a high number of collision between the gas atoms.

1.5.2 Laser sources as light amplifiers with feedback

If a medium is made laser active due to pumping, feedback can be built up by placing one mirror in front of and one mirror behind the amplifying medium, an arrangement called an **optical resonator**, see Chapter 2. If now a single photon, that is always present in daylight, appears at the input of the active medium, it propagates through the latter and is amplified and reaches the right hand mirror, it is then reflected and enters again the amplifying medium, propagates through it and is still amplified until it reaches again the end of the amplifying medium. The condition for feedback is then

$$R_1 R_2 \exp\left(2 G_0 L_{gain}\right) > 1 \tag{1.52}$$

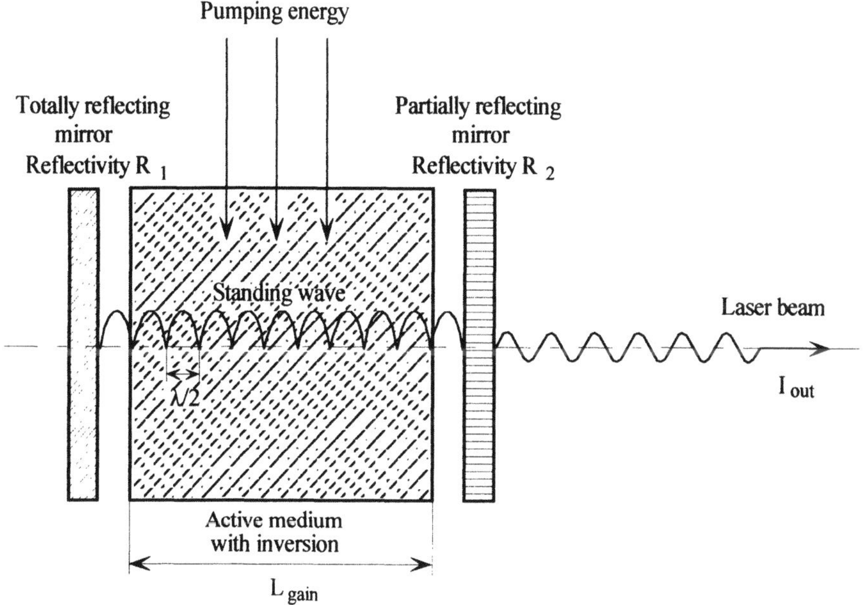

Fig. 1.18 Principal setup of a LASER.

So a light wave runs forth and back through the amplifying medium and becomes stronger and stronger. If it has then reached a certain intensity, the large signal gain has decreased to such an amount, that the losses of the system, for instance due to diffraction at the mirrors are just compensated and therefore a stable power level is established.

If one of the two mirrors is semitransparent, a part of the beam can be extracted and used for any application. Between the two mirrors the beam running forth and back leads to the building up of a standing wave with nodes on the two mirrors. So the distance of the mirrors must be a multiple of one half of the wavelength as shown in Fig. 1.18. Therefore, a very large number of wavelengths can be excited and are called the **resonance wavelengths of the resonator**. Their spacing is of course given by the distance of the two mirrors and is much smaller than the bandwidth of the gain due to stimulated emission. Of course one resonance frequency will be excited that is closest to the maximum small signal gain. If this line reaches an intensity where the gain equals the losses, then all other lines will have fallen below the limit of self excitation and cannot exist any more. The output intensity I_{out} that is finally obtained can be calculated from equation (1.50) for the large signal gain if the total losses are described by the reflectivities of the two mirrors.

$$R_1 R_2 \exp\left[2 \frac{G_0}{\left(1 + \dfrac{I_{out}}{I_S}\right)} L \right] = 1 \tag{1.53}$$

Coming back to the losses of a laser, the main source of loss in an optical resonator is of course the semitransparency of one of the mirrors, but there are also additional losses due to diffraction and imperfect reflection at the mirrors and absorption and scattering in the active medium. Some of the most important losses will be treated in detail in Chapter 2.

1.5.3 Properties of the laser beam

The most important property of light generated by a laser source is its propagation in one distinct direction parallel to the axis of the optical resonator due to the fact that all beams that show even a slight inclination towards this axis leave the resonator after some passes through the latter. Therefore, laser light is entirely different to natural light which is emitted in all directions (see Chapter 2).

Second, although natural light consists of a train of short wave packages and is essentially incoherent, laser light is a continuous wave train since there is a strict phase correlation between stimulated light and emitted light.

The artificial light of lasers is thus coherent, and causes in turn a narrow band width, referred to as monochromatic (Siegman, 1986).

The diameter of the laser beam as it is emitted at the output mirror is in principle determined by the aperture of the latter, which is assumed to be smaller than the diameter of the amplifying medium. Nevertheless, the intensity distribution over the cross section is not rectangular but shows usually a maximum on the axis and decreases with increasing radius, in principle extending to infinity caused mainly by diffraction. Although the laser beam usually leaves the output window with parallel borders, the beam shows a certain widening with increasing distance from the source caused by diffraction, although this phenomenon is rather weak.

1.6 ACTIVE MEDIA

1.6.1 Gas plasmas

In general, gases are excellent insulators without any electrical conductivity. Nevertheless, in every gas, especially in the atmosphere, the ultraviolet component of daylight carries out a number of photoionisation events, that provide a certain, although quite low density of free electrons. If now two electrodes are arranged in the gas and a certain voltage around 10000 V/cm is applied to the electrodes, the electrons are accelerated along the free path between two collisions with atoms and gain a certain kinetic energy E given in average by the mean free path λ_g and the electric field strength F, see Fig. 1.19

$$E = eF\lambda_g \tag{1.54}$$

The mean free path in a gas is inversely proportional to the gas pressure p and therefore the energy mentioned above depends mainly on F/p. If for the latter quantity a magnitude of somewhat more than 10 V/Torr cm is assumed (1 Torr = 1/760 bar = $10^5/760$ Pa), then the energy gain of the electron is larger than the ionisation energy E_i of the atoms and in consequence during the collisions electrons are liberated from the atoms leaving positively charged ions.

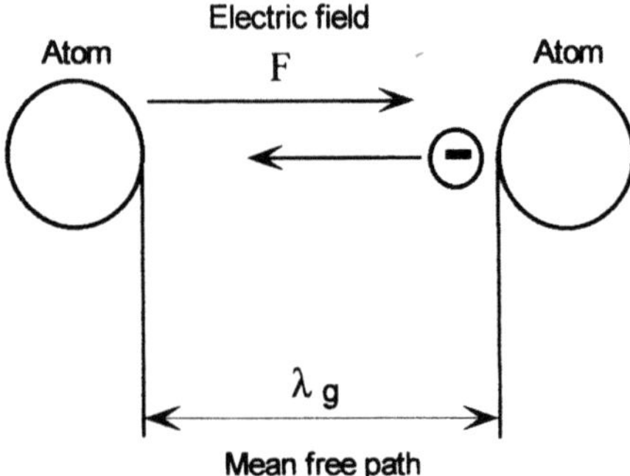

Fig. 1.19 Energy gain of an electron between two collisions with atoms due to acceleration by the electric field.

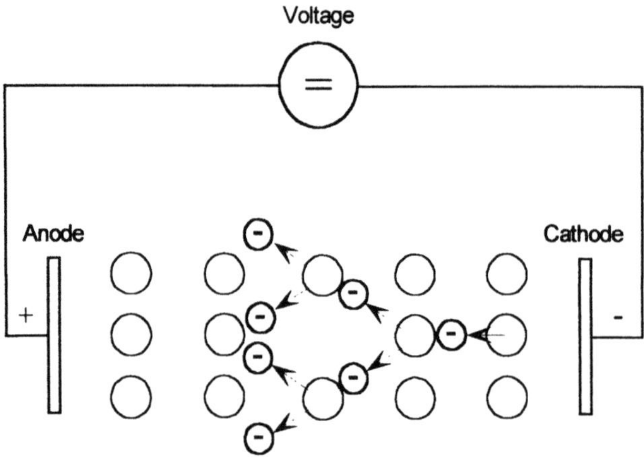

Fig. 1.20 Electron avalanche due to collisional ionisation.

The newly produced free electrons are then also subject to acceleration by the electric field towards the positive anode, and therefore an avalanche of electrons moving towards the anode is started. After a very short time a situation occurs where many gas atoms are ionised and many free electrons are present, the so called 'plasma state', characterised by an equal and high number of free electrons and positively charged ions. The process described here is called the ignition of a plasma and takes place if the voltage u across the electrodes exceeds a certain limit, which rises with the product of the pressure p multiplied by the electrode distance d, if it is considered that the energy gain along one mean free path, equation (1.54), must at least be equal to the ionisation energy. The field strength F is given by u/d. If on the other hand the product pd is decreased and approaches a value that corresponds to a mean free path, being in the order of magnitude of the electrode distance then the ignition voltage must rise also. In this case not enough atoms are on the way between the electrodes to allow the above mechanism of ignition and a new mechanism of charge carrier generation, such as field emission from the electrodes, and others, must appear. The dependence of the ignition voltage on the product of pressure and electrode distance then shows a minimum and is called Paschen's Law (Flügge, 1956).

If once a plasma has been ignited, there are in principle two main possibilities for the further development of the so-called **gas discharge**.

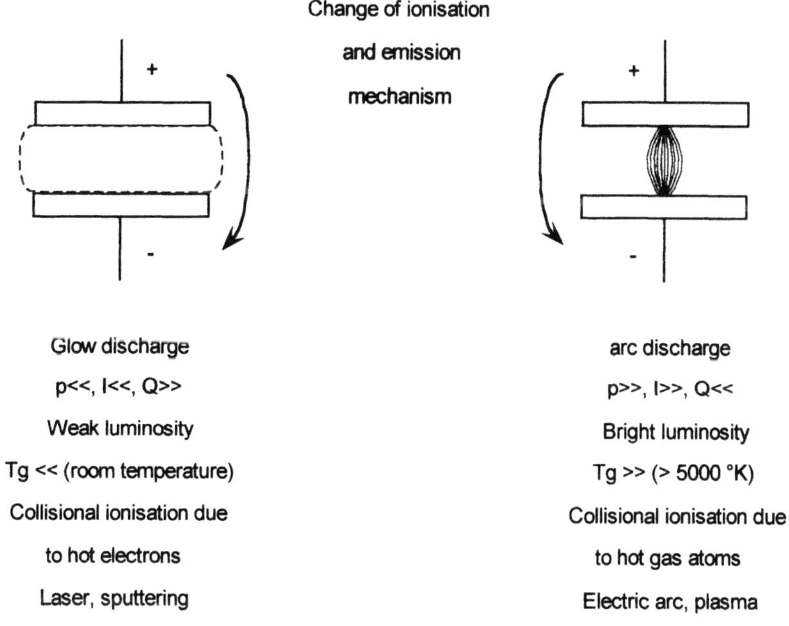

Fig. 1.21 Glow and arc discharge.

If the pressure is rather low, of the order of magnitude of a tenth of an atmosphere or lower (10000 Pa) the mean free path is rather large and therefore the electrons gain between two collisions with atoms an energy that is of the order of magnitude of the ionisation energy of the atoms, but they lose relatively little energy by collisions due to the low pressure and the low number of atoms per unit volume. Therefore, an individual electron gains more energy per unit time than it loses, which means that it heats up and reaches a relatively high average energy. However the gas atoms remain more or less cold, again due to the low number of collisions per unit time with electrons. Due to the low number of collisions and thus of excitation and spontaneous emission processes, there is only a weak light emission of the plasma and it is thus called a glow discharge, Fig. 1.21 (Flügge, 1956).

For the same reason of a relatively small number of collisions per unit time, only a very small fraction of the atoms are ionised, and thus glow discharges only show a weak electric conductivity, which nevertheless allows a supply of energy to the plasma, which is mainly converted into kinetic energy of free electrons. Due to the high average energy that the electrons obtain, they are excellently suited for excitation of the upper laser level and the generation of inversion in a gas, which makes a glow discharge most attractive for lasers.

As mentioned above the electrons gain an average energy in the order of magnitude of the ionisation energy, which is usually higher than a few Electron Volts. An electron energy of this order of magnitude is always sufficient to carry out transitions between atomic energy levels and to excite the upper laser level since the energy steps in the electron shell of the atoms are of the same order of magnitude. Since the molecular transitions show energy differences that are smaller by one order of magnitude, the electrons in a glow discharge are of course also very well suited to excite these levels.

If a glow discharge is ignited between two electrodes with a lateral extension that is larger than their distance - a situation that applies very often to high power gas lasers - the glow discharge is initially distributed more or less evenly over the volume between the two electrodes, making it most appropriate for laser operation, since it provides the generation of a more or less uniform inversion in a large volume that can be utilised by a beam with similar volume.

Nevertheless if the current across the glow discharge is increased beyond a certain limit, too much energy is supplied to the atoms and the gas heats up, thus reducing the mean free path and in consequence enhancing the number of ionising collisions per unit time. At the same time the current density at the electrodes becomes so high that new emission processes such as thermal emission take place where electrons are liberated from the electrodes due to a high temperature allowing the electrons to overcome the attraction of the metal atoms. Since these new mechanisms for charge carrier production appear first at those locations on the electrodes where the highest current density is obtained, probably due to some inhomogeneity of the surface, the electrical conductivity is enhanced. This further increases the current density and leads to a redistribution of the current across the electrode surface and to a concentration in a channel with higher temperature. This process of concentration of the current flow to a narrow channel shows of course an essential instability, since better conductivity attracts more current which in turn increases the power input per unit volume and therefore the conductivity. Due to this instability, after a very short time the current flow has constricted to a narrow channel, where the conditions have changed totally, so that the gas heats up considerably and the gas temperature becomes equal to the electron temperature. A plasma of this kind is called an **arc discharge** since the self magnetic field of the conducting channel leads to magnetic forces on the plasma that contract it in a lateral direction and give it finally an arc-like shape (Cobine, 1958).

Due to the concentration of the energy supplied to the plasma to a much lower volume than in the case of a glow discharge, the ionisation and emission processes become much more efficient, since the overall energy balance is more favourable due to the smaller surface of the arc discharge compared to the glow and therefore not only the gas temperature rises strongly to e.g. >5000 K, but also the voltage across the discharge drops to a much lower magnitude, which is in principle given by the order of magnitude of the ionisation potential. This voltage is ultimately necessary to accelerate the electrons that are leaving the cathode to an energy allowing them to carry out ionisation. It is, of course, a lower limit for the arc voltage composed of the voltage across a small layer at the cathode, and a voltage drop across the bulk of the plasma that shows a very high conductivity near to electric conductors, the latter being determined by the current flow across the arc. It should be mentioned, that the diameter of the arc adjusts itself to an equilibrium between electric energy supplied to the arc by current flow and radiative energy loss that plays an important role due to the high temperature in the plasma mentioned above and depends on the magnitude of the surface of the arc channel.

Further, it should be stressed that under the conditions of atmospheric pressure, glow discharges show a very strong tendency to collapse to arc discharges, since in the case of atmospheric pressure the high number of atoms per unit volume causes high current densities and in consequence strong heating of the plasma.

Going into the details of plasmas in general, the main process of ionisation is collision between particles that becomes stronger and stronger if either the electron temperature T_{el} or the gas temperature T_g rises, thus increasing either the average kinetic energy of the electrons or of the gas atoms. These collisional ionisations must be balanced by the loss of charge carriers due to recombination, given by the densities of electrons n_{el} and ions p_{el} and by losses to the electrodes or to the walls where also recombinations take place. If the plasma is mainly confined by the electrodes, a situation that applies to high power gas lasers, and if the regions near to the electrodes are excluded for the moment, then a balance of electrons liberated per unit time and volume and recombining with ions can be written

$$n_{el}^2 = const \; n_{el} \, n_g \exp\left(-\frac{E_i}{kT}\right) \qquad (1.55)$$

This is called Saha's equation: It allows us in principle to determine the number of charge carriers. Together with the electron **mobility** that gives the average speed of the electrons obtained by applying an electric field, the conductivity of the plasma can be estimated. So together with the energy balance of the individual electron, equation (1.54), the main properties of the plasma as they are relevant for lasers can be described.

Due to the strongly different properties of the electrodes and the plasma, thin transient regions separate them and serve for a matching, for instance of the conditions of current conduction in the electrodes with pure electronic charge transport to those in the plasma with electronic and ionic conduction. First, the region near the cathode will be considered: In the bulk of the plasma electrons and ions transport current, although the latter contribute to a comparatively small amount.

Nevertheless, the external current that is carried in the cathode merely by electrons, is larger than the electronic current in the plasma, since a considerable part of the electrons emitted by the cathode due to the emission mechanism of the glow discharge is absorbed by ions reaching the cathode and recombining with electrons. Only a few electrons can thus leave the cathode region. Therefore the latter must be multiplied by ionisation processes that take place in the cathode layer. For that, additional energy must be supplied to that region by a relatively high potential difference, the so called **cathode voltage drop**. These ionisation processes lead to an enhanced ion current reaching the cathode, which absorbs a large proportion of the electrons flowing in the latter electrode due to the necessary recombination. Thus only a small part of the electron current delivered by the cathode remains to feed electron emission of the cathode and must thus be multiplied by ionisation in the cathode layer (Fig. 1.22).

Similar phenomena can be observed in the region near the anode, that must match the plasma current condition to those of the anode, again resulting in a certain voltage drop, the **anode drop**.

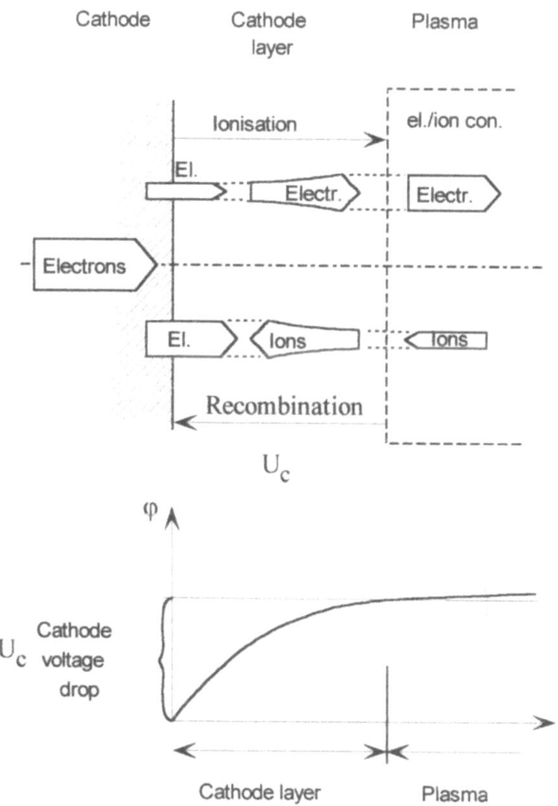

Fig. 1.22 Ionisation and recombination in the cathode layer.

Of course, for the situation at the walls, which are usually made from insulating material, similar considerations apply, although due to the insulating character no currents can appear with the exception of rf discharges, where dielectric displacement currents appear. Due to this absence of currents in the walls, equal electron and ion numbers must reach the wall per unit time, leading to full recombination. Nevertheless, due to the conditions in the bulk of the plasma mentioned above, wall charges build up, that lead again to a certain voltage drop. So at the interface between the plasma and electrodes or walls, thin layers appear where the matching between the properties of the plasma and the walls takes place.

Although in the bulk of the plasma, thermal equilibrium is obtained for a stable discharge, in the layers mentioned above no thermal equilibrium is obtained. Particles that are not in thermal equilibrium with the plasma, such as those emitted by the electrode, adapt to thermal equilibrium with the plasma during crossing of the layer mentioned above or vice-versa. Similar considerations apply also to the evaporation process, where particles leaving hot metal surfaces must adapt to thermal equilibrium in a vapour cloud building up above the boiling surface. The transition to thermal equilibrium takes place while crossing a thin layer, the so called **Knudsen layer**. The thickness of the above layers must obviously be at least one mean free path, since an influence on the particles emitted by the electrodes or reaching the electrodes can only be obtained if they carry out at least one single collision. This minimum thickness is usually obtained in arc discharges, where due to the high pressure and temperature so many collisions take place, that one mean free path is sufficient. In the case of glow discharges however, due to the low density of charge carriers, the extension of the surface layer is much larger and can even be in the order of magnitude of 1 mm.

1.6.2 Semiconductors

Generation of charge carriers and mobility

In some crystalline materials, such as the 4th group of the periodic system, generation of negative and positive charge carriers takes place by ionisation due to proliferation of energy, either thermal or optical. Since these materials are very poor conductors without an external energy supply and become highly conductive if energy of the nature mentioned above is fed to the atoms, these crystals are called **semiconductors**. The electrons thus liberated can move freely in the crystal but are also subject to recombination with carriers of the opposite charge, thus leading to the emission of radiation, the so called **recombination radiation**.

These materials are interesting candidates for the generation of laser light, since electric energy supplied by a current flow may result under certain circumstances in the emission of radiation. To understand the semiconducting properties of certain crystals, one must look at the influence of the condensation of gaseous atoms to a solid body, where the initially distinct energy levels of the free atoms are broadened into energy bands, which become wider and wider the nearer the atoms come to each other during the condensation process. This widening of the energy levels is caused by a rising mobility of the electrons that are initially bound to the atoms but can **tunnel** through the potential walls between the atoms which become lower and narrower as the atoms approach each other. This mobility leads to an irregular movement of the electrons in the crystal that is associated with a certain kinetic energy, being added to the energy of the electrons on their orbits in the atoms. So the initially distinct energy levels of the single atoms spreads up to energy bands separated by gaps where the structure finally obtained depends on the distance between the atoms in the crystal and thus on the crystal structure.

Each energy level of an atom can hold a certain limited number of electrons due to Pauli's law. Therefore, for a given density of electrons at the absolute zero of temperature all electrons occupy the lowest possible energy levels and these are therefore filled up to a certain level, the so-called **Fermi level** E_F. (Fig. 1.23).

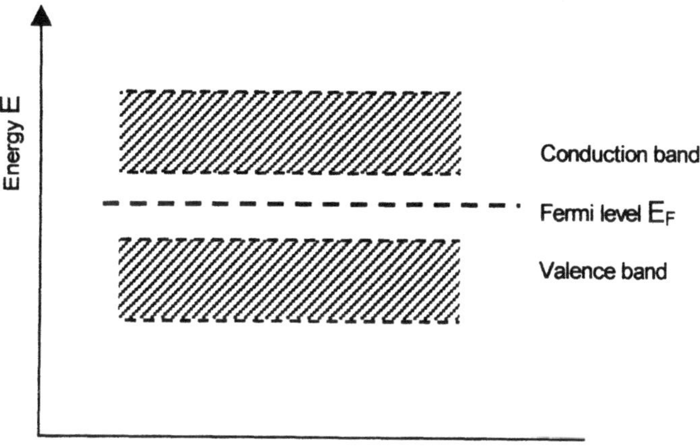

Fig. 1.23 Energy bands and Fermi energy of a semiconductor.

By combining the concept of energy bands and gaps and the Fermi level, semiconductors can now be understood as crystals where the Fermi energy level is situated in the gap between two energy bands. In this case the behaviour mentioned above can easily be understood:

The band below E_F is now called the **valence band** and the band above is called the **conduction band**. At the absolute zero of temperature all electrons remain in the valence band, where all possible energy levels are fully occupied. If energy is supplied either by raising the temperature or by irradiation, the electron energy becomes high enough to allow the particles to cross the gap and to reach the conduction band, where each electron leaves an empty space in the valence band, called a **hole**. These holes are associated with a positive charge due to the missing negative charge of the electron that has left and can freely move around in the crystal by exchanging electrons with neighbouring atoms.

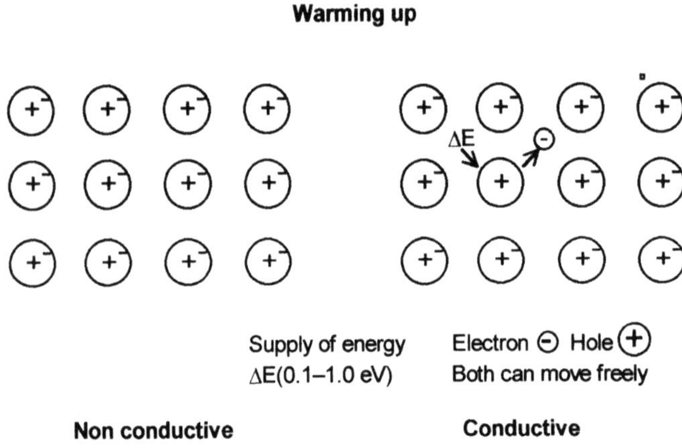

Fig. 1.24 Electrons and holes in intrinsic semiconductors.

The electron excited to the conduction band finds there a lot of available energy levels and can also move around in the semiconductor. Therefore two kinds of charge carriers are generated by energy supply, namely positive holes and negative electrons. These can both move freely and independently in the crystal, thus giving the semiconductor, that is initially insulating, a certain conductivity. The latter conduction is called **intrinsic** conduction since it is provided by the pure crystal itself without any impurities. (Fig. 1.24)

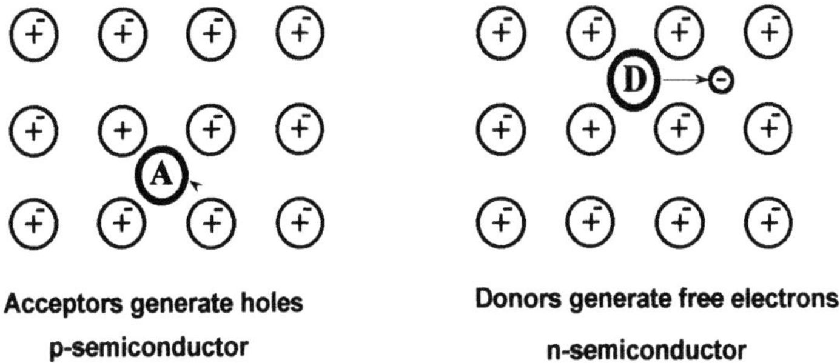

Acceptors generate holes
p-semiconductor

Donors generate free electrons
n-semiconductor

Fig. 1.25 Semiconductor with p and n impurities.

So far only monocrystals with a regular lattice and without any impurities have been considered. If certain impurities are added, these can either act as electron **donors** if they have weakly bound electrons or they can act as **acceptors** if they strongly attract free electrons. In the first case quasi free electrons are generated in the conduction band and therefore the semiconductor is referred to as **n type**, and in the second case, the **p type** semiconductors, positively charged holes are generated. (Fig. 1.25).

1.7 REFERENCES

Born, M. and Wolf, E. (1989) *Principles of Optics,* 6th edn, Pergamon Press,

Cobine, J.D. (1958) *Gaseous Conductors*, Dover Publications

Duffin, W.J. (1990) *Electricity and Magnetism*, 4th edn, McGraw Hill

Feynman, R.P. (1965) *The Feynman Lectures on Physics*, **Vol. III**: *Quantum Mechanics*, California Institute of Technology

Flügge, S. (editor) (1956) **Encyclopedia of Physics**, **Vol. XXII**: *Gas Discharges II*, Springer

Klein, M.V. and Furtak,T.E. (1986) *Optics*, 2nd edn, John Wiley & Sons Inc.

Siegman, A.E. (1986), *Lasers*, Mill Valley: University Sciences Books,

Wichmann, E.H. (1971) *Berkeley Physics Course*, **Vol. IV**: *Quantum Physics*, McGraw Hill

2

Optics, resonators and beams

D. Schuöcker and K. Schröder

2.1 THE KIRCHHOFF FRESNEL INTEGRAL

If a certain area A of arbitrary curvature and shape is illuminated with linearly polarised light with electric field strength E, all points emit spherical waves according to Huygen's law. The electric field strength generated at a point P distant from the illuminated area results from the superposition of all these spherical waves. If an infinitely small area dA is considered, and the electric field strength in this area is approximately constant and equal to E, the contribution to the field at P is proportional to $E.dA$, since the electric field strength determines the light emitted per unit area. This contribution reduces with increasing distance ρ from the emitting surface element dA, due to conservation of energy, as the wave energy distributed across the spherical wave front remains constant, while the radius increases during propagation. The field strength at point P is reduced by a factor $1/\rho$ since the light intensity is given by the square of the electric field strength. There is also a phase difference between the waves arriving at P, and those at the origin dA, given by $ik\rho$, due to the propagation time between dA and P according to equation (2.1). Finally, the electric field strength generated at P due to the emission of the area element dA also depends on the angle φ between the beam from dA to P, and the area element dA. The maximum contribution is obtained when the beam is perpendicular to the area element. An angle φ greater than zero between the vector normal to dA and the direction between dA and P, reduces the area which is seen from the point P.

This reduction follows the well known cosine law, which together with the assumptions made above yields:

$$dE(P) = const \frac{1}{\rho} E(A) \exp(-ik\rho) \cos\varphi \, dA \qquad (2.1)$$

i.e. the infinitesimal contribution of the illuminated area dA to the wave at point P. A more sophisticated wave theoretical treatment leads to the term $-i/2\lambda \ (1+\cos\varphi)$ which replaces the term *const x cosφ* in equation (2.1). Finally all the contributions of the individual elements must be summed to obtain the contribution of the whole illuminated area. As a result the following integral can be derived:

$$E(P) = -\frac{i}{\lambda} \iint_A \frac{1}{\rho} E(A) \exp(-ik\rho) \frac{(1+\cos\varphi)}{2} \, dA \qquad (2.2)$$

This integral is called the **Kirchhoff Fresnel** integral for the analytical description of diffraction. It is of crucial importance for the mathematical analysis of optical elements and their application in optical resonators.

2.2 FOURIER TRANSFORMATION BY FOCUSING

According to geometrical optics, application of the reflection law yields equal angles between an incident beam and the reflecting surface, and the reflecting surface and the reflected beam. This shows that a concave spherical mirror focuses light beams that are initially parallel to the axis of symmetry to a focus halfway between the mirror and its centre of curvature. The focal length of the mirror is thus one half of the radius of curvature R.

$$f = R/2 \qquad (2.3)$$

So far, geometrical optics yields a point like focus, which means that the intensity becomes infinitely large, an unrealistic result, since diffraction, leading to a slight bending of the beams of geometrical optics, has not been taken into account. Therefore the use of the Kirchhoff Fresnel integral that has been derived above is necessary.

In the following treatment the coordinates of a point on the focusing mirror are (x, y, z) and those for the focal plane are $(x_F, y_F, z_F = 0)$, showing that the origin is situated in the focal plane. To simplify the analysis, **paraxial beams** with a small distance from the axis and also a small inclination towards the latter are assumed. As in this case $z < R/2$ and $cos\ \varphi \approx 1$, the following approximate expression can easily be obtained for the distance ρ between a point on the mirror and a point in the focal plane:

$$\rho = \frac{R}{2} - 2\frac{xx_F}{R} - 2\frac{yy_F}{R} \tag{2.4}$$

This distance is mainly given by the focal length of the mirror $R/2$, but shows slight variations, that are small compared to $R/2$, but large compared to the wavelength λ. It is therefore ultimately necessary to consider the full expression for ρ as given by equation (2.4) for the calculation of the phase shifts in the Kirchhoff Fresnel integral, since in this case a variation of one half of the wavelength leads to the extinction of the contribution of the respective element. Nevertheless, for the calculation of the magnitude of the electric field the distance ρ can be approximated by $R/2$, thus simplifying equation (2.4) considerably, and leading to the following expression for equation (2.2):

$$E_F = -\frac{2i}{\lambda R}\exp\left(\frac{-iR\pi}{\lambda}\right)\int\limits_{x=-\infty}^{x=+\infty}\int\limits_{y=-\infty}^{y=+\infty} E(x,y)\exp\left(ix\frac{4\pi}{\lambda R}x_F\right)\exp\left(iy\frac{4\pi}{\lambda R}y_F\right)dxdy \tag{2.5}$$

If it is now assumed that the distribution of the electric field strength on the focusing mirror depends on x and y in such a way that it can be split in a product of two functions, each depending on only one coordinate, the integral in equation (2.5) can in consequence be split into a product of two integrals. Each of them depends only on the coordinates in the same direction, so that either x and x_F, or y and y_F are contained.

$$E_F = -\frac{2i}{\lambda R}\exp\left(\frac{-iR\pi}{\lambda}\right)\int\limits_{x=-\infty}^{x=+\infty} E_x(x)\exp(ix\xi_F)dx\int\limits_{y=-\infty}^{y=+\infty} E_y(y)\exp(iy\eta_F)dy \tag{2.6}$$

where

$$E(x,y) = E_x(x)E_y(y) \tag{2.7}$$

and

$$\xi_F = \frac{4\pi}{\lambda R}x_F; \quad \eta_F = \frac{4\pi}{\lambda R}y_F \tag{2.8}$$

In equation (2.6) the two integrals represent the Fourier transformations mentioned in Chapter 1. The field distribution in the focal plane results from Fourier transformations of the distribution on the mirror. The time domain of a typical Fourier transformation corresponds to the field distribution in the x or y direction on the mirror, and the frequency domain corresponds to the distribution in the focal plane. As a direct consequence of this an initially wide beam is focused to a small but finite spot, whose diameter depends not only on the wavelength and the focal length of the mirror, but also on the diameter of the initial beam. This result is of crucial importance in the application of lasers to material processing.

2.3 FOCUSING A GAUSSIAN DISTRIBUTION

As mentioned in Chapter 1, the principal shape of a gaussian distribution remains unchanged during Fourier transformation although its width is subject to change. The latter distribution shows rotational symmetry and can be written as follows, if the dependence on z is neglected due to the paraxial assumption mentioned above and if a constant phase is assumed across the mirror surface (not equal phase wave):

$$E_{00}(x,y) = E_0 \exp\left(-\frac{x^2 + y^2}{w^2}\right) \tag{2.9}$$

In this expression w is a nominal beam radius at which the electric field strength is reduced a by factor $1/e$ compared to the maximum. Therefore the Kirchhoff Fresnel integral, equation (2.2), must be soluble for this kind of distribution, since it can easily be split into two independent functions, each for one variable. According to the well known transformation rules for Fourier transformation, the following result is obtained for the field distribution in the focal plane:

$$E_F = -\frac{2\pi i w^2}{\lambda R}\exp\left(-\frac{iRx}{\lambda}\right)E_0\exp\left[-\left(\frac{\pi w}{\lambda f}\right)^2\left(x_F^2 + y_F^2\right)\right] \qquad (2.10)$$

Comparing equations (2.99) and (2.10) the factor $\pi w/\lambda f$ can be identified as the reciprocal value of the beam radius in the focal plane w_0

$$w_0 = \frac{\lambda R}{2\pi w} = \frac{\lambda f}{\pi w} \qquad (2.11)$$

The above transformation is only a good description for focusing beams with a beam divergence corresponding to a parallel beam in geometric optics. For highly converging or diverging beams the beam waist after focusing is shifted in the optical axis, giving an arbitrary, but not minimum beam radius at a distance f from the focusing element.

From equation (2.11) it can be concluded that the beam radius w_0 in the focal plane depends on the radius of the initial beam w, the focal length $f = R/2$ and the wavelength λ. The dependence on the diameter of the gaussian beam before focusing is also of practical importance, since in laser material processing the beam is widened artificially by a telescope before focusing, thus yielding a sharper focus and a higher intensity.

Equation (2.10) also shows that the amplitude of the field strength is enhanced in the focal plane. This takes into account the increase in the intensity due to the smaller focal spot and the necessary energy continuity. Figure 2.1 shows an initial gaussian distribution and its shape after focusing.

As at the focus, the field distribution at very large distances from focus, theoretically at infinite distance, is also obtained by a Fourier transformation. In the mathematical treatment the same approximations used above, namely, that the transverse coordinates x and y are small compared to z , which are valid in the focal plane, can also be applied to a plane at infinite distance from the source of the wave.

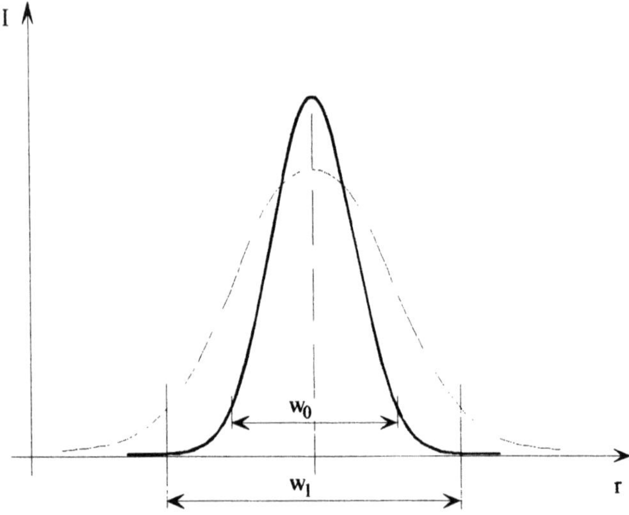

Fig. 2.1 Gaussian beam before (dashed line) and after (solid line) focusing.

2.4 GAUSSIAN BEAMS

If a light beam with a gaussian amplitude distribution with constant phase illuminates a concave mirror, the Kirchhoff Fresnel Integral determines the distribution in the focal plane. If it is now assumed, that a second mirror with the same shape is arranged opposite the first, and at equal distance from the focal plane, the field distribution generated on this mirror by the field in the focal plane can be calculated with precisely the same derivation. This gives a mirror image of the initial gaussian distribution with constant phase across the mirror surface (Fig. 2.2).

It can be seen that the focusing action is followed by spreading of the beam, yielding perfect symmetry of the beam with respect to the focal plane. In this vicinity, the beam radius is nearly equal to the beam waist w_0 and increases linearly with increasing distance from the focus for large distances, since the wave then approaches a spherical shape.

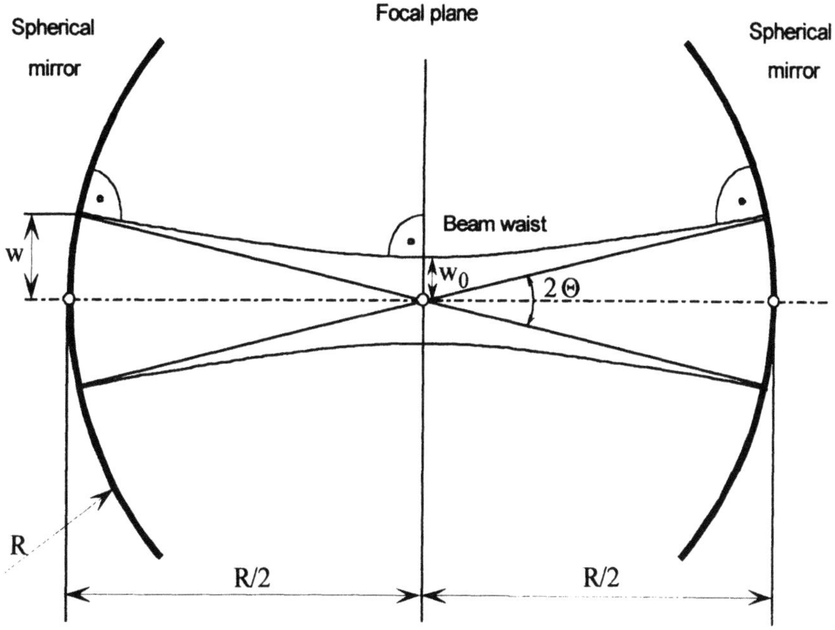

Fig. 2.2 Gaussian beam between two mirrors.

The following expression describes well the nearly constant beam radius near the focus and the linear rise far away from the focus and is obtained from a detailed mathematical treatment of the gaussian beam (Kogelnik, 1965):

$$w = w_0 \sqrt{1 + \left(\frac{z}{z_R}\right)^2} \qquad (2.12)$$

The parameter z_R, the so called **Rayleigh length**, describes the extension of the zone where the beam remains highly focused. It can be determined from the results obtained for the focusing action of a concave mirror that relates the beam radius at the mirror to the beam waist in the focal plane:

$$z_R = \frac{w_0^2 \pi}{\lambda} \qquad (2.13)$$

According to the approximations of geometrical optics, the beam edges are two straight lines through the intersections of the beam edge with the mirrors and meeting at the focus. The beam edges approach these lines more and more with increasing distance from the focus since then the geometrical optics approach fits more and more. The intersection with the beam edges at the mirrors is caused by a small error introduced by neglecting the lateral coordinates x, y, x_F, y_F in the denominator of the Kirchhoff Fresnel Integral, equation (2.6), whereas the precise treatment yields a somewhat wider beam on the mirror (see dashed line in Fig. 2.1). From equation (2.12) it can easily be seen that the opening angle Θ (**divergence**) is described by the following expression:

$$\Theta = \frac{\lambda}{\pi w_0} \tag{2.14}$$

The radius R_F of the wavefronts of the gaussian beam changes from infinity in the focus to a value that rises linearly with the distance z from the focus and can be calculated from equation (2.15) for the beam edge, since the wavefronts are always perpendicular to the beam edges (see Fig. 2.3).

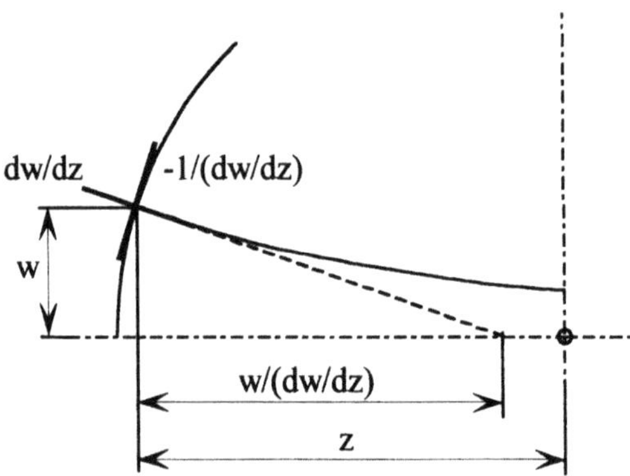

Fig. 2.3 Wavefront and beam edge of a gaussian beam.

The use of the paraxial approximation yields:

$$R_F = z_R \left(\frac{z_R}{z} + \frac{z}{z_R} \right) \tag{2.15}$$

The above expression shows a local minimum at $z = z_R$, where $R_F = 2z_R$. The wavefront radius at the focusing mirror coincides well with the radius of curvature of the latter as initially assumed. It should be mentioned that the initial field distribution focused by the mirror and leading to the generation of a gaussian beam does not correspond to a plane wave with constant phase, but is also generated by a gaussian beam. Nevertheless, under paraxial beam conditions and for flat mirrors, it is a good approximation for a plane wave. Moreover, the latter cannot exist with finite beam radius and becomes a gaussian beam if the beam radius is finite. Equation (2.14) yields for the product of divergence Θ and beam waist w_0:

$$w_0 \Theta = \frac{\lambda}{\pi} \tag{2.16}$$

This product is called the beam parameter product and depends only on the wavelength λ, thus remaining constant no matter which focusing means have been used. It becomes larger for the higher modes to be discussed later.

2.5 HIGHER ORDER TRANSVERSE MODES

An optical resonator, usually consisting of a focusing mirror and a plane mirror, provides feedback of radiation amplified by the active medium. The condition for feedback is that the amplitude and phase distribution of the electric field strength on one mirror is fully reproduced after one complete round trip in the resonator. It is a necessary condition for this reproduction, that the beam profile generated in the resonator remains in principle unchanged with the exception that the diameter of the beam is allowed to change. In consequence, all field distributions that remain unchanged by focusing and thus by Fourier transformation, give rise to beams that can be excited in optical resonators. As mentioned above a gaussian beam clearly fulfils these conditions. It can also easily be shown that a gaussian distribution multiplied by x or y remains in principle unchanged by Fourier transformation. A field distribution of that kind shows a node at the origin followed by a maximum and finally a steady decrease that converges towards zero. An intensity distribution of this kind is called a TEM_{10} radial mode of rectangular symmetry, if it is assumed that the distribution in the *y* direction is purely gaussian.

$$E_{01}(x,y) = E_0 \exp\left(-\frac{x^2 + y^2}{w^2}\right) x . const_1 \qquad (2.17)$$

If the above distribution is valid in both directions then the mode is called a TEM_{11} mode, or if a maximum appears only in the y direction then it is called a TEM_{01} mode. The theory of Fourier transformation shows that in general all the **eigenfunctions** can be formed from the product of a gaussian distribution multiplied by the so called Hermite polynomials of nth order (H_n), which are simply polynomials of nth order. These modes show n minima on the respective axis (Hodgson and Weber, 1992). So in general a TEM_{mn} mode shows *m* minima on the x axis and *n* minima on the *y* axis and is of course, symmetric with respect to the *x* and y axes, showing rectangular symmetry:

$$E_{mn}(x,y) = E_0 \exp\left(-\frac{x^2 + y^2}{w^2}\right) H_m(x) H_n(y) const \qquad (2.18)$$

The diagrams on the next page show some typical laser mode structures. As an example Fig. 2.4 shows the field distribution of a TEM_{11} mode with rectangular symmetry.

The superposition of rectangular TEM_{10} and a TEM_{01} modes results in a rotationally symmetric distribution with zero field at the origin (Fig. 2.5) surrounded by a ring of maximum field strength. This field distribution resembles a ring doughnut and is therefore referred to as a **Doughnut mode** $TEM_{01}{}^*$.

These and similar modes are very often generated by high power lasers above 5 kW, such as lasers with coaxial electrodes and a hollow cylindrical plasma, or lasers with a so-called **scraper mirror**, which is a ring-shaped mirror that extracts a certain part of the beam power out of the resonator.

Of course, there are also higher modes obtained under rotational symmetry. In this case a TEM_{mn} mode describes a distribution of *m* minima in the radial direction and *n* periods distributed evenly along a circle.

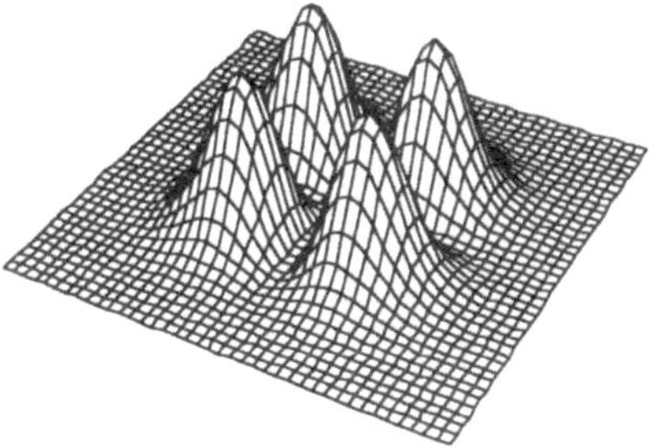

Fig. 2.4 Mode pattern for a TEM_{11} mode in rectangular symmetry.

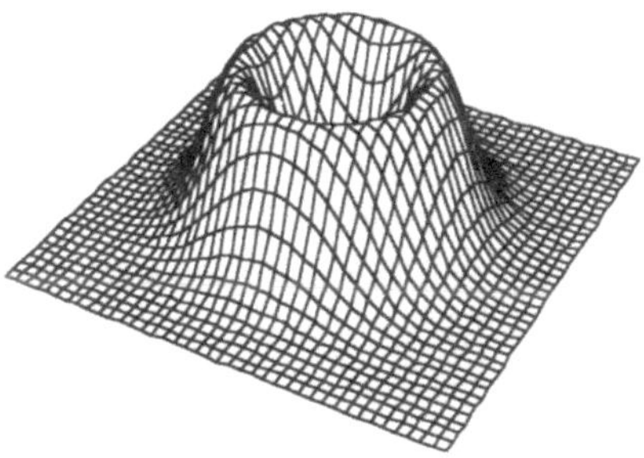

Fig. 2.5 Doughnut Mode.

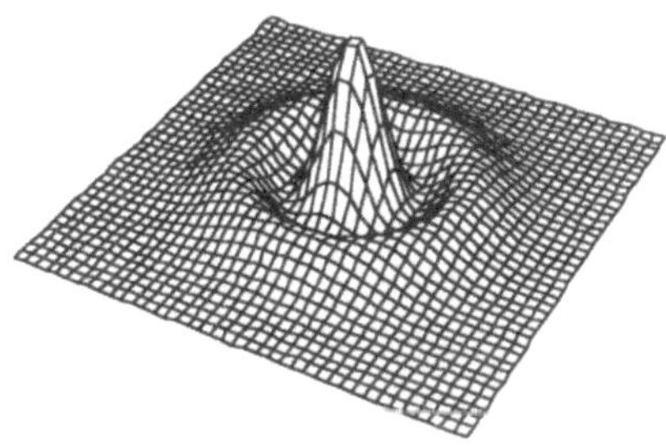

Fig. 2.6 TEM_{10} Mode.

Figure 2.6 shows a mode of this kind with one radial minimum, a so called TEM_{10} mode. The latter distribution shows not only a radial minimum but also a central maximum at the origin, which makes an important difference to the modes of rectangular symmetry. Since all the modes of rectangular symmetry form a matrix with infinite dimensions, they can be shown to form a basis for the various beam modes and can thus be used to synthesise all other modes by linear combination. Therefore, in particular the modes of rotational symmetry can be obtained from a linear combination of an infinite number of modes of rectangular symmetry.

2.6 DIVERGENCE OF HIGHER ORDER TRANSVERSE MODES

Higher order modes contain several relative maxima of field strength as for example the TEM_{11} mode that shows four maxima (Fig. 2.4). These maxima can be regarded as more or less independent spots with a diameter that is smaller than that of the fundamental mode TEM_{00}, (compare TEM_{00} and TEM_{11} in Fig. 2.4). Due to their smaller spot radius the divergence is larger than that of the TEM_{00} mode, according to equation (2.16) for the relationship between beam radius and divergence.

The higher the order of the transverse mode, the more maxima appear in the x and y directions, the smaller becomes the radius of the individual maxima and spots, and therefore the larger becomes the divergence of the whole beam. If these spots were totally independent without any phase relation, the divergence would increase almost linearly with the mode order in one direction. Since there is a phase relationship, the divergence does not increase that fast. The detailed theoretical treatment yields the following equation for the divergence angle in x direction for higher transverse modes with m maxima on the x axis, where Θ_0 means the divergence of the fundamental mode

$$\Theta_m = \Theta_0 \sqrt{2m+1} \tag{2.19}$$

Higher modes show a higher divergence, which means that they spread more rapidly than the fundamental mode. If now an arbitrary field distribution is given, for instance at the output mirror of a laser, then it can be regarded as a linear combination of the fundamental mode (the TEM_{00} mode) and a number of higher modes TEM_{nm} .

Since modes with higher m and n are subject to a higher divergence, at a certain distance from the initial field distribution the relative contribution of the higher modes becomes smaller and smaller, since their energy is distributed over a larger cross section. In the far field at infinite distance, a gaussian beam TEM_{00} is obtained. This is of great practical importance and is used in laser material processing where the beam generated by a laser source with imperfect mode is guided along a path that is artificially elongated in order to improve the mode quality and thus the sharpness of focusing.

2.7 FOCUSING ARBITRARY BEAM MODES

The effect of a long beam path on the selection of the fundamental mode component in the field distribution and the weakening of the higher modes should not be confused with the effect of focusing. In the focal plane all modes contained in the initial field distribution before focusing are reproduced, although the focal radius becomes larger and the values of m and n increase. Due to this property of the focusing process, higher modes become less important in the focal plane which leads to a similar effect as in the far field, namely a selective propagation of the fundamental mode. Special cases are Doughnut or similar modes, since they can be regarded at least to a first approximation as a gaussian mode, where the maximum is shifted along the r axis, thus yielding due to rotational symmetry a toroidal mode that resembles a ring mode, (see Fig. 2.15). The field distribution in this case is

$$E_{01}{}^* = E_0 \exp\left[\frac{(r-a)^2}{w^2}\right] \tag{2.20}$$

with $r = \sqrt{x^2 + y^2}$

If this field distribution is subject to focusing, the field distribution in the focal plane is given by the Fourier transformation. It is obvious from equation (2.5) that the shift of the maximum by a distance a results in a phase shift, if in the Fourier transformation integral a substitution of a new integration variable u is carried out, where $u = r - a$.

$$F\{E_{01}{}^*\} = E_0 \exp(-ika)F\{E_{00}\} \qquad (2.21)$$

Besides the phase shift mentioned above, the distribution obtained in the focus is again gaussian, where the diameter of the focus is given by the width w of the initial gaussian distribution. So, finally, focusing of a Doughnut mode or similar ring shaped mode yields nothing else than a focused gaussian distribution. This is of great practical importance for high power lasers since they usually produce a ring shaped mode, which is not so disadvantageous due to the focusing properties mentioned above, although the Rayleigh length differs from its value for the two gaussian beams.

2.8 STABLE RESONATORS

2.8.1 Hemispherical resonators with mirrors of infinite size

Optical resonators are in general formed by two mirrors, usually a curved one of concave shape with radius of curvature R, and a plane one at distance L. This is called a **hemispherical resonator**. In a resonator of this kind only those beams with a field distribution which is reproduced during a full round trip can be maintained. The condition for full reproduction of the field distribution, is that the distributions of the amplitude as well as phase remain unchanged after a full round trip. The latter phase distribution is fully determined by the shape of the wavefronts and therefore the above condition for the reproduction of the field distribution is precisely the same as a reproduction of the three dimensional shape of the wavefronts with the radius R and the amplitude distribution along the wavefront, characterised by the beam radius w in the case of a pure gaussian distribution.

If a gaussian beam starts at the concave mirror with a wavefront of radius R_1 and a beam radius w_1 (Fig. 2.7), then due to the focusing action of the mirror it assumes at the plane mirror a somewhat smaller radius w_2 and wavefront radius R_2. During reflection at the plane mirror, the beam radius and the radius of curvature of the wavefront remain unchanged, only the direction of propagation becomes reversed. The reflected beam then propagates towards the curved mirror, which is symmetrically arranged with respect to the beam and reverses its direction.

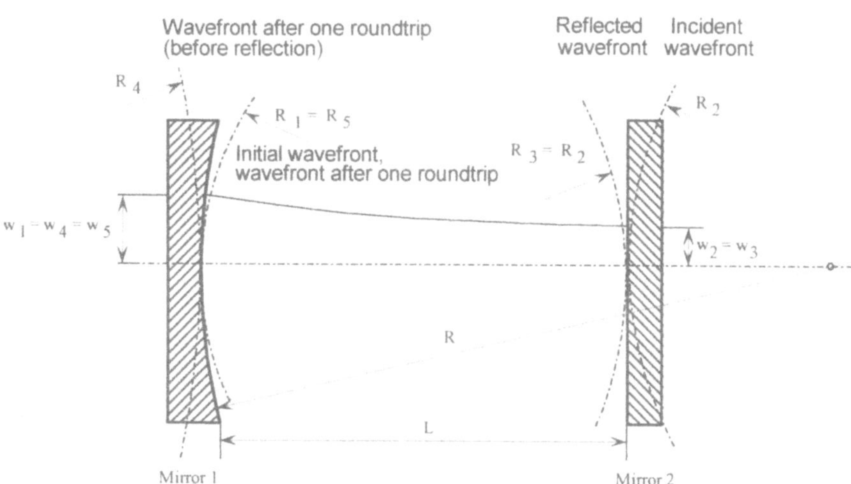

Fig. 2.7 Hemispherical resonator.

Due to the widening of the beam at the curved mirror caused by diffraction, the beam radius w_4 is assumed and the wavefront radius becomes R_4. Reflection at the curved mirror leaves the beam radius unchanged, but the wavefront curvature changes again, e.g. from convex to concave, according to equation (2.11). So after reflection, the new wavefront radius is R_5 and full reproduction of the wave is only achieved if the latter is equal to the wavefront radius R_1. Additionally, the beam radius after reflection at the concave mirror $w_4 = w_5$ must be equal to that at the start of the wave w_1. In order to evaluate the above condition, first the change of the beam radius and wavefront radius during free space propagation between the mirrors must be described. The beam radius w_1 and the wavefront radius R_1 of a gaussian beam correspond to a Rayleigh length z_R and a distance z from the beam waist given by equations (2.12) and (2.15). From the simultaneous solution of these equations, the following expressions result:

$$z_w = \frac{z_1^2 R_1}{R_1^2 + z_1^2} \tag{2.22}$$

$$z_R = \frac{z_1 R_1^2}{R_1^2 + z_1^2} \tag{2.23}$$

To simplify the treatment the characteristic length z_1 has been defined:

$$z_1 = \frac{w_1^2 \pi}{\lambda} \tag{2.24}$$

With equations (2.12), (2.15), (2.22) and (2.23) the beam radius w_2 and the wavefront radius R_2 at the plane mirror, that have been altered due to free space propagation of the gaussian beam, can be calculated.

$$w_2 = \sqrt{\left(\frac{\lambda}{\pi}\right) \frac{R_1^2 z_1^2 + L^2\left(R_1^2 + z_1^2\right) - 2Lz_1^2 R_1}{R_1^2 z_1}} \tag{2.25}$$

$$R_2 = \frac{R_1^2 z_1^2 + L^2\left(R_1^2 + z_1^2\right) - 2Lz_1^2 R_1}{L^2\left(R_1^2 + z_1^2\right) - z_1^2 R_1} \tag{2.26}$$

During reflection at the plane mirror, the beam radius remains unchanged and the direction of propagation changes. After the latter reflection, the beam propagates towards the curved mirror and assumes there a beam radius w_3, given by equation (2.12) for a distance from the beam waist $z = 2L - z_w$. After reflection it propagates as a mirror image of the beam, that would have existed without the mirror.

If $z = 2L - z_w > 0$ the beam waist is situated inside the resonator and an initially e.g. concave wavefront will arrive as a convex wave after one roundtrip, since it has passed the beam waist. The radius of curvature can be calculated from equation (2.15) where the sign must be chosen according to the rules for signs of radii of curvature. A negative sign indicates an initially concave wave and a positive sign a convex wave arriving after one roundtrip.

If $z = 2L - z_w < 0$ the beam waist is situated outside the resonator and an initially concave wave front would arrive again as a concave wave after one roundtrip, since it does not pass the beam waist. The wavefront radius is given by equation (2.15) for $z = 2L - z_w$, where the sign does not change.

$$w_4 = \sqrt{\left(\frac{\lambda}{\pi}\right)\frac{R_1^2 z_1^2 + L^2\left(R_1^2 + z_1^2\right) - 2Lz_1^2 R_1}{R_1^2 z_1}} \tag{2.27}$$

$$R_4 = \frac{R_1^2 z_1^2 + 4L^2\left(R_1^2 + z_1^2\right) - 4Lz_1^2 R_1}{2L\left(R_1^2 + z_1^2\right) - z_1^2 R_1} \tag{2.28}$$

Reflection at the spherical mirror leaves the beam radius unchanged, indicating that the beam radius w_4 equals the beam radius w_5 after reflection: $w_4 = w_5$. For a full reproduction of the wave, the latter beam radius must be equal to the initial one, therefore the first necessary condition for reproduction after one round trip is: $w_1 = w_5$. According to equation (2.11), the wavefront radius changes during reflection at the spherical mirror. The resulting dimensionless wavefront radius R_5 is:

$$R_5 = \frac{R_4}{-2\dfrac{R_4}{R} + 1} \tag{2.29}$$

For a full reproduction of the wave, the initial wavefront radius must be equal to that obtained after reflection at the spherical mirror. It is therefore a second necessary condition, that $R_1 = R_5$. Using equation (2.27) the condition $w_1 = w_5 = w_4$ yields:

$$L\left(R_1^2 + z_1^2\right) = z_1^2 R_1 \tag{2.30}$$

Comparing this result to equation (2.22) it can be concluded that $L = z_w$ indicating that the plane mirror is situated in the beam waist where the phase front of the beam is also plane. Therefore the plane mirror matches the phase front.

Substituting equation (2.28) in (2.29) and using (2.30) for the reproduction of beam radii, the condition for the radii of curvature $R_1 = R_5$ can be written as:

$$R_1 = R \tag{2.31}$$

From this it can also be concluded that at the curved mirror the phase front must correspond to the curvature of the spherical mirror. Equation (2.30) can also be used to calculate the beam radius at the curved mirror in terms of the length L of the resonator and the radius of curvature R of the spherical mirror:

$$w_1 = \sqrt[4]{\left(\frac{\lambda}{\pi}\right)^2 \frac{LR^2}{R-L}} = \sqrt[4]{\left(\frac{\lambda L}{\pi}\right)^2 \frac{(R/L)^2}{R/L-1}} \tag{2.32}$$

Equation (2.32) is plotted in Fig. 2.8.

To describe optical resonators the so called **g parameters** are often used. In a resonator of length L for each mirror with the radius of curvature R_i the dimensionless g parameter is defined by:

$$g_i = 1 - \frac{L}{R_i} \tag{2.33}$$

The spherical mirror with radius of curvature R therefore is described by:

$$g_1 = g = 1 - \frac{L}{R} \tag{2.34}$$

For the plane mirror with $R_2 = \infty$ this becomes:

$$g_2 = 1 - \frac{L}{\infty} = 1 \tag{2.35}$$

Using the g parameter, equation (2.32) can be transformed to the following expression which is useful in considering the existence of reproducible waves in the resonator.

$$w_1 = \sqrt{\frac{\lambda L}{\pi}} \sqrt{\frac{1}{g(1-g)}} \tag{2.36}$$

The square root in equation (2.36) possesses real solutions only when $0 \leq g \leq 1$. Three cases within this regime are of special interest, since they are limits for the hemispherical resonator:

1. **$g_1 = 0$** $(R = L)$ The centre of curvature of the spherical mirror is situated on the plane mirror, a **hemiconcentric resonator**. Equation (2.36) yields $w_1 = \infty$, which corresponds to the fact that in this case spherical waves develop with their origin at the centre of the plane mirror, filling the whole volume between both mirrors (Fig. 2.8).

2. **$g_1 = 1$** e.g. $R = \infty$, or that the spherical mirror is in fact also a plane one, in the case of a so called **Fabry Perot** Resonator with parallel planar mirrors, Equation (2.36) again yields $w_1 = \infty$, since each beam leaving one mirror at an arbitrary angle different from zero runs zigzag through the mirror indefinitely and therefore fills the whole volume between the mirrors which extends to infinity. (Fig. 2.8).

3. **$g_1 = \frac{1}{2}$** . That means that $L = R/2$ and that the focus of the spherical mirror is on the plane mirror, the so called **hemiconfocal resonator** can easily be compared with the results obtained for the treatment of the spherical mirror by diffraction theory. In this case the wavefront fits precisely the curved and the plane mirror, the wave at the latter just being obtained by focusing the distribution that appears at the curved mirror. This gives a beam waist at the plane mirror with a radius inversely proportional to that at the curved mirror, according to equation (2.11). (Fig. 2.8).

All cases shown by Fig. 2.8 correspond to $0 \leq g_1 \leq 1$ and $g_2 = 1$ especially the three specific resonators treated above, yielding perfectly reproducing waves and thus allowing the maintenance of waves with constant amplitude for infinite time. Such resonators are called **stable resonators**.

In general this is the case if $0 \leq g_1 g_2 \leq 1$. It will be shown later, that for $g_1 g_2 \geq 1$ an entirely different situation arises, where losses cannot be avoided, since a part of the beam continuously leaves the resonator. This is called an **unstable resonator**.

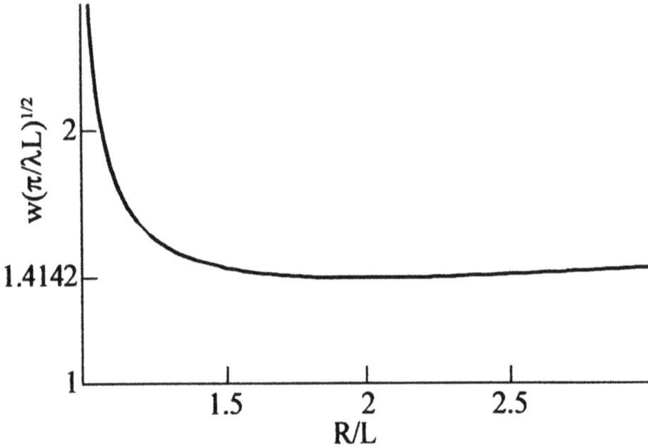

Fig. 2.8 Beam radius at the spherical mirror as a function of R/L.

2.8.2 Resonators with finite mirrors

So far only stable resonators with infinitely large mirrors have been treated, a case that cannot be realised in practice. Resonators of this kind have absolutely no losses and therefore for instance a gaussian beam can be maintained indefinitely without any amplification, which again justifies the name stable. In practice, the mirrors must be of finite size, for instance with radii a_1 and a_2.

To understand the main properties of these resonators, it is assumed that the concave mirror still has infinite size and the plane mirror has a finite radius of a_2. In this case reflection at the plane mirror causes cutting of a considerable part of the field distribution at this mirror, leading to a loss of beam energy depending on the ratio of the beam radius w_2 and the radius of the mirror a_2.

Of course, the beam reflected by the plane mirror with finite size no longer shows a gaussian distribution and therefore the distribution at the concave mirror changes too. These changes take account of the energy losses at the plane mirror and thus yield reduced amplitudes. So full feedback is no longer possible and the amplifying effect of the active medium is necessary to maintain a stable beam. As a first approximation it can be assumed that a gaussian beam is also formed in this case, but with a reduced diameter at the plane mirror, thus compensating losses at the edge of the mirror to a certain extent. Nevertheless due to the infinite extension of a gaussian beam some losses remain called **diffraction losses**. They lead to the appearance of a light wave behind the plane mirror due to diffraction. Since the plane mirror is usually semitransparent in order to extract some part of the beam for practical applications, there are also **transmission losses**. The diffraction losses mentioned above depend on the ratio between beam radius and mirror radius w_2/a_2. According to equation (2.36), the beam radius depends in turn on the wavelength and on the length of the resonator. Therefore the diffraction losses are determined by:

$$\frac{\sqrt{\lambda L}}{a_2} \tag{2.37}$$

The dimensionless constant N given by:

$$N = \frac{a_2^2}{\lambda L} \tag{2.38}$$

is called the **Fresnel Number** and characterises the losses of a resonator by diffraction, where an increasing Fresnel number leads to decreasing diffraction losses.

Higher order transverse modes with their higher diameter and volume lead of course to increased diffraction losses for a given Fresnel number (see Fig. 2.9). This figure shows that for a medium value of the Fresnel number near unity (1) the diffraction losses for the fundamental mode TEM_{00} are relatively small but much larger for all higher order transverse modes. Therefore a Fresnel number of $N = 1$, or thereabouts, ensures that only the fundamental mode is excited in the laser. This situation is favoured for most applications, since it ensures optimum focusing ability and maximum intensity.

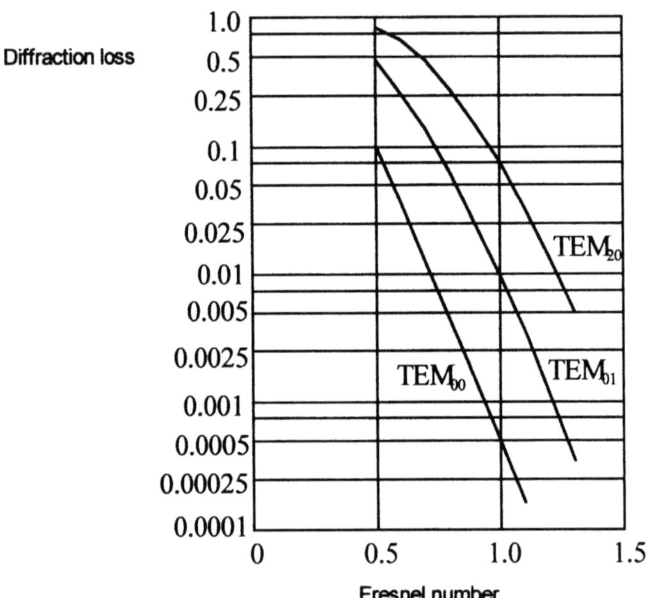

Fig. 2.9 Losses of gaussian and higher order modes.

2.9 UNSTABLE RESONATORS

Optical elements, such as the spherical mirrors treated above can be described under the conditions of geometrical optics by the so-called **beam matrix**. The latter describes an incoming beam by a vector. The first element is the distance of the beam from the axis. The second element is the gradient of the beam with respect to the axis (see Fig. 2.10).

The beam matrix for a focusing mirror is written as follows, if the transformation of a spherical wave is described by Eq. (2.29):

$$\begin{pmatrix} 1 & 0 \\ -2/R & 1 \end{pmatrix} \tag{2.39}$$

For a plane mirror $g_2 = 1$ and the beam matrix is:

$$\begin{pmatrix} 1 & 0 \\ 0 & 1 \end{pmatrix} \tag{2.40}$$

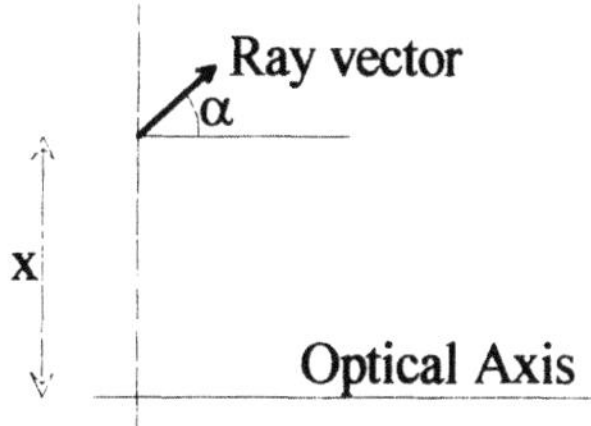

Fig 2.10 Definition of the ray vector $\mathbf{r} = \begin{pmatrix} x \\ \alpha \end{pmatrix}$

For free space propagation, the beam matrix is:

$$\begin{pmatrix} 1 & L \\ 0 & 1 \end{pmatrix} \tag{2.41}$$

According to this formalism, subsequent optical elements are described by multiplying their beam matrices, yielding a final matrix for the whole system. In the case of a laser resonator consisting of a focusing mirror and a plane mirror, the matrices for free space propagation, the plane mirror, free space propagation again and finally the spherical mirror, must be multiplied together giving a final matrix that describes the whole resonator:

$$\begin{pmatrix} 1 & 0 \\ -2/R & 1 \end{pmatrix}\begin{pmatrix} 1 & L \\ 0 & 1 \end{pmatrix}\begin{pmatrix} 1 & 0 \\ 0 & 1 \end{pmatrix}\begin{pmatrix} 1 & L \\ 0 & 1 \end{pmatrix} = \begin{pmatrix} 1 & 2L \\ -2/R & 1-4L/R \end{pmatrix} \tag{2.42}$$

Due to its dimension, the final matrix has two **eigenvectors**. If $0 \leq g_1g_2 \leq 1$, as in the case of the stable resonator, the two eigenvectors are complex, which means that no reproducible beams are possible. Conversely if $g_1g_2 > 1$ or $g_1g_2 < 0$, as in the case of the unstable resonator, the two eigenvectors are real, which means that in principle two beams are reproducible, although the magnitude will change according to the **eigenvalue** associated with each eigenvector.

Analysis of the matrix in equation (2.42) shows that the two real eigenvectors of the beam matrix of the unstable resonator correspond to two real eigenvalues, of which the first is smaller than one and the second is larger. If the second of these eigenvectors is now considered in detail, a beam described by this vector generates after one round trip, a beam where the radius and the gradient are larger than that of the initial beam and the proportionality constant is given by the appropriate eigenvalue. The next round trip again yields a beam that shows the same proportionality to the initial one. So all these beams produced from the one which corresponds to the eigenvector mentioned above, generate a spherical wave in the resonator that propagates in the direction of the initial beam. In the opposite direction a second spherical wave must propagate, since the initial beam is reproduced during one full round-trip, although its radius and its gradient are magnified. This second spherical wave has of course a different radius of curvature. Concerning the eigenvalue that is less than one, this leads after several round trips to a beam that runs along the axis of the resonator. This is the trivial solution for all resonators treated here having of no practical importance.

A good example of the behaviour of an unstable resonator, as mentioned above is the **confocal resonator**, consisting of one concave and one convex mirror with the same focal point, The diameter of the convex mirror is smaller than that of the concave mirror (Fig.2.11).

It is obvious that the condition for this kind of resonator is that:

$$L = \frac{R_1}{2} - \frac{R_2}{2} \tag{2.43}$$

A beam starting from the convex mirror and crossing the axis at the focus is reflected by the concave mirror in a horizontal direction, due to the common focus point of both mirrors. If the initial radius of the beam, that is, the distance from the axis, is not too large, then the beam reaches the convex mirror and is again reflected as a beam going through the focus.

After round trips, the wave running from the convex mirror to the concave mirror is spherical with the common focus point as its origin.

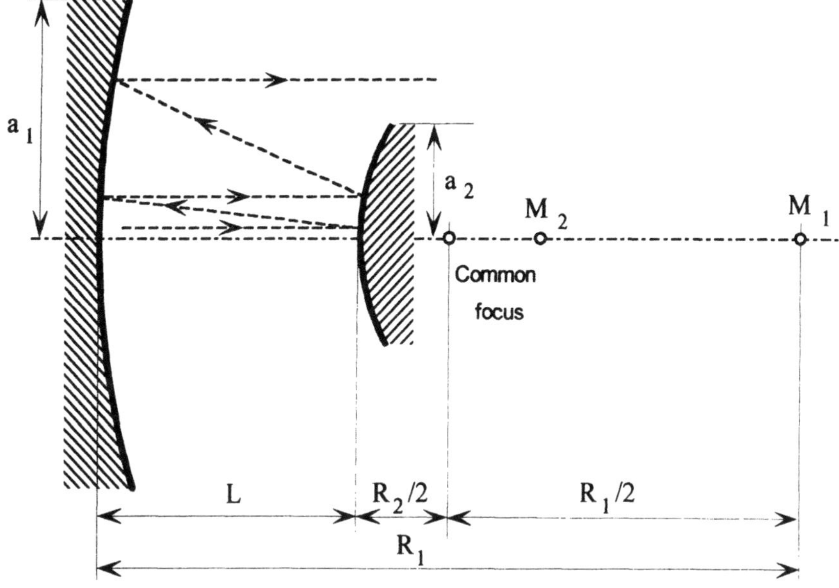

Fig. 2.11 Confocal unstable resonator.

Of course, the wavefront of this wave neither fits the surface of the convex mirror, nor that of the concave mirror, a situation that is entirely different from the case of the stable hemispherical confocal resonator, where the wavefront fits to the surface of the two mirrors. Since all beams running back from the concave mirror are parallel to the axis, the wave that propagates towards the convex mirror is a plane wave. This behaviour corresponds precisely to that mentioned above on the basis of the real eigenvalues of the beam matrix, that is larger than one. Since the convex mirror has a radius a_2 that is smaller than that of the concave mirror a_1, a fraction of the plane wave generated by the larger concave mirror leaves the resonator, thus generating a ring shaped beam. Considering that the radius of the beam obtained at the larger and concave mirror depends on the radii of the two mirrors and their separation, it becomes obvious that these quantities also determine the outer radius of the ring shaped beam produced by the resonator. A straightforward calculation shows that the magnification of the beam at the second concave mirror, which passes through the focus and starts at the convex mirror with radius r_2, is given by the following expression, assuming a paraxial beam and slight curvature of the mirror:

$$r_1 = \frac{R_1}{R_2} r_2 \tag{2.44}$$

This magnification determines the transmission of the output beam across the edge of the smaller convex mirror and is most important for the unstable resonator. Simple considerations yield a relation between output power and total power given by the ratio relation of output beam cross section and total cross section:

$$T = \frac{\left(\dfrac{R_1}{R_2}\right)^2 - 1}{\left(\dfrac{R_1}{R_2}\right)^2} = \frac{R_1^2 - R_2^2}{R_1^2} \tag{2.45}$$

Since the resonator basically shows a steady loss during each round trip, it is impossible to maintain a stable wave without amplification, which justifies the name of the resonator. This is also the major advantage of this kind of resonators, since they can extract some part of the beam power present inside the mirror without any transmissive optics, which are associated with severe heating problems due to residual absorption in the case of high power beams. Unstable resonators do not need any transmissive optics and can therefore be made from metal mirrors with 100% reflectivity and which can be cooled in a very efficient way by flowing water through thin capillaries. In practice, unstable resonators are very often designed with a so called **scraper mirror**, a ring shaped mirror oriented at an angle to the beam axis, consisting of two or more mirrors, leaving an inner part of the beam undisturbed while reflecting an outer ring of the beam.

A further advantage of the unstable resonator is that the mode volume is usually wider than in the case of stable resonators, that are in general designed to be as narrow as possible to yield a low Fresnel number and thus produce favourably a fundamental mode. This wider mode volume of the unstable resonators fits much more to the bulky active medium of the usual high power lasers such as CO_2 lasers. It is a disadvantage of the unstable resonator, that the field distribution of the beam extracted shows a sharp rise at the inner radius and a maximum near to the latter followed by a monotonic decrease in radial direction.

This complicated field distribution in the output plane is composed mainly from the toroidal gaussian mode obtained by a shift in radial direction, but also from a large number of higher modes. Although the ideal ring mode contained in the field distribution can be focused to a gaussian beam in the focus and also has a low divergence, the higher modes contained in the beam disturb the gaussian mode and the focus and lead to a higher divergence.

A detailed analysis shows that the beam parameter product depends on the magnification of the unstable resonator and is always larger than for a comparable stable resonator. The M^2 parameter which is defined as the value $(\pi/\lambda)w_0\Theta_0$ is shown in Fig. 2.12 as a function of the magnification of the unstable resonator. The smallest possible value of $M^2 = 1$ can be derived from equation (2.16) and is obtained for a gaussian beam.

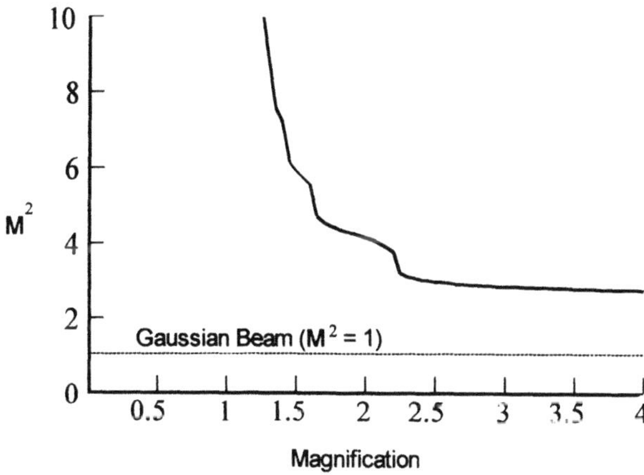

Fig. 2.12 M^2 of an unstable resonator as function of the magnification.

2.10 CHARACTERISATION OF THE BEAM QUALITY

To ascertain the parameters necessary to specify a laser beam, a gaussian beam is considered first. Since the gaussian beam represents the ideal case with respect to the beam parameter product, as defined by equation (2.16), it is permissible to define the beam quality with respect to this case.

Considering a gaussian beam, it can be concluded that the position of the beam waist, the beam radius w_0 at the waist, the divergence angle Θ and the orientation of the optical axis are sufficient to specify the beam. The position of the beam waist is defined by the position in the direction of propagation z_0 and the transverse coordinates x_0 and y_0 of the point of maximum intensity.

To evaluate these numbers they must be derived from quantities which can be found by measurement. The best method is the measurement of overall power or energy. When small detectors or apertures are used, the power per unit area[A], termed power density or intensity, or the energy per unit area, respectively, can be measured as well. Using more sophisticated measuring systems the transverse power density distribution $I(x,y)$ can be scanned automatically. It must be pointed out in this context that a complete characterisation of the laser beam would require also the measurement of the phase distribution, which is a very complicated procedure. Instead of the phase distribution, the intensity distribution is usually measured at several distances from the beam waist which allows in many cases an approximate description of the beam propagation. A more complicated situation must be faced, however, when the beams of unstable resonators shall be characterised. In this case the parameters for beam quality characterisation are only of restricted value.

Assuming that a system for measuring the transverse power density distribution is available, a procedure to derive the parameters mentioned above must be found. Instead of measuring the beam waist directly, the position of the beam and the beam radius must be evaluated at different distances until the beam waist position and radius can be calculated from

---

[A] Unfortunately in the literature there is no uniform nomenclature for the radiometric quantities so the different terms can be quite confusing. The power per unit area at the detector surface which is displayed by the power meter is actually the irradiance. In the field of laser physics, however, this quantity is often called 'intensity' or 'power density'. Since 'power density' is the standardised term it should be preferred, although it is not very consistent, as 'density' usually is related to the unit volume but not to the unit area. In addition in many publications the term 'intensity' is also used for the radiant power per solid angle. In this chapter the terms 'power density' and 'intensity' shall be used synonymously for the irradiance at the detector surface.

equation (2.12). It must be considered that this procedure only makes sense for beams from stable resonators, beams from unstable resonators must not be treated in this way. The beam position and the beam radius in a given plane, however, are very general quantities and can be derived independently of the resonator properties.

Lasers which are used in material processing are usually designed in order to produce a symmetric beam with respect to the optical axis. For this reason in most cases it is sufficient to characterise the beam by one beam waist and one beam radius. In a more general case, however, it is possible that the beam behaves differently in the transverse directions (astigmatism). In this situation the beam characterisation gets more complicated (ISO/DIS 11146, 1995) In the following, beams with the same properties in all transverse directions will be considered exclusively.

2.10.1 Position of the beam

To define the position of a laser beam, the centre of the transverse intensity distribution must be found. The co ordinates x_0, y_0 of this point are obtained by calculating the centre of gravity or the first moments of the power density distribution function, respectively:

$$x_0 = \frac{\iint x I(x,y) \, dx \, dy}{\iint I(x,y) \, dx \, dy} \tag{2.46}$$

$$y_0 = \frac{\iint y I(x,y) \, dx \, dy}{\iint I(x,y) \, dx \, dy} \tag{2.47}$$

This definition is of course trivial for symmetric distributions, e.g. for a gaussian beam, where the centre of gravity and the position of the maximum power density are identical. For real world laser beams, however, it is possible that the beam position deviates from the point of maximum power density. In this situation it is important to use the above definition in order to get reliable results for the beam radius and the beam divergence.

2.10.2 Beam radius

The beam radius w of the gaussian beam as introduced by equation (2.9) is at first only a scaling factor without any physical meaning. At a radius of $r = w$ the electric field has reduced to a multiple of $e^{-1} = 0.37$ of the maximum field strength E_{00} at the centre. Since the electric field cannot be measured directly it is not of great relevance. Because of the importance of powers and power densities the power included within the circle $r = w$ related to the total power shall be estimated next. For this calculation it must be considered that the power density is proportional to the square of the electric field. Additionally the integration over the area is simplified by substituting r^2 for the expression (x^2+y^2) in equation (2.9) and by replacing the area element $dxdy$ by $2\pi r dr$. This ratio v can then be calculated:

$$v = \frac{\int_0^w E_{00}^2 e^{-\frac{2r^2}{w^2}} 2\pi r dr}{\int_0^\infty E_{00}^2 e^{-\frac{2r^2}{w^2}} 2\pi r dr} = 1 - e^{-2} = 0.865 \tag{2.48}$$

This simple calculation yields the important result that 86.5% of the power of the laser beam is located within the beam radius w. This fact is generalised and used as the definition for the beam radius of real laser beams. For a given power density distribution function the beam position must be calculated first. Afterwards, the beam radius $w_{86.5}$ can be calculated by looking for the integration boundary within which 86.5 % of the total beam power is located.

In addition to this very practical approach to evaluate the beam radius there is also a second definition which uses the second moment of the energy density distribution. The beam radius owing this definition can therefore be understood as the standard deviation of the power density distribution function from the centre point.

In order to get a unique beam radius for a gaussian beam the second moment is normalised in the following way:

$$w = \sqrt{\bar{r}^2} = \left[2\frac{\int_0^\infty r^2 I(r)2\pi r\,dr}{\int_0^\infty I(r)2\pi r\,dr} \right]^{\frac{1}{2}} \tag{2.49}$$

For beams without symmetry properties two beam diameters in orthogonal directions must be evaluated in a similar way (ISO/DIS 11146, 1995).

Although the beam radius $w_{86.5}$, which is defined on basis of the beam power, is easier to interpret from the laser user's point of view, the second definition equation (2.49) is standardised and should therefore be used. To get an understanding of the difference between the beam radii according to these two definitions, some examples are given in Table 2.1

Table 2.1 Beam radii for different beam modes according to the two definitions

Mode	$w_{86.5}$ [mm]	$\sqrt{\bar{r}^2}$ [mm]	Difference %
TEM$_{00}$	5.000	5.000	0.0
TEM$_{10}$ (polar)	5.000	5.266	5.0
TEM$_{01}$*	5.000	5.341	6.4
Ring aperture $\dfrac{r_1}{r_a} = \dfrac{1}{3}$	5.000	5.619	10.0

According to the definition, the beam radii $w_{86.5}$ and $\sqrt{\bar{r}^2}$ are identical for gaussian beams. Consequently the difference between the two beam radii is small also for gaussian-like beams. For a TEM$_{10}$ mode (Fig. 2.6) and a TEM$_{01}$* mode (Fig. 2.5) the error amounts to approximately 5%, which can still be tolerated as measuring errors within this range must be expected too. When the beam produced by a ring aperture is considered, which is produced by an unstable resonator according to Fig. 2.11, the difference already exceeds 10%. This fact indicates that the characterisation of beam quality relates in the first place to beams produced by stable resonators. Beams from unstable resonators behave in a more complicated way and the use of beam radius and other quantities is more restricted than for gaussian like beams.

2.10.3 Beam divergence

A second important quantity to characterise a laser beam is the beam divergence. As shown in detail for the gaussian beam, at distances larger than the Rayleigh length a linear increase of the beam radius with distance can be observed. When a real beam is considered, the beam radius $w_{86.5}$ (or $\sqrt{\bar{r}^2}$) must be measured for at least two distances from the beam waist well above the Rayleigh length (far field). The beam divergence Θ can then be calculated according to:

$$\Theta = \frac{w(z_2) - w(z_1)}{z_2 - z_1} \tag{2.50}$$

For more accurate measurements the beam radius should be evaluated for more than two distances, which provides the possibility to make a linear regression and thus to check the linearity.

Finally the condition for the far field measurement shall be described more clearly. From equation (2.12) the relation $z \gg z_R$ can be obtained for the regime where the divergence of the beam is defined. This condition can be rewritten as:

$$z \gg z_R = \frac{\pi w_0^2}{\lambda} \quad \text{or} \quad \frac{w_0^2}{\lambda z} \ll \frac{1}{\pi} \approx 1 \tag{2.51}$$

This equation introduces the dimensionless number

$$N = \frac{w_0^2}{\lambda z}$$

the so called Fresnel number, which always appears when diffraction phenomena are described. A Fresnel number less than one guarantees that the far field is reached and that the beam divergence can be observed.

For the beams of unstable resonators this number is of great importance too. In this case the beam shows in the near field ($N > 1$) a strong change of the transverse power density distribution with distance. In the far field, on the contrary, the power density distribution becomes independent of distance, except for a scaling factor which increases linearly with distance according to a constant beam divergence.

2.10.4 Beam parameter product

The importance of the product $w_0\Theta$ has already been outlined. This product reaches its minimum λ/π for the gaussian beam and is always greater for real laser beams. Since both the divergence during free space propagation and the minimum focal spot diameter are affected by this number, it is one of the main requirements for laser source development to reduce this product. A fundamental way to reduce the beam parameter product is the selection of a short laser wavelength since the theoretical minimum is proportional to the wavelength λ. From this point of view the Nd:YAG laser ($\lambda = 1.06$ μm) should provide a ten times smaller beam parameter product than the CO_2 laser ($\lambda = 10.6$ μm). Due to technical problems (thermal lensing, fibre coupling, etc.), however, it is much more difficult to reach the theoretical minimum of this number with the Nd:YAG laser than with the CO_2 laser. For this reason the CO_2 laser shows usually (especially high power lasers) a smaller beam parameter product than the Nd:YAG laser.

2.10.5 K number and M^2

In order to specify the beam quality with respect to the theoretical minimum, the K number and the M^2 are defined, respectively:

$$\frac{1}{K} = M^2 = \frac{\pi}{\lambda} w_0 \Theta \tag{2.52}$$

According to equation (2.16) this expression equals 1 for gaussian beams. For real beams it must get greater than one, meaning that K is always less than one and M^2 is always greater than 1 respectively.

The increase of the M^2 value or the reduction of K has a strong effect on beam properties which influence processing results with lasers. When focusing action is considered, the beam divergence behind the lens is obtained by the ratio arctan$(w/f) \approx (w/f)$, where w means the beam radius at the lens and f the focal length. With the corresponding beam waist radius or focal spot radius, respectively, it is obtained:

$$\frac{1}{K} = w_0 \frac{w}{f} \frac{\pi}{\lambda} \quad or \quad w_0 = \frac{1}{K} \frac{\lambda f}{\pi w} \tag{2.53}$$

This equation yields the important result that the focal spot diameter increases linearly with $1/K$ or M^2, respectively. It must be mentioned, however, that this relation yields a very optimistic result for the focal spot radius, since real optical systems usually cause an additional decrease of the beam quality.

From equation (2.53) the focal spot area increases with the square of 1/K and thus the power density in the focal spot is proportional to K^2. Since many laser processes depend sensitively on the power density a high beam quality, meaning a K factor close to 1 is essential. As the power density depends on the square of the K factor but only linearly on the beam power it often is more advantageous to increase the K factor than the beam power.

2.11 REFERENCES

Kogelnik, H. (1965) *Imaging of Optical Modes - Resonators with Internal Lenses*, Bell Syst. Tech., **Vol. 44**, No. 3, pp. 455-494

Hodgson, N. and Weber, H. (1992) *Optische Resonatoren*, Springer.

Optics and optical instruments - Lasers and laser related equipment - Test methods for laser beam parameters: Beam widths, divergence angle and beam propagation factor, **ISO/DIS** 11146 (1995).

3

Carbon dioxide lasers

D. Schuöcker

3.1 INTRODUCTION

The efficient use of high power lasers for material processing depends on several properties of the laser beam. First of all the intensity must reach at least a million Watts per cm^2 to achieve secure melting and evaporation of the workpiece, which are both necessary for nearly all production processes with lasers such as cutting, ablation, welding and coating. Secondly, the beam waist must have a diameter as small as a few tenths of a millimetre and the extension of the focused region, the so called Rayleigh length, must be at least several millimetres long to be able to obtain narrow processing paths and to achieve a reasonable depth of processing. To obtain this narrow beam waist and a low divergence, the radial beam mode must come as close as possible to the fundamental Gaussian distribution. Thirdly, the wavelength of the laser beam must be matched to the absorption properties of the workpiece material to avoid high reflectivity and to achieve an efficient production process. Therefore, for the treatment of metals, wavelengths up to one micrometre (μm) are most appropriate. Concerning the treatment of plastics, glass and similar materials, the wavelength of 10 μm is excellently suited. Finally, the overall power of the laser beam must be as large as possible, since for the treatment of thick workpieces the beam must be focused to a low divergence, what means in turn, that the beam waist diameter becomes rather large and therefore the necessary intensity mentioned above requires a high beam power.

The above requirements are mainly met by the CO_2 lasers invented by Patel (1964), which use a mixture of CO_2, N_2 and He gases for the conversion of electric energy, which is usually supplied by DC or RF currents, to infrared radiation with a wavelength of 10.6 μm. These lasers combine excellent beam mode with nearly unlimited beam power, in which context it should be mentioned that lasers with a CW power of 45 kW are currently available on the market. The reason for this high power capability of the CO_2 laser is on the one hand its relatively large efficiency of theoretically more than 30% for the conversion of electric energy to radiation power and on the other hand the possibility of a very efficient cooling by removing the hot CO_2 gas mixture from the electrode system, where it is heated, cooling it by heat exchangers and recirculating it to the electrode system where it converts electric energy to radiation.

CO_2 lasers can be operated in the fundamental mode or in a doughnut mode, which is also very appropriate for material processing as shown in Chapter 2. It should be mentioned that CO_2 lasers are nowadays the most commonly used lasers for industrial material processing and have become rugged and reliable machines that are very well suited for the operation in the rough environment of a workshop. It can be estimated that more than ten thousand CO_2 lasers with a beam power above 1 kW are employed around the world. So these lasers have a major importance for production technology and it is assumed that they will keep this position for many years.

3.2 EXCITATION, EMISSION AND RELAXATION MECHANISMS

3.2.1 The CO_2 molecule and its vibrations

The conversion of electric energy into radiation is performed by the CO_2 molecule (Witteman, 1987). It consists of a central carbon atom that is tied to two oxygen atoms, one on each side of the carbon, see Fig. 3.1. The three atoms are held together by elastic forces and can therefore carry out oscillations around their equilibrium positions. The first main possibility is transverse oscillation, where the carbon atom oscillates perpendicularly to the main axis of the molecule. The latter oscillations are called bending oscillations. A certain energy is associated with these oscillations, in which the lowest possible energy is a 0 oscillation and the higher energies are referred to as 1 or 2 and so on.

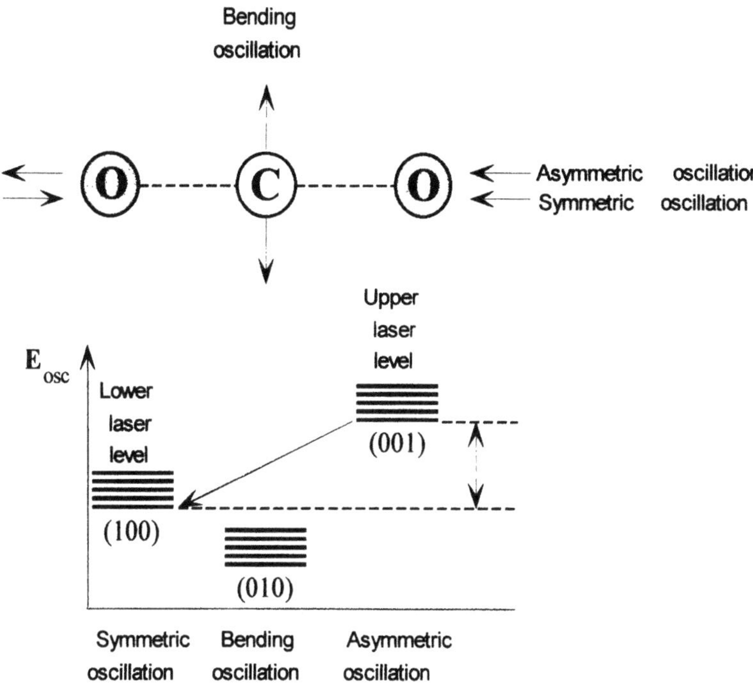

Fig. 3.1 Vibrational-rotational states of the CO_2 molecule.

A second main possibility are longitudinal oscillations, where the carbon atom remains in its equilibrium position and the two oxygen atoms move either to opposite sides, called a symmetric longitudinal or stretching oscillation, or to one and the same side, called an asymmetric longitudinal oscillation. These two kinds of oscillations are quantised and according to their energy content are referred to as 0, 1 or 2 and so on. All three kinds of oscillations can take place at the same time and therefore the complete oscillation status is described by a triplet of numbers, the first accounting for the symmetric stretching oscillation, the second for the bending oscillation and the third for the asymmetric oscillation. For example an 001 oscillation is solely asymmetric with lowest energy, used as the upper laser level. A 100 oscillation is the lowest energy symmetric oscillation and is used as the lower level of the CO_2 laser, see Fig. 3.2.

It should be mentioned that the lifetime of the lowest energy asymmetric oscillation is somewhat higher than that of the symmetric oscillation, so that the CO_2 molecule itself is well suited for laser operation.

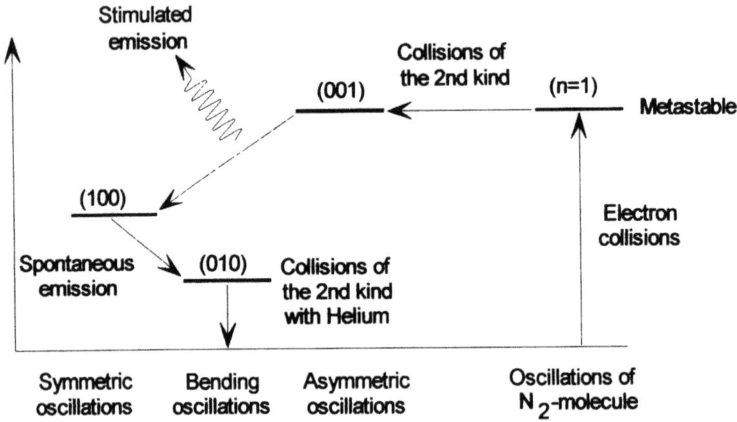

Fig. 3.2 Transition scheme of the CO_2 laser.

3.2.2 Selective excitation of the upper laser level

The building up of inversion between the upper and the lower laser level of the CO_2 laser can be strongly enhanced if selective excitation of the upper laser level is achieved. This is usually achieved by the use of an auxiliary gas, namely Nitrogen. The molecules of this gas consist of two atoms that can carry out longitudinal vibrations, in which the lowest energy state shows an energy that is very close to the energy of the upper laser level of the CO_2 laser and is nearly metastable (Fig. 3.2). So if a mixture of Nitrogen and CO_2 is subject to a gas discharge with many free electrons moving with high speed, Nitrogen molecules are excited to their first vibrational energy and remain there for a very long time and are therefore acting as energy accumulators. If the excited Nitrogen molecules collide with CO_2 molecules in the ground state, that is the latter do not carry out any molecular vibrations, then the excitation energy of the Nitrogen vibration is transferred to the CO_2 molecule by a collision of the second kind as mentioned in Chapter 1, thus exciting CO_2 to the upper laser level and leaving the Nitrogen molecule without vibrational energy. So a selective excitation of the upper laser level of CO_2 takes place enhancing inversion considerably.

3.2.3. Light emission and relaxation of the CO_2 molecule

If the CO_2 molecule that occupies the upper laser level is hit by a photon with a wavelength of 10.6 µm corresponding to a quantum energy of roughly 0.1 eV, which is the energy difference between the upper and the lower laser levels, stimulated emission of a second photon takes place and the CO_2 molecule changes to the lower laser level, the symmetric oscillation 100. The molecule then shows a fast relaxation to the bending level 010. The latter level is unfortunately nearly metastable and therefore relaxation to the ground state of the molecule is difficult. Therefore, the bending vibration is called the **bottleneck** of the CO_2 laser. A solution of the problem is the use of a second auxiliary gas, namely Helium (Fig. 3.2), which shows a very high heat conductivity. It can thus absorb very effectively the excitation energy of CO_2 molecules and subsequently carries this energy to the walls of the laser or to the heat exchanger, where this energy is removed. The CO_2 laser molecule is then left in the ground state and can take part in the excitation and emission process again. The bending vibration with lowest energy leads to a second problem of CO_2 lasers. Its energy is very close to zero, and can therefore be excited by thermal collisions if the temperature of the gas mixture exceeds 100°C. In this case no relaxation of the lower laser level can be obtained and therefore inversion breaks down due to a high thermal occupation of the lower laser level. Therefore a temperature above 100°C must strictly be avoided to achieve a high inversion and optical amplification. It is the main problem of high power CO_2 lasers to keep the gas temperature below this limit, and therefore a lot of effort must be made to cool down the laser gas. This is usually achieved by a gas transport system that moves the hot gas away from the electrodes where it is heated, and circulates it through a water cooled heat exchanger before returning it to the electrodes with a sufficiently low temperature.

3.2.4 Fine structure of the CO_2 laser spectrum due to molecular rotation

The CO_2 laser molecule shows three axes of symmetry (Fig. 3.1), one that connects the centres of the three atoms and two others perpendicular to these. Therefore, the CO_2 molecule can carry out rotations around these axes. These three rotations around the axes of symmetry are also associated with a certain kinetic energy, which is quantised. The two rotations around the axis perpendicular to that connecting the centres of the atoms are degenerate, meaning they have the same energy. The typical difference of the energy levels is 0.01 eV and is thus ten times smaller than the energy of the vibrations. So the combined vibrational rotational states of the molecule show energy levels roughly given by the vibrations, but having a fine structure due to the rotational levels, which lead to a certain spreading of the main energy levels, thus yielding laser wavelengths between roughly 9 and 11 μm, corresponding to certain pairs of vibrational rotational states. It should be mentioned that due to the influence of a rising pressure on the line width of the energy levels, a high gas pressure leads to an overlapping of the various vibrational rotational energy levels, thereby forming a continuous energy band that extends from 9 to 11 μm allowing continuous tuning of the laser frequency (Witteman, 1987).

Due to the above mentioned excitation, emission and relaxation processes of the laser, an appropriate mixture of usually 10% CO_2, 10% Nitrogen and finally 80% Helium with a total pressure of around one tenth of an Atmosphere, in which a glow discharge with a large volume due to a low pressure and appropriately shaped electrodes is maintained, leads automatically to the generation of strong inversion which can be utilised for the amplification of infrared radiation with a´wavelength around 10 μm. In the next section the influence of various parameters such as gas pressure and temperature, mixture, electric field strength, and current will be discussed in detail, showing optimum parameter choices for laser operation.

3.3 PARAMETRIC BEHAVIOUR OF THE CO_2 LASER

3.3.1 Gas pressure and electric field strength

First the influence of total gas pressure and field strength will be discussed, since the ratio of these quantities is of main importance for reliable operation of a CO_2 laser. The energy gain by an electron in the glow discharge of a CO_2 laser between two collisions with a gas particle is given by the product of the mean free path λ_g and the field strength F times the elementary charge e as explained in Chapter 1 section 1.6.

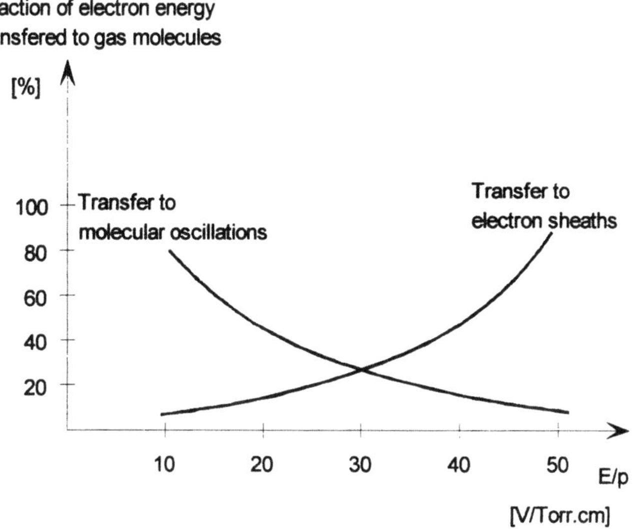

Fig. 3.3 Influence of electric field and gas pressure on the performance of a CO$_2$ laser.

Since λ_g rises with decreasing gas pressure p, the quotient F/p determines the kinetic energy of the electron. If this energy is near to the ionisation potential of the atoms, many of the electron-atom collisions will lead to ionisation and therefore a large part of the electron energy is used for the latter process and only a low part remains for excitation of Nitrogen molecules with their much lower excitation energy. So the number of ionisation processes will strongly outnumber excitation processes (Fig. 3.3).

The gain of the laser will then be rather low. If the quantity F/p is lower, then many more excitations of the Nitrogen molecules will take place and only a few ionisation events will appear, which means the necessary regeneration of free electrons will become too weak to maintain the electron density against losses by recombination and therefore the gain will be rather low due to a reduced number of free electrons (Witteman, 1987). Finally an optimum situation with high electron number and strong excitation of the Nitrogen molecules and thus strong inversion and optical gain will appear if F/p shows a medium value in the order of magnitude of 10 V/cmTorr. The latter quantity has the most important influence on the efficient operation of a CO$_2$ laser. With an optimum value for F/p, typical values for the small signal gain and the saturation intensity for $p = 100$ mbar are $g_0 = 0.8\%/\text{cm}$ and $I_S = 500$ W/cm^2 (Breining and Pfeiffer, 1994).

3.3.2 Influence of gas pressure

A high gas pressure of course leads to a high number of gas particles that take part in the light amplification process, and therefore a high output power per unit volume will be obtained. A good example is the so called TEA laser (Walter, 1986) that operates typically at atmospheric pressure and is able to yield a pulse beam power of several Megawatts from a volume of some 10 cm^3. A high gas pressure allows continuous tuning which is interesting for applications in spectroscopy. Nevertheless, a high gas pressure leads to stability problems of the glow discharge. Usually, a glow discharge can only be maintained up to a pressure of a tenth of an atmosphere, since a higher pressure leads to a higher current density, as more charge carriers are available per unit volume and therefore the heating per unit volume becomes stronger and also the heating of the electrodes is enhanced. Therefore the emission mechanism at the cathode and the ionisation mechanism in the bulk of the discharge change to more efficient mechanisms such as thermo field emission and thermal ionisation, thus increasing the current density even more. This process starts in a random way along a certain path through the discharge and leads to a constriction of the discharge to the region of initial enhancement of conduction thus leading finally to the transition of the glow discharge to an arc discharge with a very small cross section of the discharge channel and a high temperature. This is entirely inappropriate for CO_2 laser action.

3.3.3 Influence of the gas mixture

The radiation power output per unit volume of the laser plasma depends on the density of the CO_2 molecules. So the more CO_2 is available, the higher is the laser power per unit volume. The same is true for the content of Nitrogen molecules, since at least one Nitrogen molecule is necessary to excite one CO_2 molecule. So usually the percentages of CO_2 and of Nitrogen are nearly equal or at least similar. Nevertheless, the Helium content is also of major importance for the performance of the laser, since it is necessary to carry out a total relaxation of the CO_2 molecules in order to allow a new excitation emission and relaxation sequence. In addition, Helium helps to stabilise the plasma against thermal instabilities as mentioned above due to its excellent heat conduction properties. So the more Helium is contained in the gas mixture the higher is the energy input per unit volume that can be achieved without the onset of instabilities and the higher becomes the inversion and subsequently the optical gain. So both the Helium, CO$_2$ and Nitrogen content should be as high as possible and therefore an experimental optimum has been found with around 70 to 80% Helium with the remainder covering Nitrogen and CO$_2$ (Witteman, 1986).

3.3.4 Gas temperature

The very important influence of the gas temperature has already been discussed in the preceding paragraph. It can only be re-emphasised that the gas temperature must definitely remain below 100°C to avoid an interruption of the excitation, emission and deexcitation cycle due to bottlenecking of the bending energy state.

3.3.5 Voltage and current

The voltage applied to the plasma and the current conducted through the plasma are limited by several conditions. First for the ignition a voltage given by PASCHEN's law (see Chapter 1 section 6) and thus dependent on electrode distance and gas pressure must be applied. After ignition has been completed, usually after a very short time, the voltage across the plasma normally drops down and is then dependent on the current through the plasma, when Ohm's law is more or less obeyed. If a voltage source with a low internal resistance is used, the current through the discharge rises as long as the voltage across the plasma plus the voltage across all resistive elements between the source and the plasma equals the driving source voltage. Finally the voltage remaining across the plasma is determined by the electric properties, especially the resistance, of the whole circuit and must then correspond to the optimum field strength determined by the gas pressure according to the preceding paragraph. The same is true for the current flowing through the discharge, which is also determined by the electric properties of the whole electric circuit, so that a higher current leads to an improved performance of the laser since more electric power is then fed to unit volume of the active medium. Nevertheless, if the current density becomes too large, instabilities, mainly at the electrodes but also in the bulk of the plasma, lead to a transition from the beneficial glow discharge to the unwanted arc discharge. A typical practical example is 10 W/cm^3.

3.4. ENERGISING AND COOLING THE PLASMA

3.4.1 Energy supply

Energy supply to the CO_2 plasma is, with some minor exceptions, usually carried out by electric power, either with DC current, low frequency currents from 50 Hz to a few kHz, RF current up to a frequency of about 100 MHz, or finally with microwaves.

DC current has been used for a long time following the invention of the CO_2 laser, since it is easy to realise, simple to operate and correspondingly rather cheap. Nevertheless, the maximum current density for stable operation of the plasma is rather low, corresponding to only 10 W/cm^3 power input per unit volume of the plasma, thus imposing a relatively low limit for the radiation power output per unit volume of DC excited CO_2 lasers.

In order to improve this number in the late 1970s alternative excitation schemes with low frequency currents were developed such as the Silent laser developed by MITSUBISHI (Tabata, 1984) and with RF currents (Hügel, 1986). In the late 1980s very high frequencies up to about 100 MHz were used especially for lasers with a very low electrode spacing (Opower, 1991). At the same time successful operation of CO_2 lasers with microwave excitation was demonstrated by Uhlenbusch (1992) at the University of Düsseldorf .

The main advantage of the use of alternating currents, especially in the RF domain, is that a direct electric contact between the plasma and the electrodes can be avoided by the use of so called dielectric electrodes, where an electric insulator separates the metallic electrodes from the plasma. In this case the electrode-plasma system acts electrically as two capacitors connected on each side of the plasma (see Fig. 3.4), where the plasma can mainly be regarded as a combination of a resistor and an inductive coil, the latter representing the inertia of the charge carriers in the plasma, that prevents a sudden rise of the current. The two condensers of course stand for the two dielectric electrodes that allow dielectric displacement currents of an alternating nature but inhibit DC current flow. The electric equivalent circuit explained above clearly shows that an alternating current supplied to the system leads to the build up of a certain voltage across the plasma thus providing ignition. If the amplitude is sufficiently high, it allows energising of the plasma and thus its maintenance and excitation of the laser levels.

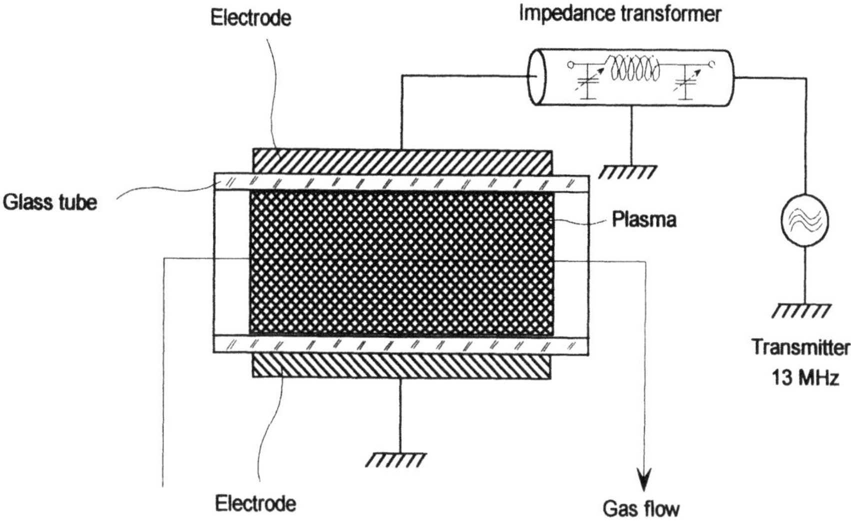

Fig. 3.4 RF excited CO_2 laser plasma.

If the electrode plasma system is considered in more detail, it is clear that surface layers build up between the plasma and the dielectric electrodes, as explained in a preceding section. It can be shown that the thickness of these layers decreases with increasing frequency of the alternating current (Schröder, 1990) and therefore the use of very high frequencies in the range of 100 MHz allows very thin surface layers to be achieved, which is very important if electrodes with a very small separation as described later are used.

It is the main advantage of the use of alternating currents and in consequence of dielectric electrodes, that direct contact between the plasma and the metallic part of the electrodes is avoided. Thermal instabilities due to a changing emission mechanism at the electrodes, that appear if a certain current density is reached, can no longer take place. However a further rise in current density finally leads to thermal instabilities in the bulk of the plasma. Nevertheless, the current density where these instabilities appear is shifted to much higher values and the corresponding power input per unit volume rises to 50 W/cm^3 (Hall and Baker, 1984), so a much higher radiation power per unit volume of the plasma can be obtained with alternating currents. In addition to the much more complicated construction of lasers excited by alternating currents and the correspondingly increased costs, there is also another severe disadvantage.

Particularly in the case of RF currents, radio frequency sources such as transmitters can only deliver their full power to the electrode/plasma system, if the RF impedance due to resistance and reactance (capacitors and inductors) is precisely equal to the resistance of the generator (Walter, 1989). Since the latter impedance is usually resistive, the reactive part of the electrode/plasma impedance must be compensated by a matching network, usually a combination of two capacitors and one inductor (see Fig. 3.4). Besides the fact that the use of a matching network adds an additional complication and is expensive, the electrical properties of the plasma depend on the power supplied to it, and the matching network must therefore be changed if the laser power is altered, leading to severe practical problems. Nevertheless, due to the rising output power of industrial lasers during the recent years, RF excitation has been used more and more and seems to be the only possibility for the construction of ultra high power CO_2 lasers with a beam power of 100 kW, as they are envisaged for the next ten years. Although most of the lasers energised by alternating currents use RF transmitters, low frequencies are also used by industrial laser devices, particularly of Japanese manufacture, where the main advantage is the cheapness of the energy sources and the avoidance of dangerous emissions, that can be a problem if RF currents are used, although a total shielding of the laser and the use of frequencies that are reserved for industrial use (13.56 and 27.12 MHz) have proved reliable and totally safe in operation. The main advantage of the use of microwave energy sources is that cheap microwave generators as magnetrons and klystrons (Skolnik, 1970), are available since they are very commonly used for microwave ovens. Nevertheless, no industrial laser is available up to now, that uses this kind of excitation.

3.4.2 Cooling

Usually in practice only less than 20% of the electric power delivered to the laser plasma is converted to radiation and therefore more than 80% of the energy is lost, leading to heating of the laser gas and the electrodes, and therefore cooling down to a temperature that keeps the gas below 100°C is definitely necessary. Two major possibilities have been developed for heat dissipation in the case of high power lasers.

Diffusion Cooling

The first and most simple solution is the use of water cooled electrodes with a very narrow spacing and a large surface. In this case the laser gas is cooled by heat conduction due to the high temperature gradients between the hot plasma and the cooled surface of the electrode. Since in this case of diffusion cooling the electrode separation is only a few millimetres, the usual extension of the plasma surface layers would be much too large for DC excitation. The surface layers are not appropriate for laser operation due to their low electron temperature which does not allow excitation of the laser levels (see Fig. 3.5). Therefore, this kind of laser can only be excited by RF currents with very high frequencies of about 100 MHz that keep the layers thin enough.

Due to the small electrode separation it is also impossible to use a freely propagating Gaussian beam, since the latter would suffer from very strong losses due to 'friction' with the electrodes. Instead a beam is used that impinges on the electrodes at a very small angle, thus allowing total reflection leading to a zigzag propagation between the electrodes that guide the beam. This is called a waveguide laser (Hall and Baker, 1994). These lasers generate a plasma with a geometry where the separation between the two electrodes is very low compared to the large dimensions parallel to the electrode surfaces. An elliptical beam cross section is best suited for an efficient use of the plasma. Thus, lasers of that kind usually employ resonators with elliptical mirrors, in which the resulting elliptical beam can easily be converted to a rotational symmetric geometry, as it is best suited for focusing and laser material processing. Lasers of this kind have been developed at the DLR in Stuttgart (1988) (Nowack et al, 1991) and are now produced by a German company, where a CW beam power up to 2 kW can be obtained, very well suited for cutting and similar applications.

Convection cooling

A second possibility of cooling is the use of a gas transport system, which moves the hot laser gas away from the electrode system and circulates it through a heat exchanger, usually water cooled, and finally brings it back to the electrode system with a much lower temperature. This gas circulation is provided by mechanical blowers, such as ROOTS blowers (Fig. 3.6), or turbo-blowers (Swysen and Auer, 1990), driven by high frequency motors.

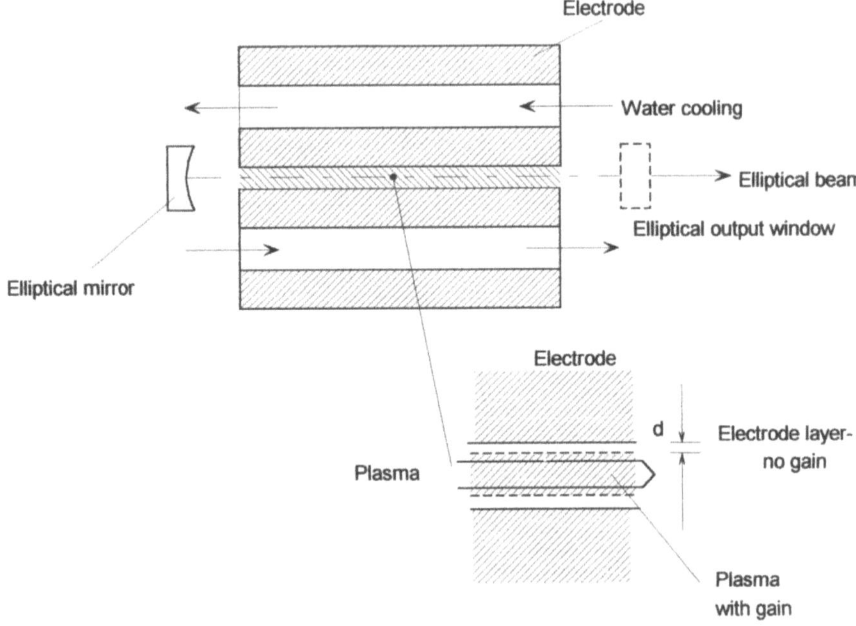

Fig. 3.5 Electrode layer in diffusion cooled laser.

A practical example of one of the strongest turbo-blowers is a system that operates at a frequency of 300 Hz, generating a gas speed of nearly 2000 km per hour, although the gas pressure is only one tenth of an Atmosphere. In this case the flow speed remains just below the speed of sound.

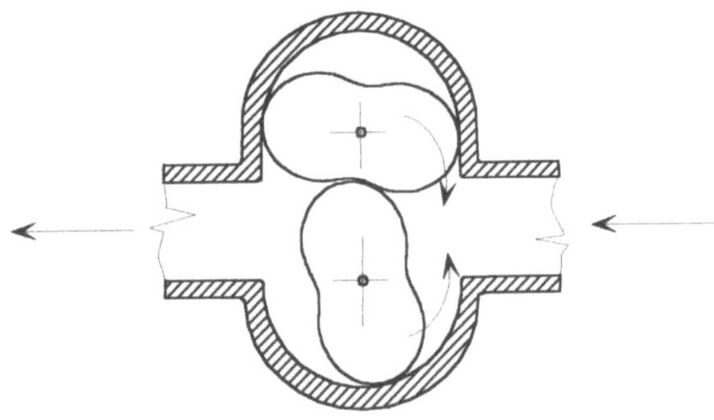

Fig. 3.6 Schematic layout of a blower.

The heat removal capability of a gas transport system that consists of the electrode system, gas ducts, heat exchangers and blowers, depends on the mass flow per unit time, in turn given by the maximum pressure difference handled by the blower, divided by the flow resistance of the system. This flow resistance is limited and determined mainly by the geometrical dimensions of the electrode system. In the case of a narrow and lengthy flow channel through the discharges, a high flow resistance appears that makes efficient cooling a difficult task thus limiting the beam power. This situation occurs for example in longitudinal flow lasers that will be treated below (Witteman, 1987) (Fig. 3.7).

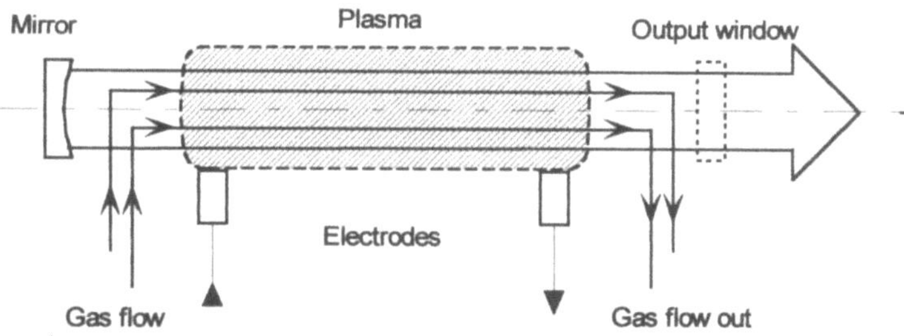

Fig. 3.7 Longitudinal flow CO_2 laser.

In contrast, transverse flow lasers, where the gas crosses the discharge section perpendicularly to the optical axis, show a low flow resistance and thus efficient cooling, which results in a high beam power. These lasers utilise a wide and short flow channel through the electrode system and will be treated somewhat later (Witteman, 1987) (Figs. 3.8 and 3.10).

There is also a significant influence of the laminar or turbulent nature of the gas flow, since in the latter case the flow channel is jammed and therefore the flow resistance becomes rather high, thus inhibiting efficient gas cooling. Nevertheless, turbulent gas flows are very well suited for laser operation, since they lead to a homogenisation of the plasma in terms of temperature, electron density and density of excited particles, due to a more or less random motion of all particles through the plasma.

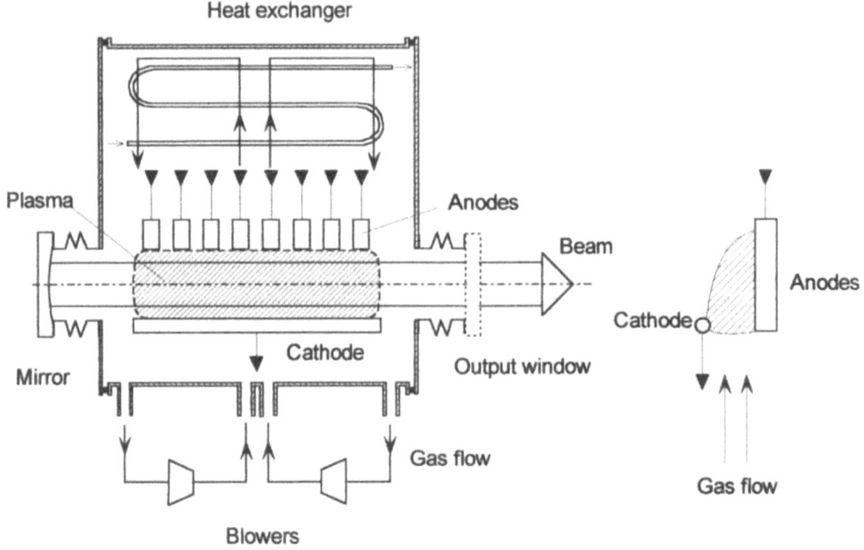

Fig. 3.8 Transverse flow CO_2 laser.

Therefore, the optical gain is homogeneous across the whole plasma volume, which is very well suited for laser amplification, since it avoids a separation of the amplifying volume into parts where maximum beam intensity and in consequence saturation appears, whereas in other parts of the amplifying volume the excited particles remain practically unused, since there is no or only negligible beam intensity.

3.5 MAIN TYPES OF ELECTRODE SYSTEMS AND PLASMA SHAPES, OUTPUT POWER

The most simple plasma geometry is given by the shape of the beam, namely a slim cylindrical shape, where the circular cross section of the plasma is matched to the cross section of usual laser beams. For instance in the fundamental mode, the length is as large as possible to allow beam amplification across a long distance (Witteman, 1987) (Fig. 3.7). A plasma with a shape like this can be ignited and maintained between two ring-shaped electrodes at both ends of the plasma column, where the ignition and operation voltage depends on the distance between them, which limits the length of the discharge as mentioned above. In practice a maximum of 20 to

30 kV is used, thus limiting the discharge length for instance in DC discharges to a few tens of centimetres. To prevent the plasma from diffusing into the surrounding volume, a discharge of this kind is usually contained in a hollow cylindrical glass or ceramic tube. Cooling must then be performed by a fast longitudinal gas flow through the tube and therefore these lasers are called **longitudinal flow lasers**. The flow resistance is comparatively high in this geometry and therefore the radiation output power is somewhat limited, e.g. in the case of DC discharges to a few hundred watts. Due to the total rotational symmetry of the plasma and furthermore to the low Fresnel number (see Chapter 2) associated with it, longitudinal flow lasers can easily be operated in the fundamental mode and generate a very high beam quality, which means the beam is a sharp tool very well suited to materials processing. Nevertheless, the total beam power is limited, and usually therefore several discharge columns with longitudinal flow cooling are optically connected in series to provide sufficient beam power. In practice most lasers (Fig. 3.9) use 8 discharge columns that are arranged in two lines each with four discharges, thus forming a U-shaped beam path, obtained by the use of two bending mirrors in addition to the two resonator mirrors, one 100% reflecting and one partly reflecting.

Especially with this kind of laser utilising RF excitation, a cw beam power up to more than 20 kW has been obtained by German companies. In this case the electrodes are plates extending along the beam axis and forming a capacitor, where the curvature of the electrodes matches to the cylindrical shape of the glass tube that confines the plasma column (see also Fig. 3.4).

A second possibility makes use of electrodes extending along both sides of the plasma column parallel to its axis, where the electrode separation is in principle not much larger than the lateral extension of the plasma and therefore much smaller than in the case of the longitudinal flow laser (Fig. 3.7). This leads to a widely reduced ignition and operation voltage. The gas flow is then conducted through the plasma perpendicular to the axis of the discharge and of the beam and therefore this kind of laser is called **transverse flow** (Witteman, 1987). Obviously, the flow channel through the electrode system is rather wide and short and therefore the flow resistance is quite low, allowing a very efficient cooling of the plasma, leading in turn to nearly unlimited beam power. Lasers of this kind with a length of less than one metre, which can generate a beam power of 8 kW and more, are available on the market. Nevertheless, these lasers suffer from poor beam quality, that means the presence of higher radial modes. This prevents efficient focusing, since the combined action of transverse current flow through the plasma and transverse gas flow leads to a blowing away of the plasma from the main current path (Fig. 3.8), finally resulting in a more or less triangular plasma cross section.

Due to this shape of the plasma, the anode that collects the current delivered by the cathode, must show a considerable extension in parallel to the current flow, because otherwise excited charge carriers would be lost. The anode extends in a direction parallel to the plasma column over a considerable length. It must be divided into small segments to avoid constriction of the current flow and distribute the current evenly over the full length of the plasma column. This complicates the construction of transverse flow lasers. It is by no means obvious that the triangular shape of the plasma cross section is matched to the rotational symmetry and circular cross section of a fundamental mode beam. Complex modes of higher order are therefore usually built up as mentioned above.

Nevertheless, by using circular apertures, rotationally symmetric beams can be forced, but then the beam must be guided zigzag through the plasma to make efficient use of the excited volume. These beam paths are obtained by the use of so-called **folded resonators**. These consist of the two main mirrors, one totally reflecting and one partly reflecting, and several bending mirrors (Fig. 3.10). Although the beam power of the transverse flow laser is very high as mentioned above, its beam quality in terms of the maximum intensity that can be obtained by focusing, is lower than in the case of longitudinal flow lasers. Nevertheless, the most powerful lasers available on the market with a total beam power near 50 kW are built up from several transverse flow modules optically in series.

So, in principle there are two main possibilities for CO_2 lasers. The longitudinal flow laser with its high beam quality but limited beam power and the transverse flow laser with its high beam power but limited beam quality. A further problem is the relatively low efficiency for the conversion of electric energy to beam energy that is theoretically near 40% but practically below 20% and becomes even lower if RF excitation is used.

A compromise between these two competing geometries of high power CO_2 lasers is offered by the coaxial or ring gap laser, that was first realised by Hall (1986). These lasers use hollow cylindrical and coaxial electrodes that are similar to coaxial cables. With these electrodes a hollow cylindrical plasma is formed.

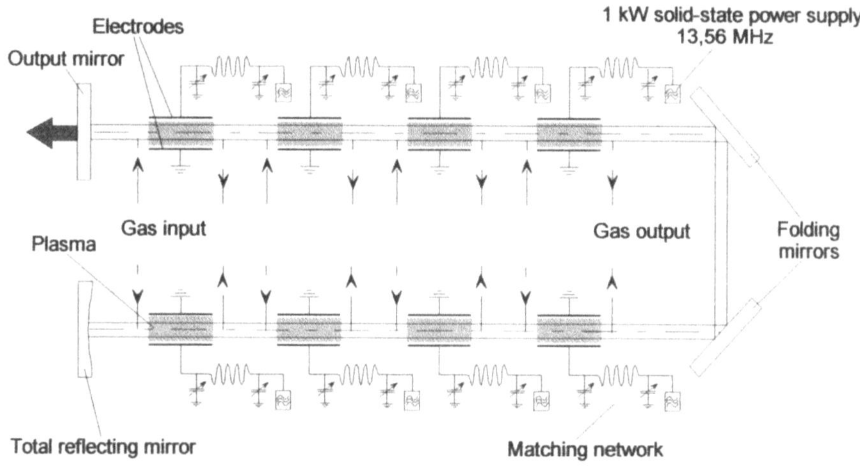

Fig. 3.9 Radio frequency excited CO_2 laser.
(Department of High Power Beam Technology-IST, TU Vienna 1986).

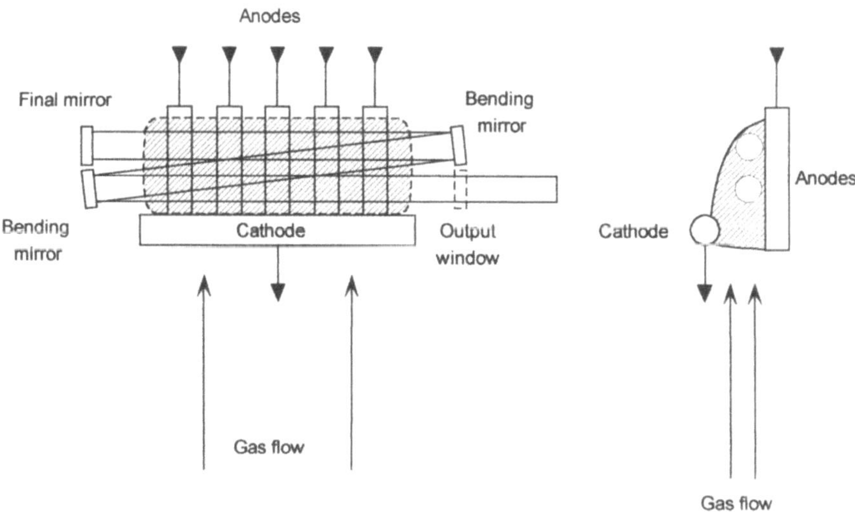

Fig. 3.10 Cross section of the transverse flow CO_2 laser and folded resonator
geometry.

Since the surface of the inner electrode is much smaller than that of the outer one, a considerable constriction of the current takes place at the inner electrode, which would lead to instabilities and arcing in the case of DC excitation. Therefore coaxial lasers can only be realised with dielectric electrodes and in consequence with RF excitation. The main advantage of this electrode geometry is then a total rotational symmetry of the plasma that is by far superior to a conventional cylindrical plasma column of the same outer diameter, since the latter leads to a more or less rectangular shape of the amplifying region if it is energised by condenser plate-like electrodes, that are usually used for RF excitation. This leads to considerable deviation of the beam from rotational symmetry, and can even degenerate to two half moons. Cooling of the coaxial laser for high power is best carried out by a longitudinal gas flow, first proposed by the author (Schuöcker and Schröder, 1994). An optimum situation is obtained if the hot gas flowing out on one side of the electrode system, is reversed and conducted back along the outer side of the electrode, where a heat exchanger is arranged (see Fig. 3.11).

The gas is then cooled down during its way through the heat exchanger and arrives at the other end of the electrode system where it is again reversed and enters the electrode system with an appropriately low temperature. This kind of closed loop gas transport has the important advantage of a very short flow length that reduces the flow resistance and allows a high flow speed with a given blower.

Moreover, due to a comparatively low cross section of the electrode system (Fig. 3.11), the gas flows rather fast where it is heated and due to the much larger cross section of the heat exchanger, the gas flows much slower where it is cooled. Therefore a very efficient energy balance results. In consequence, the fast flow coaxial laser with direct recirculation along the outer electrode is capable of a very high beam power per unit volume.

Theoretical estimations (Schuöcker and Schröder, 1994) show, that with a volume of a few metres length and a diameter well below one metre, a beam power of about 100 kW CW can in principle be obtained. So, the coaxial laser with fast gas flow is superior to all other types of high power CO_2 lasers in terms of beam power and volume. Concerning the resonator, simply two mirrors, namely a totally reflecting ring-shaped mirror and a circular partly transmissive mirror, can be used to obtain a hollow cylindrical beam, that resembles a doughnut mode and shows total rotational symmetry and can thus be focused to a very small focal spot, well suited for the use of material processing.

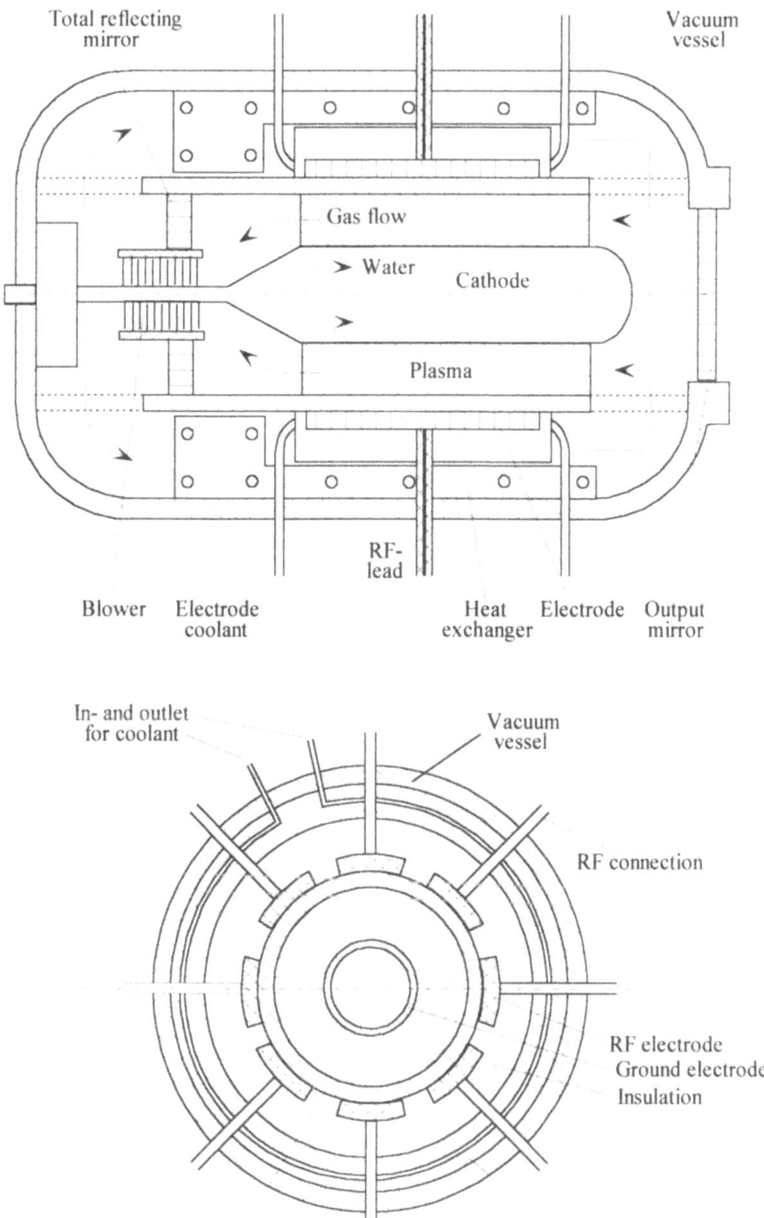

Fig. 3.11 CO_2 laser with coaxial electrodes, RF excitation and fast closed axial flow with integrated heat exchanger (above longitudinal section, below cross section, not to scale).

Of course, there are also other possibilities as for instance a zigzag beam that is generated by a folded resonator, usually yielding lower beam power but even enhanced mode quality.

A fast flow coaxial laser has been developed at the TU Vienna/Austria with a beam power of 6 kW and a beam quality characterised by a spot diameter of 0.5 mm and a maximum intensity above 4×10^6 W/cm^2, that is thus very well suited for deep penetration welding and surface treatment, the main application for lasers with a beam power above 3 kW (Fig. 3.12, 3.13 and 3.14).

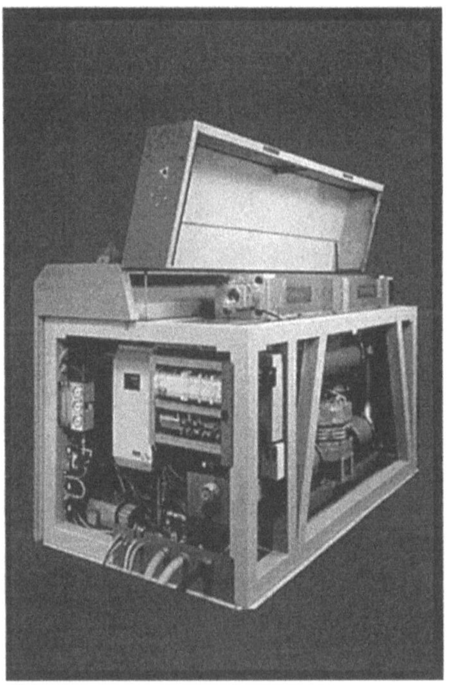

Fig. 3.12 Overall view of a 6 kW coaxial laser: The cover of the laser head is open, right side view into blower and heat exchanger compartment (Department of High Power Beam Technology, Technical University of Vienna 1994).

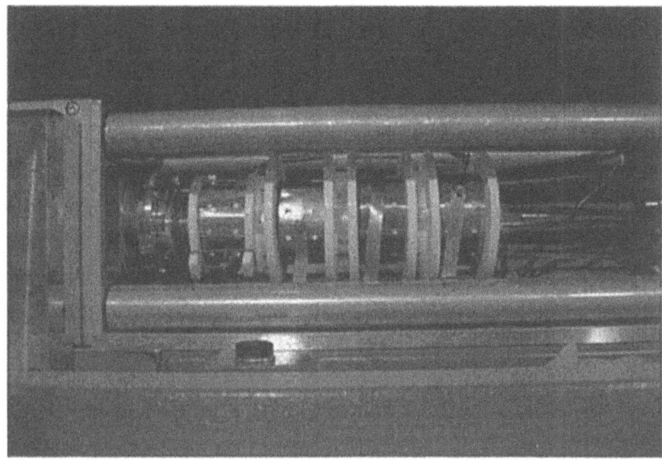

Fig. 3.13   One of two plasma columns of a 6 kW coaxial laser, energised by 32 kW RF power at 13 MHz and cooled with a gas flow with a speed of nearly 2000 km/h (Department of High Power Beam Technology, Technical University of Vienna 1994).

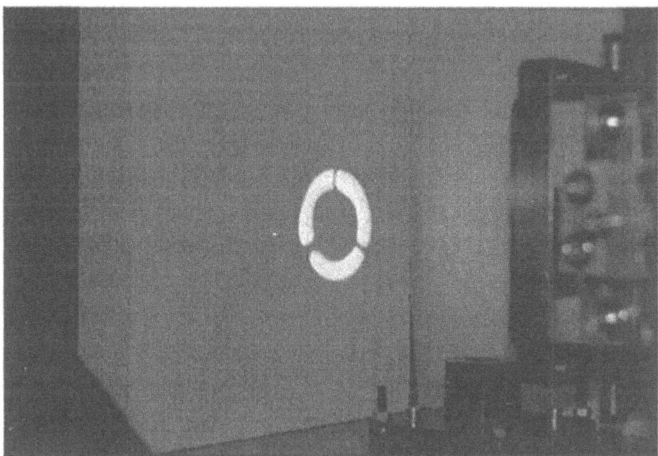

Fig. 3.14   'Burn in' of the beam of the 6 kW coaxial laser at moderate beam power (Department of High Power Beam Technology, Technical University of Vienna 1994).

3.6 OPTICAL RESONATORS AND BEAM QUALITY

The output beam power of CO_2 lasers depends mainly on the electric power input per unit volume. The plasma volume can be much larger in the case of RF excitation than in the case of DC excitation, and optimum use of the excited plasma volume by the laser beam is achieved if the laser beam reaches all parts of the excited plasma. To achieve this optimum use of the excited plasma volume, the optical resonator must provide an appropriate beam configuration with a mode volume that ensures the volume covered by the beam mode comes as close as possible to the excited plasma volume. Moreover, the optical resonator determines the beam geometry and thus the beam mode and quality.

Two main types of resonators as treated in Chapter 2 are used in CO_2 lasers, namely stable resonators for lower and medium power up to several kW and unstable resonators for high beam powers. Stable resonators must employ partly reflective optics as transmissive windows, leading to the limitation of beam power, since only a low beam power avoids an unacceptable heating of the transmissive optics due to residual absorption.

Various construction schemes are used for stable resonators such as the U-shaped resonators, consisting of a spherical totally reflecting mirror, two folding mirrors and a final partly transmissive plane window, usually used in fast longitudinal flow lasers. These lasers usually consist of an optical series connection of several plasma modules, and are thus narrow and lengthy, which leads to a low Fresnel number and preferable generation of a fundamental mode (see also Fig. 3.7). Due to the low Fresnel number a very high beam quality in terms of the participation of the fundamental mode is obtained.

As an example, Fig. 3.15 shows the intensity profile in the focal plane of a longitudinal flow laser at a beam power of 1 kW, using the U-shaped resonator as described above. The intensity profile comes relatively close to a Gaussian mode and shows a beam quality number (see Chapter 2) of $K = 0.5$, which means it is very well suited for cutting, since it has a small focus diameter that yields the desired narrow kerf and a high concentration of power which yields a high processing speed. It also exhibits a low divergence, allowing it to cut even relatively thick materials.

For transverse flow lasers, so called folded resonators consisting of a spherical totally reflecting mirror, several bending mirrors and a final partly transmissive plane mirror, are used to generate a beam that runs several times zigzag through the plasma (see Fig. 3.10).

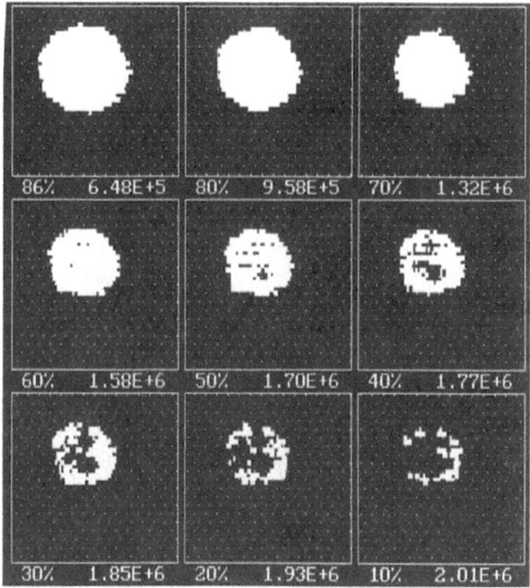

Fig. 3.15 Intensity profile in the focal plane (f = 125 mm) of an DC excited longitudinal flow laser with a stable resonator at a beam power of 1 kW (window size: 0.5 x 0.5 mm^2).

Due to the absence of rotational symmetry, the beam quality is considerably lower, meaning the fundamental mode is accompanied by higher order radial modes. As an example, Fig. 3.16 shows the intensity profile in the focal plane of a transverse flow laser with a beam power of 3 kW, using a folded resonator as described above. The intensity profile is far away from a Gaussian mode and has a beam quality number of K = 0.1, meaning it is not very well suited for cutting, but has proved very suitable for welding and surface treatment. For welding in particular, a small focus size and a low divergence are not the most important aspects, but rather a high peak intensity, that allows the onset of evaporation, ultimately necessary for a deep penetration of the beam into the depth of the workpiece through a vaporised channel, and a high total power, that determines welding depth and speed.

As mentioned above elliptical optics are used in the case of the diffusion cooled laser.

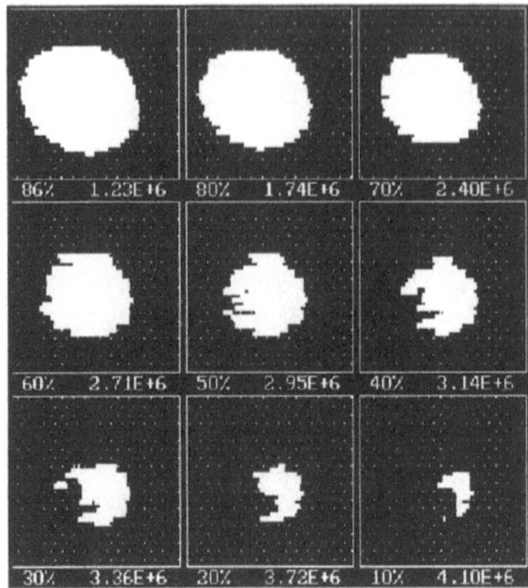

Fig. 3.16 Intensity profile in the focal plane (f =125 mm) of a DC excited transverse flow laser with a stable resonator and a beam power of 3 kW (window size: 0.5 x 0.5 mm^2).

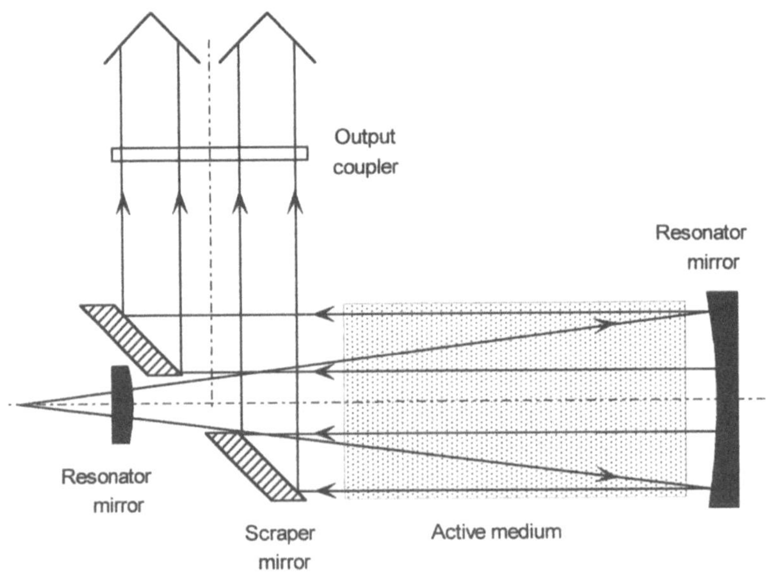

Fig. 3. 17 Unstable resonator with scraper mirror.

For the generation of high and ultra-high beam powers up to several tens of kilowatts, unstable resonators are used, since they do not need transmissive optics that would be heated too much, but can be constructed solely with metallic mirrors.

The usual design for CO_2 lasers is two 100% reflecting end mirrors, one spherical and one plane, at the ends of the beam path in the resonator, accompanied by two folding mirrors, in a U-shaped beam path.

Beam coupling is performed by a ring-shaped mirror (Fig. 8.17), inclined towards the axis of the beam and reflecting a certain part of the beam power in perpendicular to the resonator axis, allowing the extraction of internal beam power. The ring-shaped mirror is called a **scraper mirror** and is widely used in high power lasers. These resonators generate hollow cylindrical beams with a ring shaped cross section, that comes close to a Doughnut mode and can be transformed to a nearly fundamental mode by focusing. As an example, Fig. 3.18 shows the intensity profile in the focal plane of an RF excited longitudinal flow laser with a beam power of 10 kW, using a folded resonator. The beam quality number K = 0.1, indicating it is not very well suited for cutting, but quite similar to the transverse flow laser, it has proved its excellent capability of welding and surface treatment.

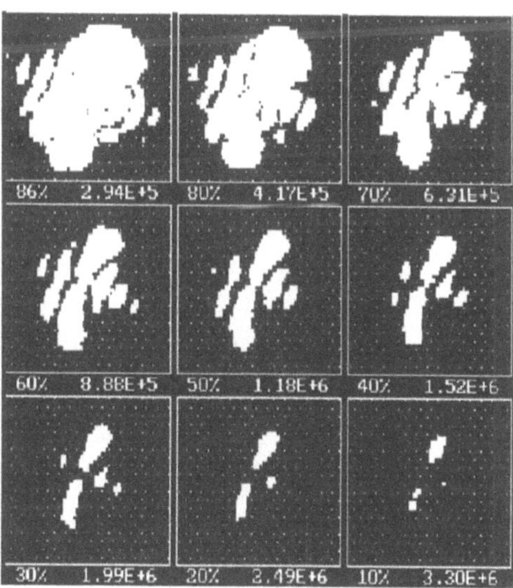

Fig. 3.18 Intensity profile in the focal plane (f =200 mm) of an RF excited longitudinal flow laser 10 kW with an unstable resonator and at a beam power of 5.5 kW (window size: 1 x 1 mm^2).

Beams generated by unstable resonators are very well suited for laser material processing, although they show some slight disadvantages with a higher beam parameter product. Since unstable resonators do not use transmissive optics as a window between the low pressure of the discharge chamber and the surrounding atmosphere since they would overheat, special means must be used. The solution is the so called **aerodynamic window** (Fig. 3.19), where an opening in the discharge chamber is covered by a very fast gas flow that prevents atmospheric gas entering the discharge chamber, but allows a more or less undisturbed transmission of the laser beam. The only disadvantages of this high power laser window are additional expenses and noise produced by the fast flowing gas.

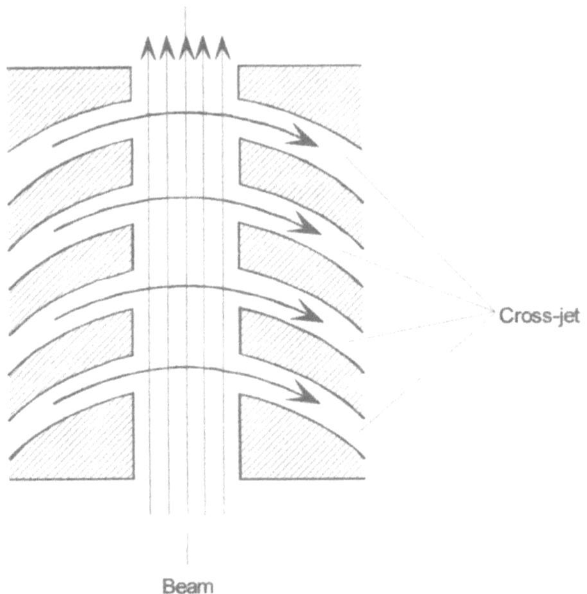

Fig. 3.19 Aerodynamic window.

In the case of the coaxial laser mentioned in the preceding section, the resonator volume is of hollow cylindrical shape, a geometry where the general distinction between stable and unstable resonators fails. So these resonators form an additional main group and are neither stable nor unstable. Nevertheless, they can be extremely simple and cost saving, since they consist only of a ring shaped totally reflecting mirror, with a concave surface being in fact a part of a **Thorus** surface, that helps to focus the hollow cylindrical beam and compensates spreading by diffraction between the resonator mirrors. The second mirror can then simply be a flat, partly transmissive and partly reflective mirror, that performs the extraction of a hollow cylindrical beam. As an example, Fig. 3.20 shows the intensity profiles at the output mirror and Fig. 3.21 in the focal plane of the RF excited longitudinal flow coaxial laser with a beam power of 6 kW developed in Vienna and described above. The intensity profile has a beam quality number of K = 0.1, indicating that it is also not very well suited for cutting, but similarly to the transverse flow laser it is most appropriate for welding and surface treatment.

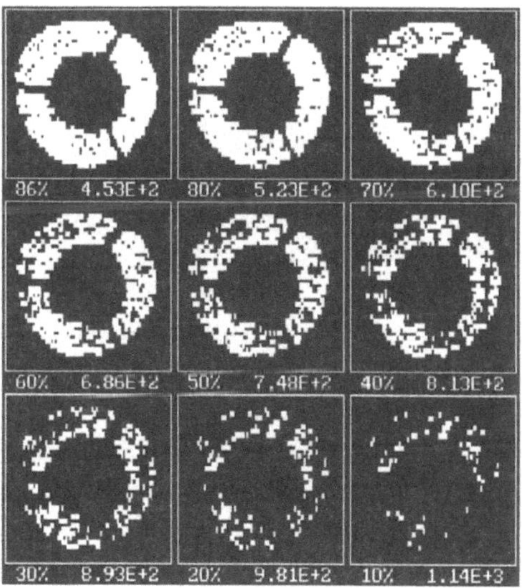

Fig. 3.20 Intensity profile at the output mirror of an RF excited longitudinal flow 6 kW coaxial laser at a beam power of 5 kW.
(window size: 40 x 40 mm^2) (Vienna 1993).

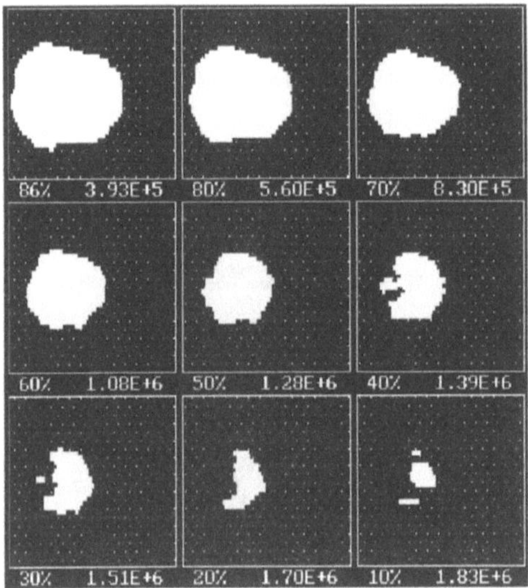

Fig. 3.21 Intensity profile in the focal spot ($f = 250$ mm) of an RF excited longitudinal flow 6 kW coaxial laser at a beam power of 4.3 kW. (window size: 1x1 mm^2) (Vienna 1993).

A further possibility is a zigzag beam running through the hollow cylindrical plasma in such a way that a majority of the plasma volume is filled, although a total coincidence between the latter and the beam volume can not be practically achieved, resulting in a reduced beam power. Nevertheless, a low Fresnel number and therefore a high beam quality, say a mode coming close to the fundamental modes, is obtained, making this kind of resonator well suited even for cutting.

3.7 SUMMARY

Table 3.1 Overview of industrial CO_2 laser sources (part 1)

	Fast Longitudinal Flow	*Fast Transverse Flow*	*RF excited Longitudinal Flow*	*Coaxial*	*Diffusion Cooled*
Electrode geometry	longitudinal DC discharge	cathode rod anode segments	helical stripe electrodes	coaxial electrodes	slab
Plasma volume	1200 cm^3			1700 W/cm^3	
Excitation frequency	DC	DC	27.12 MHz	27.12 MHz	82 MHz
Power input per unit volume	approx 16 W/cm^3			approx 30 W/cm^3	
Gas flow system	turbo blower	tangential blower	turbo blower	turbo blower	none
Gas flow geometry	longitudinal	transverse	longitudinal	longitudinal	none
Gas flow speed				approx 550 m/s (100 mBar)	
Resonator type	stable mode kit optional	stable	stable	hybrid	unstable

Table 3.1 Overview of industrial CO_2 laser sources (part 2)

	Fast Longitudinal Flow	*Fast Transverse Flow*	*RF excited Longitudinal Flow*	*Coaxial*	*Diffusion Cooled*
Resonator geometry	folded x1	folded	folded	2 mirrors	hybrid ext. beam shaping
Fresnel number	≈ 1.5			≈ 1.5	
Output power	1800 W	3000 W	10000 W	6000 W	2000 W
Pulse modes	0-5 kHz 30 µs - DC	CW	0.1-100 kHz 10 µs - CW	0-20 kHz 10 µs - CW	0-5 kHz 20 µs - CW
Beam mode	low order nearly TEM_{00} with kit	multi	low order	multi	low order
Gas flow system	turbo blower	tangential blower	turbo blower	turbo blower	none
K	0.45 0.8 with kit	> 0.1	>0.3	>0.1	>0.7
Beam diameter	14 mm	41 mm	31 mm	78 mm	<25 mm
Divergence full angle	<2.1 mrad	<3 mrad	<1.5 mrad	<1.7 mrad	<1.5 mrad
Industrial examples	Oerlikon OPL 2000	Rofin-Sinar RS 825	Trumpf TLF 12000	Wild CX 6000	Rofin-Sinar DC 020

3.8 REFERENCES

Bielesch, U., Budde, M., Fischbach, M., Freisinger, B., Schäfer, J.H., Uhlenbusch, J. and Viöl, W. (1992) *A Q-switched Multikilowatt CO_2 Laser System Excited by Microwaves* Proceedings of the SPIE: Gas Flow and Chemical Lasers, **1818**, pp. 57-60

Breining, K., Pfeiffer, W., Giesen, A. and Hügel, H. (1994) *Spatially resolved measurements in CO_2 laser active media* Proceedings of the SPIE: Tenth international Symposium on Gas Flow and Chemical Lasers, **2502**, pp. 542-547

Hall, D.R., and Baker, H.J. (1984) *RF excitation of diffusion cooled and fast axial flow lasers* Proceedings of the SPIE: Seventh International Conference on Gas Flow and Chemical Lasers, **1031**, pp. 60 - 67

Hall, D.R., and Baker, H.J. (1994) *Diffusion Cooled Large Surface area CO_2/CO Lasers* Proceedings of the SPIE: Gas Flow and Chemical Lasers, **2505**, pp. 12-19

Hügel, H.E. (1986) *RF excitation of high power CO_2 lasers* Proceedings of the SPIE: High Power Lasers and Their Industrial Applications, **650**, pp 2 - 9

Nowack, R., Opower, H. and Wessel, K. (1991) *Diffusionsgekühlte CO_2 Hochleistungslaser in Kompaktbauweise, Laser und Optoelektronik* **23**, No. 3, pp. 68-81

Opower, H. and Schuöcker, D. (1993) *Gaslaser*, Patent No. DE 3810604 A1

Patel, C.K.N (1964) *Selective excitation through vibrational energy transfer and optical maser action in N_2 - CO_2* Phys. Rev. Letts. **13**, pp. 617

Schröder, K. (1990) *Theoretical treatment of rf discharges in CO_2 waveguide lasers* J.Appl.Phys. **68** (11), pp.5528-5531

Schuöcker, D. and Schröder, K. (1994) *New strategies for the development of high power CO_2 lasers with beam powers up to 100 kW* LANE 94, Oct.12 - 14, Erlangen, Germany

Skolnik, M.I. (1970) *Radar Handbook*, McGraw Hill

Swysen, R. and Auer, M. (1990) *Radial blower for gas circulation in a compact 2 kW CO_2 laser* Proceedings of the SPIE: Lasers and Applications II, **1276**, pp. 68 - 76

Tabata, N (1989) *High Power Industrial CO_2Lasers* Proceedings of the SPIE: Gas Flow and Chemical Lasers

Walter, B. (1986) *TEA CO_2 Lasers, physical problems and technical solutions* Proceedings of the SPIE: High Power Lasers and Their Industrial Applications, **650**, pp. 52 - 58

Walter, B. (1989) *Impedance Matching of rf excited CO_2lasers* Proceedings of the SPIE, **1020**

Wittemann (1987) *The CO_2 Laser* Springer Series in Optical Sciences, **53**, Springer

Xin, J.G. and Hall, D.R. (1986) *Multipass coaxial radio frequency discharge CO_2 lasers* Opt. Commun, **58**, pp. 420-422

4

Solid state lasers

H. Weber

4.1 INTRODUCTION

4.1.1 The diversity of solid state lasers

There are three types of lasers of which the active medium is formed by solid materials:

1. pn-junctions, directly excited by an electric current;
2. colour centre lasers, that are crystals with X-ray induced defects, pumped by other lasers
3. crystals or glasses, doped with ions, and pumped by gas discharge lamps or other lasers.

Among these, only the doped crystals or glasses are usually called solid state lasers. Hundreds of crystals exist with thousands of wavelengths, covering the visible and infrared range of the spectrum from about $\lambda = 0.3$ μm up to $\lambda = 3$ μm. Some examples are given in table 4.1, (a detailed survey is compiled in Weber, 1986; and in Kaminskii, 1981). Most of them are only of scientific interest, or are used for very special applications where this particular wavelength is required.

4.1.2 Solid state lasers for industrial applications

Up to now, only a few solid state lasers are used in industrial applications, mainly in the following fields:

1. pollution control by laser spectroscopy;
2. material processing;
3. medical applications;

For spectroscopic applications tuneable lasers are required, that are laser crystals with a broad emission band. Although many crystals with this particular property exist (Budgor et al, 1985; Hammerling, P., Budgor, A.B., Pinto A., eds., 1986), only a few are used and available on the market, as shown in Table 4.2. These lasers are mainly pulsed and produce a low average output power of several watts.

For material processing, the bandwidth of the laser (or the coherence) is of lower importance. Beam quality and output power are the relevant numbers. Up to now, only a few systems are of interest for material processing. They are compiled in Table 4.3.

The ruby and alexandrite lasers are only employed for some very special applications and therefore of little interest. The titanium sapphire (Ti:sapphire) laser might become of interest because of its ability to generate ultrashort, high energy pulses useful for ablation. There are also many applications for the frequency converted Ti:sapphire laser at the second and third harmonic (0.39 µm and 0.26 µm, tuneable). Nd:YAG and Nd:Glass are the favourite solid state lasers used in material processing (about 30%). They are the most advanced systems with high reliability, and they are at the high technical standard necessary for industrial use, with the special feature that high power transmission by flexible fibres is possible. Recently Ytterbium:YAG crystals were operated successfully. Using diode pumping the efficiency can be higher than for Nd:YAG and crystal heating is lower by a factor 2–3.

Different modes of operation exist as compiled in Table 4.11. Continuous emission is mainly used for welding and cutting. Spot welding, drilling, trimming and engraving require pulsed emission in the range of 10^{-4} s down to 10^{-8} s. Ultra short pulses of 10^{-12} s are not yet used in material processing, but are under consideration in the laboratory.

4.1.3 Requirements for solid state lasers in material processing

Particular requirements depend on the material and kind of processing. These cannot be discussed here, only a rough summary is given in Table 4.4. The main parameters are power or energy and the beam quality.

Table 4.1 Selection of doped crystals used as solid state lasers

Active Ions		Host Crystal	Wavelength λ (μm)
Gadolinium	3+	$Y_3Al_5O_{12}$ (YAG)	0.3146
Terbium	3+	$LiYF_4$	0.5445
Holmium	3+	CaF_2	0.5512
Praseodymium	3+	LaF_3	0.5985
Europium	3+	Y_2O_3	0.6113
Chromium	3+	Al_2O_3 (Sapphire)	0.6943
Samarium	2+	SrF_2, CaF_2	0.6969
Neodymium	3+	Glass, $Y_3Al_5O_{12}$ (YAG)	1.06
Neodymium	3+	CaF_2, $CaWO_4$, LaF_3	0.9–1,06
Neodymium	3+	$YLiF_4$(YLF)	1.04–1.05, polarised
Neodymium	3+	$YAlO_3$(YAP,YALO)	1.06
Ytterbium	3+	$Y_3Al_5O_{12}$ (YAG)	1.03
Praseodymium	3+	$CaWO_4$	1.04
Thulium	3+	CaF_2	1.1
Thulium	3+	$CaWO_4$, $Y_3Al_5O_{12}$	1.93–2.01
Erbium	3+	$Y_3Al_5O_{12}$, $Ca(NbO_3)$	1.6
Nickel	2+	MgF_2	1.6
Cobalt	2+	MgF_2, ZnF_2	1.7–2.6
Uranium	3+	CaF_2,SrF_2	2.4–2.6
Dysprosium	2+	CaF_2	2.35
Dysprosium	3+	$Ba(Y_{1.26}Er_{0.74})F_8$	3.02

Table 4.2 Tuneable solid state lasers, commercially available for spectroscopic applications

Crystal	Dopant	tuning range (μm)
Alexandrite($BeAl_2O_4$)	Cr	0.72–0.795
Sapphire (Al_2O_3)	Ti	0.69–0.95
LiCAF ($LiCaAlF_6$)	Cr	0.72–0.84
LiSAF ($LiSrAlF_6$)	Cr	0.78–0.92

Table 4.3 Solid state lasers for material processing

System	Wavelength $\lambda(\mu m)$	Average output per unit $P_L(W)$	Output energy per pulse and unit $E_L(J)$
(Unit means one laser head with a length of the active medium of 10–20 cm)			
Neodymium:YAG	1.06	600	50
Neodymium:Glass	1.06–1.05	20	45
Titanium:Sapphire	0.79	200	1–5
Chromium:Alexandrite	0.76	150	10
Chromium:Sapphire (Ruby)	0.69	100	10

Beam quality, or preferably the beam parameter product, will be discussed later. Besides the parameters in the table, two other features are important:

1. small transient effects, that means that the beam structure (beam quality) does not vary considerably with the output power;
2. the position and diameter of the focused beam must be constant.

For all industrial applications investment costs, running costs and reliability are finally the decisive parameters, especially for the decision CO_2 or Nd:YAG laser.

Advantages compared with CO_2 lasers

1. higher absorption in metals;
2. better focusing in principle;
3. transmission by fibres;
4. short pulses ($>10^{-12}$s);
5. high peak power (ablation);
6. compact set-up;
7. high energy storage.

Table 1.4 Typical parameters of solid state lasers used in material processing

Parameter	Value
average power P_L (kW)	0. –5
energy per pulse E_L (Ws)	0. –20
beam parameter product BP = $D\theta/4$ (mm·mrad)	micro processing <0.5
	cutting <10
	fibre transmission <25
	welding <40
efficiency η_{tot}(%)	4–5
lifetime (h)	500–1000
core-diameter of fibres d_{core}(μm)	< 600
pulse width δt	cw, 10–0.1 ms, 100 ns
repetition rate f	10 Hz–100 kHz

Drawbacks in comparison with CO_2 lasers

1.   lower beam quality;
2.   low efficiency;
3.   output power limited to $\sim$ 8 kW per system;
4.   low life time (< 1000 h).

Some of these disadvantages will be removed within the next years, especially if high average power diode pumped solid state lasers are available at moderate cost.

1.5 Beam quality

There is a lot of confusion about beam parameters and beam quality, therefore the main facts are summarised in this section. In Fig. 4.1 a schematic laser system is shown with the emerging laser beam. Somewhere inside or outside the resonator will be the waist, the minimum beam diameter. Very often it is located on the plane output mirror. The waist diameter can be characterised by its second moment or by its power content value D (ISO 11146, 1992).

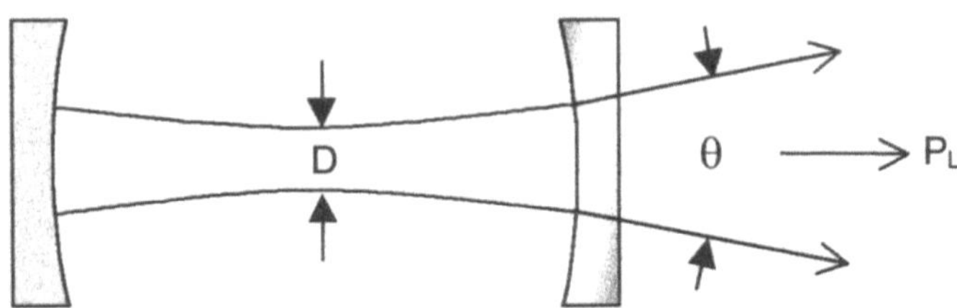

Fig. 4.1 Definition of beam quality by the product of the waist diameter D and full far field divergence θ. $BP = D\theta/4$.

D is the diameter of a circle which at the waist position will contain a certain amount of the total power, in general, 86.5%. The divergence θ (full angle) of the beam is measured in the far field, i.e. at a distance $z \gg D^2/\lambda$, or in the focal plane of a lens. It is again defined as a power content value. For the 86.5% values the relationship is:

$$BP = \frac{D\theta}{4} \geq \frac{\lambda}{\pi} \tag{4.1}$$

This relation is a fundamental one, well known as the Fourier relation, and a consequence of diffraction. The equality in the above relation holds for the fundamental mode, often called TEM_{00} or gaussian beam with a gaussian field and intensity distribution:

$$J = J_0 \exp\left[-2\left(\frac{2r}{D_0}\right)^2\right] \tag{4.2}$$

$$(BP)_{00} = \frac{D_0\theta_0}{4} = \frac{\lambda}{\pi} \tag{4.3}$$

The beam parameter product BP of equation (4.1) is identified with the beam quality and can be normalized to the fundamental mode:

$$\frac{D_0\theta_0}{D\theta} = K \leq 1 \tag{4.4}$$

with K a number characterising the beam quality.

In many beam analysing systems the inverse of K is used and denoted by M^2, the beam propagation factor:

$$M^2 = \frac{1}{K} \geq 1$$

Instead of the power content values the beam waist and the divergence can be defined by the second intensity moments, which is very convenient for gaussian or gaussian-like fields, produced by lasers with spherical, stable resonators. For the second moments a formalism exists to calculate the beam propagation in optical systems (Hodgson and Weber, 1997). The M^2 characterisation is now recommended by the International Standards Organisation ISO (ISO 11146, 1992). The numbers M^2 and K are constant in aberration free optical systems, which means spherical or parabolic optics in the paraxial approach (small angles θ). K becomes smaller in optics of low quality (beam quality decreases). It can be increased up to $K = 1$ by inserting a pinhole in the focal plane of a lens which means reduced power. For ideal optical systems:

$$D\theta = const \tag{4.5}$$

For large apertures as may occur for very small beam waists in microprocessing the angle θ has to be replaced by $\sin\theta$. If in the input/output planes of an optical system the refractive indices are different this has to be included and the above relation reads:

$$Dn \sin\theta = const \tag{4.6}$$

Some examples of beam quality are given in Table 4.5. Anyway, if a laser producer offers a system with a certain beam quality, it has to be carefully examined what it really means and the way it was measured. The beam parameter product is of major importance for all applications as will be demonstrated by two examples.

Focusing

If a laser beam is focused by a telescope, Fig. 4.2, the waist diameter is reduced and beam divergence enhanced according to the magnification of the telescope. An essential parameter is the focal length z_f. That is the distance

Table 4.5 Beam parameter products and K values of different laser systems. (For diodes the beam parameter product is different in both transverse directions)

Laser	Power P_L (W)	Wavelength $\lambda_L(\mu m)$	$D_0\theta_0/4$	$D\theta/4$ (mm·mrad	$K=1/M^2$
He:Ne	5×10^{-3}	0.633	0.2	0.2	1
Nd:YAG	15	1.06	0.34	0.36	0.94
	90			2.25	0.15
	1000			25	0.014
Nd:Glass. directly coated	20	1.05	0.34	28	0.012
external mirrors	15			10	0.034
CO_2	500	10.6	3.2	4.2	0.76
	1500			8	0.4
Low power diode	10^{-2}	0.809	0.26	0.3/0.9	0.84/0.28
High power diode	10			$0.3/1.5\times10^4$	$0.84/1.5\times10^{-5}$

in the propagation direction between the waist with the maximum intensity and half the intensity, which means that the diameter has increased by a factor of $\sqrt{2}$. The focal length reads:

$$z_f = K\pi\frac{D_{foc}^2}{4\lambda} = \frac{D_{foc}}{\theta_{foc}} \geq \frac{\pi D_0^2}{4\lambda} \tag{4.7}$$

Notice, the focal length depends on the K number as well as on the wavelength.

Examples

He/Ne-laser: $\lambda = 0.63\ \mu m$, $D_0 = 1\ mm$, $K = 1$, $z_f = 1.2\ m$;
Nd-YAG-laser:$\lambda = 1.06\ \mu m$, $D = 5\ mm$, $K = 0.94\ (P_L = 10\ W)$, $z_f = 17.4\ m$;
* $D = 5\ mm$, $K = 0.012\ (P_L = 1\ kW)$, $z_f = 0.2 m$;*
same beam focused, $D = 0.5\ mm$, $K = 0.012$, $z_f = 0.002\ m$.

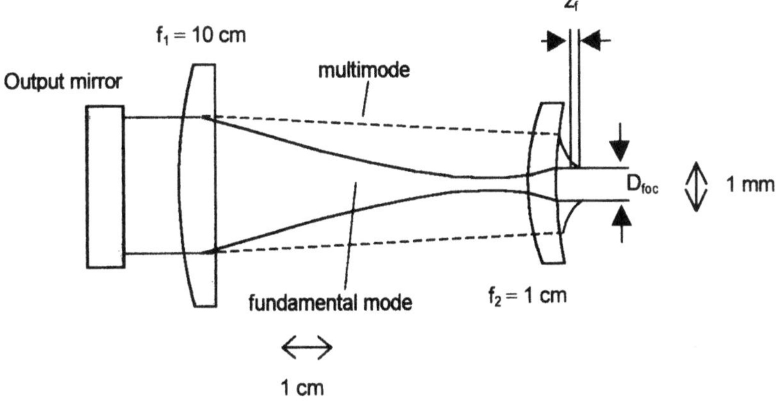

Fig. 4.2 Focusing of fundamental and higher order mode radiation by a telescope.

According to equation (4.7) the focal length has a maximum for the fundamental mode and decreases for other radiation fields. The focal length defines the positioning accuracy of optical systems necessary for precise and reliable material processing, but also the aspect ratio of laser drilled holes (among other parameters). For these reasons high z_f values and a K number near one is desirable. The striking difference between fundamental mode and multimode focusing is demonstrated in Fig. 4.2. Some numbers are compiled in Table 4.6.

Table 4.6 Typical beam parameters for the fundamental mode TEM$_{00}$ and a higher order mode TEM$_{mn}$ with $K = 0.01$ focused by a telescope with a magnification 1:10

Parameter	*Symbol and units*	*fundamental mode TEM$_{00}$*	*multimode TEM$_{mn}$*
diameter of laser output	D (mm)	4	4
fundamental mode diameter	D$_0$ (mm)	4	0.4
laser output divergence	θ(mrad)	0.34	34
beam parameter product	Dθ/4(mm·mrad)	0.34	34
divergence of focused beam	θ_{foc}(mrad)	3.4	340
focus diameter	D$_{foc}$(mm)	0.4	0.4
focal length	z$_f$ (mm)	118	1.18

Fibre coupling

It will be shown in section 4.5 that the low loss transmission of high power laser radiation by fibres requires

$$\frac{D\theta}{4} < d_{core} \cdot \frac{NA}{2} \tag{4.8}$$

where d_{core} is the fibre core diameter and NA its numerical aperture. Fibres of small core diameter are desired, which again requires a low beam parameter product (high beam quality).

Radiance

The two relevant parameters - output power P_L and beam parameter product BP can be combined to one characteristic number

$$\frac{\text{power}}{\text{beam parameter product}} = \frac{P_L}{(BP)_x (BP)_y} = \pi L_R$$

where the beam parameter products in both directions have to be considered. L_R is a well known quantity in geometrical optics: the radiance. It is defined as the output power dP_L per emitted over an area dA and solid angle $d\Omega$:

$$L_R = \frac{dP_L}{dA.d\Omega.\cos\alpha} \quad (\text{W}/\text{m}^2.\text{ster.}) \tag{4.9}$$

where α is the angle between the beam direction and the surface normal vector (Fig.4.3). For small far field divergence θ and emission perpendicular to the surface ($\alpha = 0$) the radiance L_R reads:

$$L_R = \frac{P_L}{\left[\pi D\theta/4\right]_x \left[\pi D\theta/4\right]_y} = K_x K_y \frac{P_L}{\lambda^2} \tag{4.10}$$

which for fundamental mode operation reduces to:

$$L_{R0} = \frac{P_L}{\lambda^2} \tag{4.11}$$

The radiance of some light sources is given in Table 4.7.

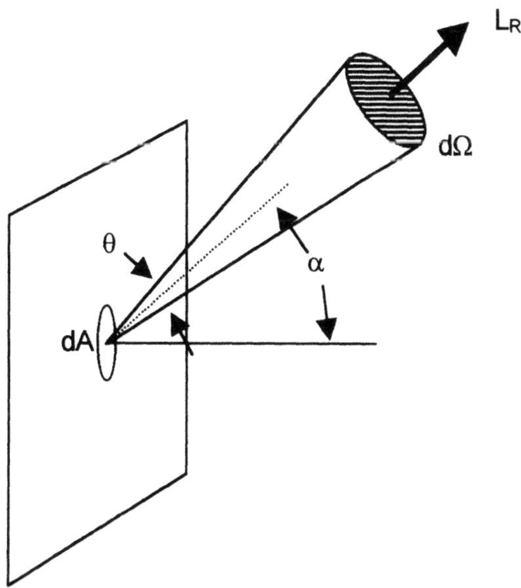

Fig. 4.3 Geometry for radiance calculations.

Table 4.7 Radiance of some light sources

Light source	Radiance L_R [W/ster·m²]
Sun P_L=3.9 x 10^{26} W	2.1 x 10^7
Single diode laser P_L= 10 mW, K_x = 0.84, K_y= 0.28, λ = 0.8 µm	3.6 x 10^9
He:Ne Laser with 10mW fundamental mode, K=1, λ= 0.64 µm	2.4 x 10^{10}
Nd:YAG Laser, 1kW output, K=0.012, λ = 1.06 µm	1.3 x 10^{11}
CO_2 Laser 500 W output, fundamental mode, K=1, λ = 10.6 µm	4.5 x 10^{12}

The radiance seems to be a quite useful number to characterise light sources. But that is only a rough characterisation because other parameters are involved such as time dependence (pulse duration), wavelength (absorption) and the intensity structure of the focal area. Furthermore, the radiance is a radiometric quantity, assuming that the different emitting areas of the light source are not correlated (incoherent source). Interference effects may occur and the radiance becomes a bit obvious for lasers (Wolf, 1978).

4.2 MAIN CHARACTERISTICS OF HIGH POWER Nd LASERS.

In the following section the basic relations of four level lasers are summarised. For the derivation of these formulas from first principles the well known text books on laser physics are recommended (Siegmann, 1986; Yariv, 1970).

4.2.1 Some fundamental facts.

The Nd $^{3+}$ ion in glass or YAG has a very complicated energy level diagram. A simplified diagram with the relevant transitions of interest for high power systems is shown in Fig. 4.4.

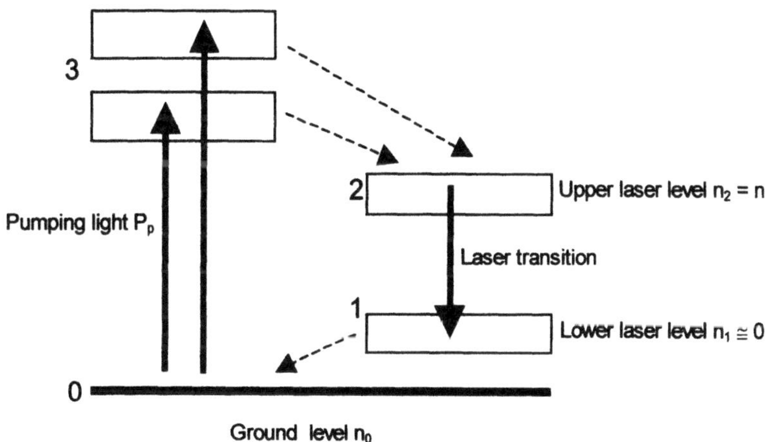

Fig. 4.4 Simplified energy level diagram for Nd^{3+} in glass or YAG.

It holds for Nd in glass or YAG, although the absolute values of the energy levels differ slightly. By absorbing pumping light from a gas discharge lamp or a laser diode, transitions from the ground level 0 to higher absorption bands 3 are induced. The lifetime of the excited ion in these levels is very short and nearly all ions decay very rapidly into the upper laser level 2 which has a long lifetime τ_l of some hundreds of microseconds. The lower laser level 1 again has a very short lifetime and the density n_1 of Nd ions is nearly zero at room temperature. By a sufficiently strong pumping power a considerable density $n_2 = n$ of ions is accumulated in the upper level and produces a small signal gain factor G_0 per transit:

$$G_0 = \exp(n\sigma_L l) = \frac{J}{J_0}, \quad J_0 \ll J_S \tag{4.12}$$

where G_0 = small signal gain factor
$\qquad \sigma_L$ = cross section of laser transition (effective area of the atom)
$\qquad n$ = density of ions in the upper laser level
$\qquad l$ = length of the active medium
$\qquad h\nu_L$ = quantum energy of the laser transition

A radiation field of intensity J_0 or a pulse of the energy density e_0 (fluence) can be amplified by the factor G_0 per transit, as long as the intensity is small compared with the saturation intensity J_S, an atomic parameter (Table 4.8). For high intensities the upper level population is depleted by induced emission.

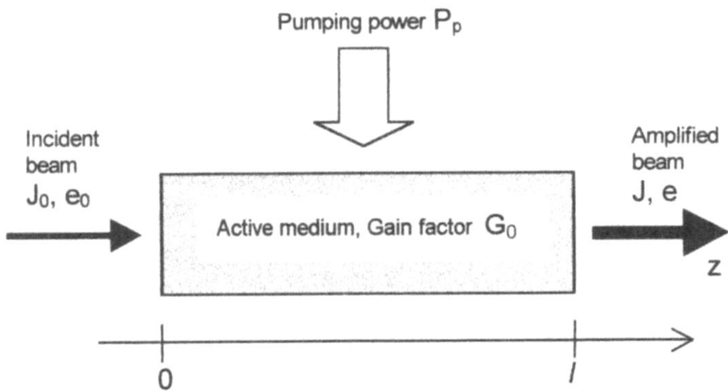

Fig. 4.5 The active medium as a light amplifier.

In the steady state:

$$n_s = \frac{n}{1 + J/J_s} \tag{4.13}$$

and the gain reduces to the saturated gain factor

$$G_s = \exp\left(n_s \sigma_L l\right) \le G_0 \tag{4.14}$$

J_s is the saturation intensity and given by:

$$J_s = \frac{h\nu_L}{\sigma_L \tau_L}$$

with τ_L the upper laser level lifetime. Similar relations hold for the amplification of laser pulses. If the pulse energy e_0 is comparable with the saturation energy density $e_s = h\nu_L/\sigma_L$ the inversion is depleted and the gain decreases. The total stored energy in the crystal $E_{st} = nh\nu_L Fl$ (F = rod cross section) can be extracted by a single pulse, if its energy density is large compared with e_s. Some numbers are given in Table 4.8.

Example

Consider a Nd:YAG-crystal with l = 100 mm and 6 mm diameter. By pumping a small signal gain factor G_0 = 5 is produced and can be measured. Equation (4.12) delivers a density of upper level ions of n = 4.6 x 10^{17} cm^{-3}. An incident beam P_L = 1 kW, intensity J_o = P_L/F = 3.5 kW/cm^2 reduces the inversion density to n_s = 2.1 x 10^{17} cm^{-3}, resulting in a saturated gain of G_s = 2.1.

4.2.2 Excitation

The active medium is excited by the pumping power P_P, which is the usable part of light emitted by a gas discharge lamp (pulsed, cw) or an array of laser diodes. In Fig. 4.6 a,b, the absorption spectrum of Nd:YAG is given, and in Fig. 4.7 that of Nd:Glass. These have to be compared with the emission spectrum of the commonly used Xenon or Krypton lamps as shown in Fig. 4.8. The emitted light can only be used for Nd excitation in part. Broad ranges of the emission spectrum are useless or even undesired, because they produce heat. The diode laser with λ_P = 0.809 μm (Fig. 4.6a) fits the Nd absorption spectrum much better.

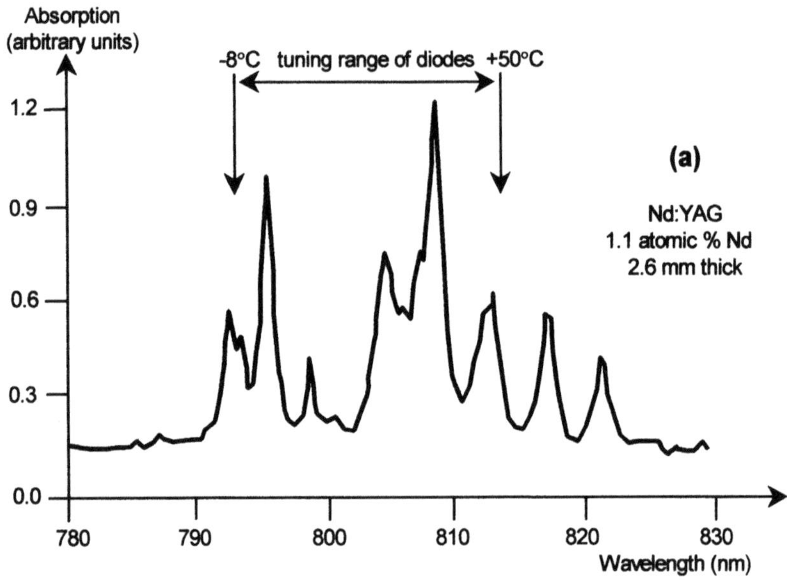

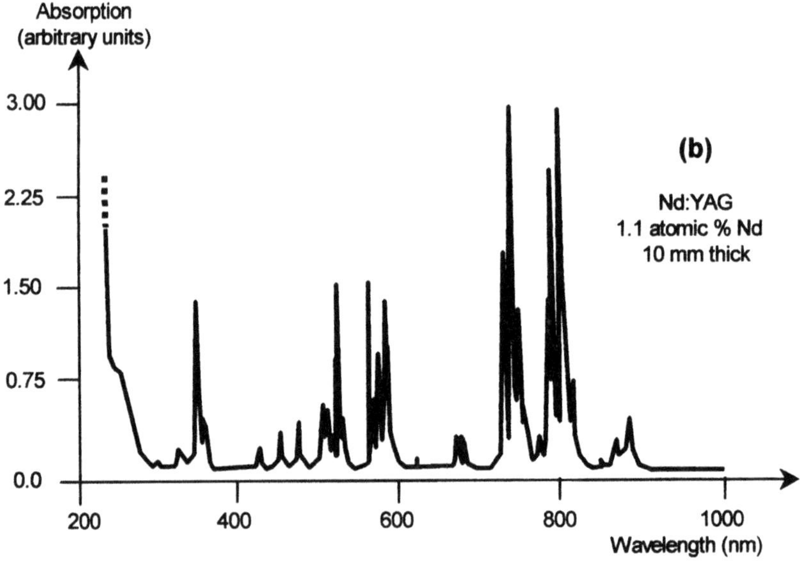

Fig 4.6 Absorption spectrum of Nd:YAG.
(a) wavelength range λ_p = 780–830 nm for diode pumping.
(b) wavelength range λ_p = 200–1000 nm for lamp pumping.

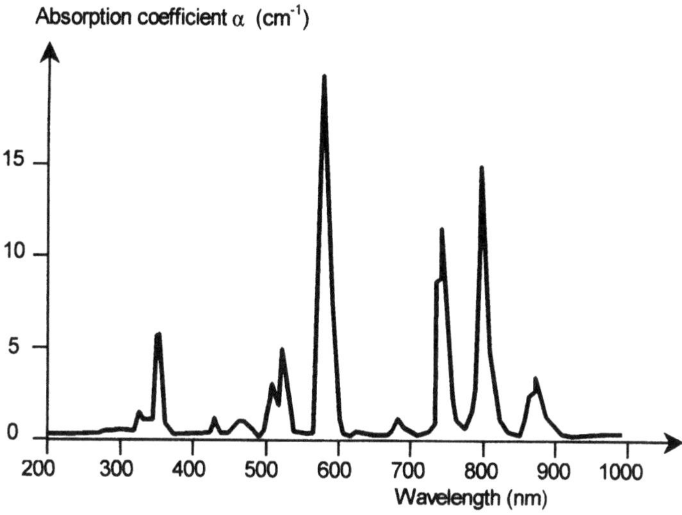

Fig 4.7 Absorption coefficient $\alpha = n_0\sigma_p$ of Nd:glass in the wavelength range $\lambda_p = 300\text{--}1000$ nm for lamp pumping. (Hoya data sheet 1987, $n_0 = 5.8$ ions/cm^3.

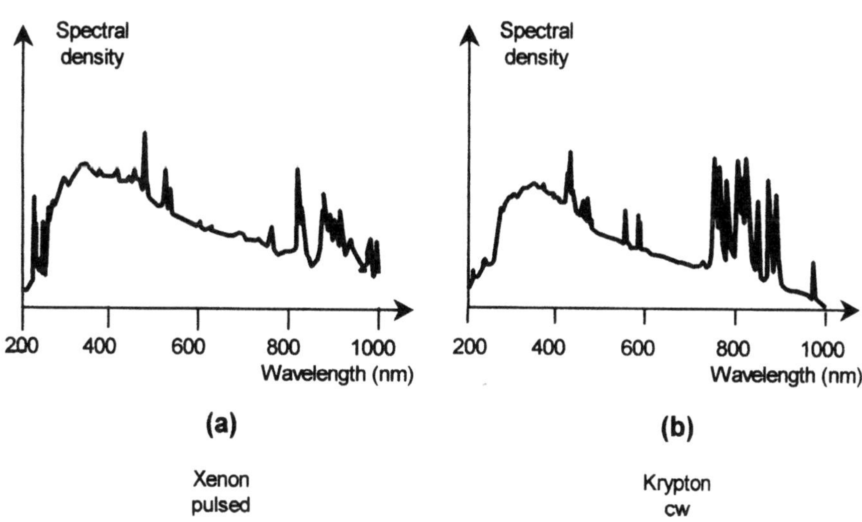

Fig 4.8 Emission spectra of gas discharge lamps (Iffländer, 1990) in the wavelength range $\lambda_p = 200\text{--}1000$ nm.

Table 4.8 Some parameters of Nd:YAG, Yb:YAG and Nd:Glass.

Parameter	Nd:YAG	Yb:YAG	Nd:Glass
typical doping $n_0 (cm^{-3})$	10^{+20}	10^{+20}-10^{+21}	$3 \times 10^{+20}$
cross section σ_L (cm^2)	3.5×10^{-19}	1.8×10^{-20}	3×10^{-20}
upper laser level lifetime τ_L (s)	2.3×10^{-4}	1.2×10^{-3}	3.8×10^{-4}
quantum energy $h\nu_L$ (Ws)	1.4×10^{-19}	1.94×10^{-19}	1.89×10^{-19}
saturation intensity $J_S (kW \cdot cm^{-2})$	1.62	9.4	129
saturation energy density $e_S (Ws \cdot cm^{-2})$		0.37	10.9
thermal conductivity k $(Wm^{-1}{}^{\circ}C^{-1})$	10.3	10.3	0.67
thermal expansion α $(^{\circ}C^{-1})$	6.9×10^{-6}	6.9×10^{-6}	13.8×10^{-6}
Poisson's ratio	0.25	0.25	0.27
Young's modulus E(GPa)	300	300	53.7
index of refraction n_r at $\lambda = 1.0$ μm	1.82	1.82	1.52
$\delta n_r / \delta T$ $(^{\circ}C^{-1})$	1.0×10^{-5}	1.0×10^{-5}	6.8×10^{-6}
rupture stress σ_{max} (MPa)	280	280	50
maximum heating power per length $P_{H,max} / \ell$ (kW/m)	14	14	0.7
quantum defect $h\nu_P - h\nu_L$ (Ws)	5.5×10^{-20}	1.5×10^{-20}	5.5×10^{-20}
pumping wavelength for diodes λ_P (μm)	0.809	0.911	

The energy of the pumping photons $h\nu_P$ is always larger than the energy of
the emitted laser photons $h\nu_L$. Again heat is produced. The ratio of both
energies is called the quantum defect:

$$\eta_Q = \frac{\nu_L}{\nu_p} \leq 1 \tag{4.15}$$

A quantum defect η_Q near to 1 is desirable for low heating of the crystal
and high efficiency. The power which excites Nd ions into the upper laser
level is called excitation power P_{excit} and is more or less proportional to the
electrical power P_E of the lamps. The ratio of both is the excitation efficiency

$$\eta_{excit} = \frac{P_{excit}}{P_E} \tag{4.16}$$

In the steady state without induced emission the number of excited ions per second is equal the number of decaying ions per second, which delivers the relation:

$$\frac{P_{excit}}{h\nu_L Fl} = \frac{n}{\tau_L} \tag{4.17}$$

from which the upper level population density and the small signal gain factor of equation (4.12) result in:

$$G_0 = \exp\left(\frac{\eta_{excit} P_E}{J_s F}\right) \tag{4.18}$$

The small signal gain factor increases exponentially with the pumping power, but not so the saturated gain factor which depends on the pumping power and the intensity.

4.2.3 The cw laser oscillator

Output power

The laser oscillator is the amplifier with two mirrors as shown in Fig. 4.9. In most cases one mirror has a reflectivity of 100%, the other one has to be adapted to the special conditions. For the following discussion $R_1 = R$ and $R_2 = 1$ is assumed. Between the mirrors two travelling waves of intensity J', J appear. The output power reads:

$$P_L = (1 - R)FJ^+_{(z=L)} \tag{4.19}$$

The two waves are amplified by the active medium and attenuated by internal losses. Losses (characterised by the loss factor V) are produced by scattering, diffraction. Additional losses occur by the reflectivities of the mirrors. In steady state the total losses must be compensated by the gain, which requires per round trip:

$$RVG_s G_s V = 1 \tag{4.20}$$

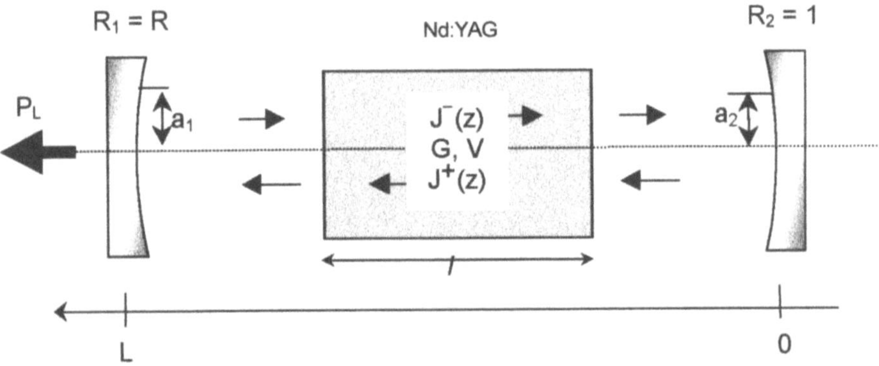

Fig. 4.9 Laser oscillator with high reflectance mirror R_2 and output mirror R_1. Two travelling waves J^+ and J^- exist inside the resonator. G: saturated gain factor, V: loss factor.

where G_s is the saturated gain factor. With equation (4.14) the well known threshold condition is obtained:

$$\left| \ln\left(V \sqrt{R} \right) \right| = n_s \sigma_L l \tag{4.21}$$

In the steady state this equation must hold for any pumping power level. Increasing pumping power will therefore not increase the population in the upper level but the internal intensity will rise, which by induced emission ensures constant population.

The intensities J^+, J^- inside the resonator are now varying with z, but the superposition of both does not depend on z very strongly as long as $R > 0.5$. A good approximation is (Iffländer, 1990):

$$J^+ + J^- = \left(1 + R \right) J^+_{(z=L)}$$

Equation (4.13) delivers for the steady state population of the upper level:

$$n_s = \frac{n}{\left[1 + \dfrac{\left(1 + R \right) J^+_{(z=L)}}{J_s} \right]} \tag{4.22}$$

The internal intensity is $J^+_{(z=L)}$ and with equation (4.19) the output power is obtained. Replacing n, n_s by the electrical power (equations.4.16/17) and approximating $(1-R)/(1+R)$ by $-\ln\sqrt{R}$ (which is fairly good for $R > 0.5$, error $< 5\%$) delivers the output power of the laser oscillator:

$$P_L = \left|\ln\sqrt{R}\right| FJ_s\left(\frac{P_E}{P_{th}} - 1\right) \qquad (4.23)$$

with

$$P_{th} = \frac{1}{\eta_{excit}}\left|\ln V\sqrt{R}\right| FJ_s \qquad (4.24)$$

P_{th} is the minimum electrical power necessary for the laser onset. It is called the threshold power. At threshold the gain factor just compensates the losses R and V. Although equation (4.23) is a rough approximation the typical behaviour of the oscillator is described and the results are confirmed by the experimental data. For the pulsed operation, the same relations can be used, by multiplying the powers P_L, P_{th} and P_E with the pumping pulse duration Δt and replacing them by the energies E_L, E_{th} and E_E, respectively. This holds as long as $\Delta t \gg \tau_L$. Small oscillations may occur at the laser onset, as shown in Fig. 4.10.

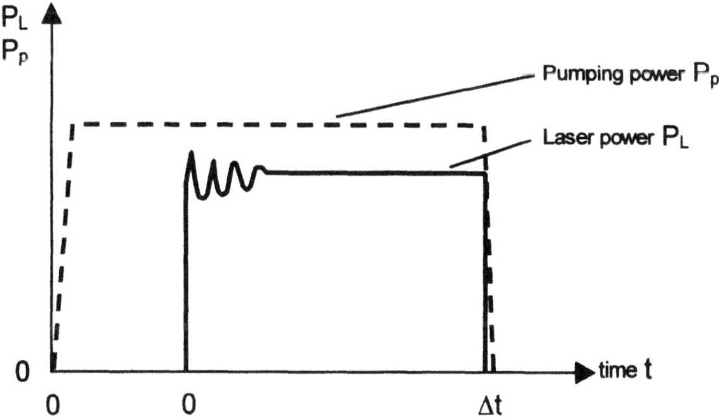

Fig 4.10 Pulsed operation of a Nd:YAG laser pumped with a rectangular pulse.

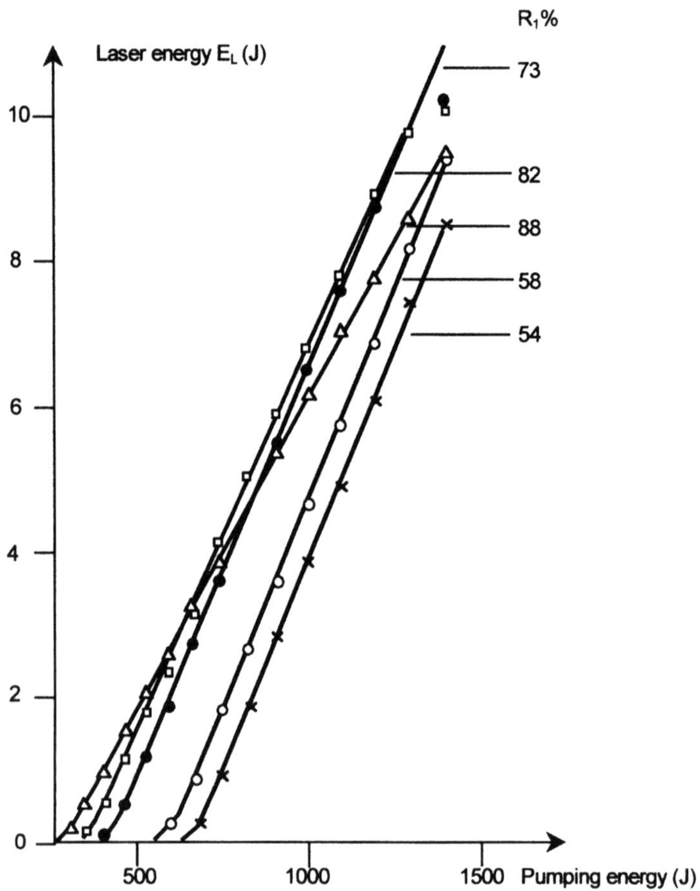

Fig 4.11 Laser output energy E_L (ordinate) vs electrical pumping energy E_E per pulse. Parameter is the reflectivity $R_1 = R$, $R_2 = 1$.

Near threshold the above equations fail in the pulsed regimes, because then the energy for building up the population n has to be taken into account (Ifflländer, 1990), but in most cases this energy can be neglected. A typical experimental result is shown in Fig. 4.11. The laser output energy increases proportional to the pumping energy, the slope depends on the reflectivity R.

The laser output power versus reflectivity for constant pumping power is plotted in Fig.4.12 and fits well the theoretical results of equation.(4.23). For the reflectivity $R \rightarrow 1$, the output power is zero, because the mirror transmission is zero. If R is below a critical value, the pumping power is below the threshold power and no laser oscillation occurs.

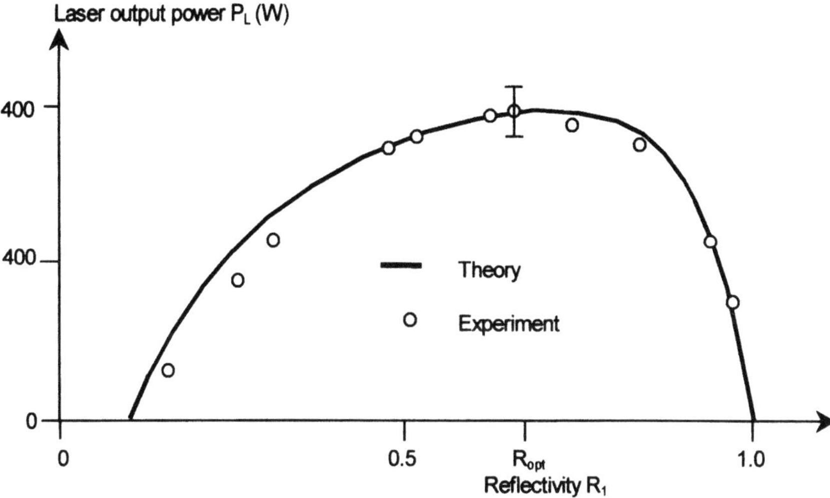

Fig. 4.12 Laser output power P_L (W) vs reflectivity R_1 for constant pumping power, $R_2 = 1$.

In between, the output power has a maximum for $R = R_{opt}$. A complete family of characteristics is given in Fig. 4.13. From equation.(4.23) a straightforward calculation results in the optimum reflectivity:

$$\ln\left(\sqrt{R_{opt}}\right) = \ln(V)\left[\left(\frac{P_E}{P_{min}}\right)^{\frac{1}{2}} - 1\right]^2 \tag{4.25}$$

P_{min} is the minimum electrical power to start the laser, if the reflectivity of both mirrors is 100%. The gain factor has only to overcome the internal loss factor V. This minimum power is:

$$P_{min} = \frac{1}{\eta_{excit}} FJ_s|\ln V| \tag{4.26}$$

The maximum output power $P_{L,max}$ for $R = R_{opt}$ is obtained from equations (4.23–25) and reads :

$$P_{L,max} = \eta_{excit}\left(\sqrt{P_E} - \sqrt{P_{min}}\right)^2 \tag{4.27}$$

The maximum output power depends of course on the electrical pumping power, but the internal loss factor V (scattering, diffraction) determines very sensitive the output power, especially at low pumping powers. The optimum reflectivity equation (4.25) depends on the pumping power, which means it has to be adapted to the actual pumping power level. But for practical reasons the lasers have a fixed reflectivity of the output mirror. At not too low pumping power, the output power depends only smoothly on the reflectivity, as can be seen in Fig.4.12.

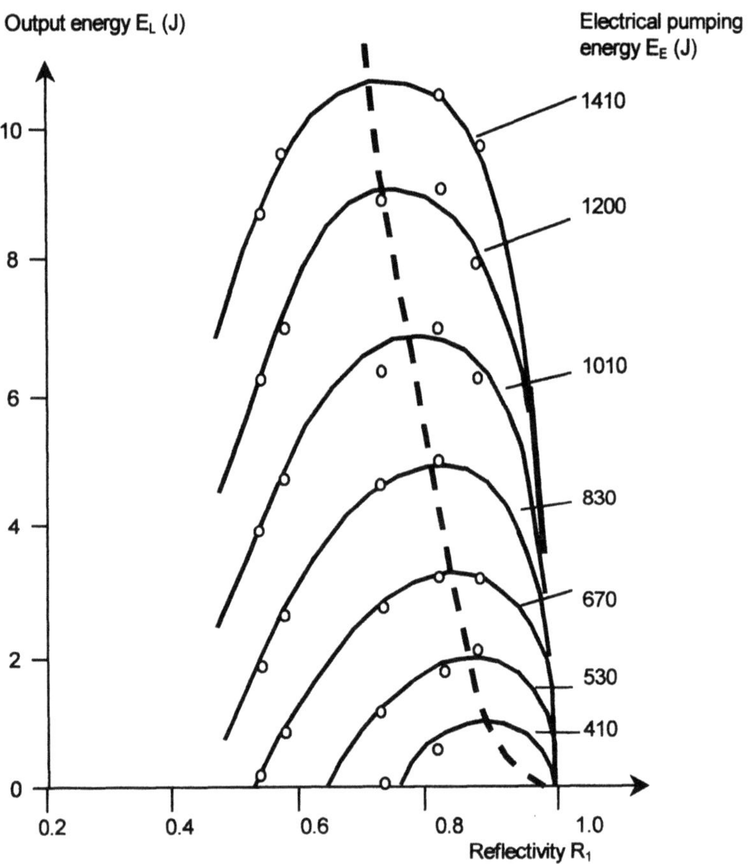

Fig. 4.13 Laser output energy E_L vs reflectivity R_1 for different electrical pumping energies E_E. $R_2 = 1$.

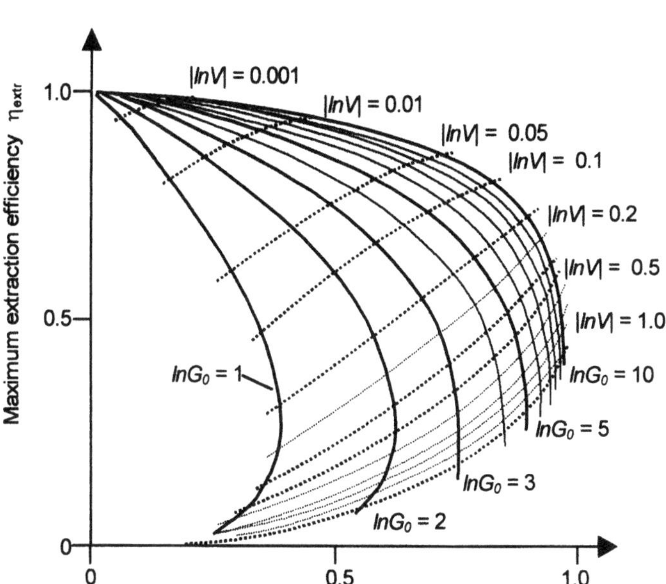

Fig. 4.14 The Schindler diagram to optimise the ideal 4 level laser system. G_0 is the small signal gain factor, equation (4.12), V the loss factor per transit, R_{opt} the optimum reflectivity, equation (4.25) and η_{extr} the extraction efficiency.
Schindler, Q.M., (1980) *IEEE J.QE* **16**, p546, © 1980 IEEE

The above relations hold for reflectivities $R > 0.5$. A more precise presentation of this optimization problem was given by Rigrod (1978) and Schindler (1980). The result is compiled in Fig. 4.14. For any small signal gain factor G_0 related to the pumping power according equation (4.18) and for any loss factor V the optimum reflectivity can be taken from this diagram.

Example

Nd rod, $l = 20$ cm, electrical pumping power $P_E = 8$ kW, cross section of rod $F = 0.5$ cm^2, 5 % losses $V = 0.95$, $\eta_{excit} = 0.07$. The minimum electrical power for both reflectivities equal one results with equation (4.26) in $P_{min} = 570$ W; optimum reflectivity equation (4.25) $R_{opt} = 0.7$; output power Eq.(4.27) $P_{L,max} = 300$ W

Efficiencies

Several efficiencies are used to characterise the laser system and the different components. The quantum defect η_Q and excitation efficiency η_{excit} have already been discussed. The total efficiency includes several other transfer efficiencies including:

- plug → power supply;
- power supply → lamp output;
- lamp output → cavity;
- cavity → active medium;
- active medium → absorption level;
- absorption level → laser level;
- laser level → radiation field of the resonator;
- resonator → output power.

A rough representation of these efficiencies is given in Fig. 4.15. They strongly depend on the particular experimental setup, the mode of operation (cw, pulsed) and the mode structure (fundamental mode, multimode).

Slope efficiency is defined by $\eta_{slope} = \delta P_I/\delta P_E$, which is the increase of output power per electrical input power and is from equations (4.23–24):

$$\eta_{slope} = \eta_{excit}\left[\frac{\ln\left(\sqrt{R}\right)}{\ln\left(V\sqrt{R}\right)}\right] \tag{4.28}$$

For the optimal reflectivity equation (4.25) and at high pumping powers, the slope efficiency is equal to the excitation efficiency.

Total efficiency is the the ratio of laser output power to electrical power of the lamp

$$\eta_{tot} = \frac{P_L}{P_E} \tag{4.29}$$

which approaches the slope efficiency for high pumping powers.

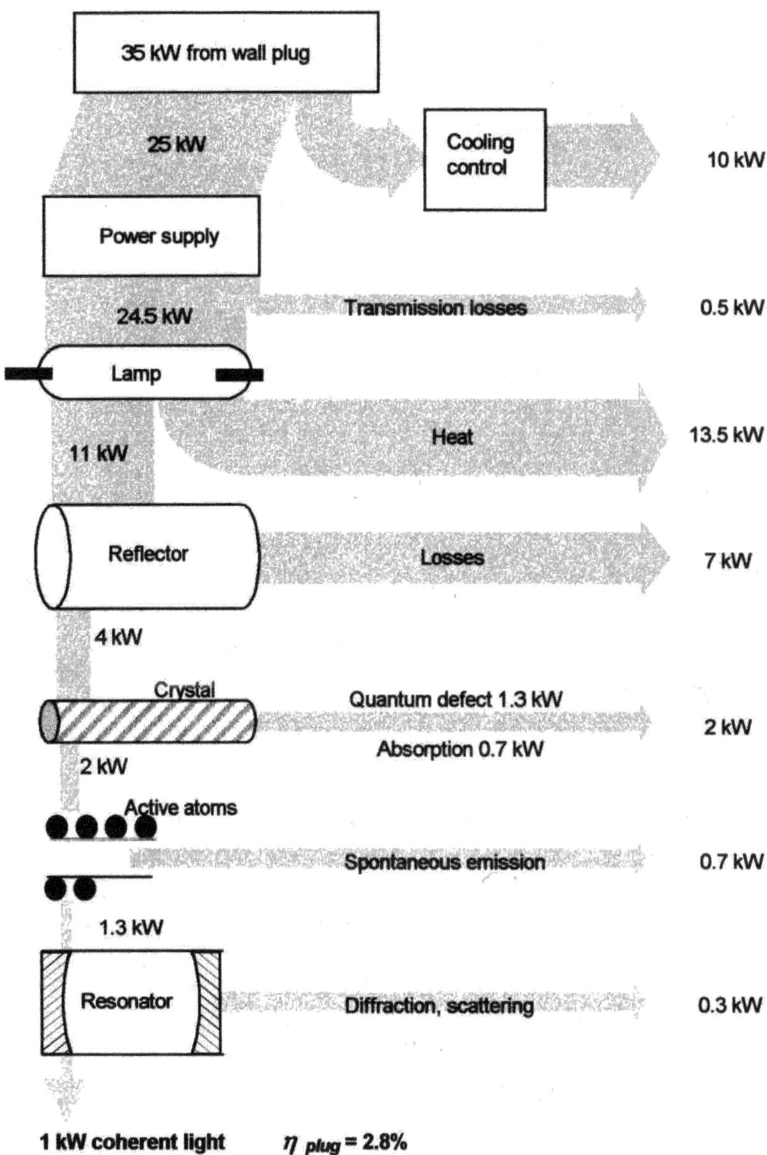

Fig 4.15 Energy flow of a 1 kW lamp pumped Nd:YAG laser.

Table 4.9 Efficiencies of typical multimode Nd:YAG lasers

(The numbers vary considerably, depending on the particular design (lamps, cavities, crystal, flow tubes). For pulsed systems the efficiencies are generally higher than for cw operation.)

Efficiency (%)	Equation	lamp pumped	diode pumped
η_{excit}	4.16	7–8	30–50
η_{slope}	4.28	4–6	30–40
η_{tot}	4.29	4–5	10–30
η_{extr}	4.31	50–70	50–80
η_{plug}	4.40	2–3	5–10
η_Q	4.15	30–50	76
η_H	4.32	5–10	5–10

Plug efficiency. A certain amount of the total electrical power is used for the control and cooling units. This is considered by the plug efficiency, a number which characterises the total power consumption of the laser system:

$$\eta_{plug} = \frac{P_L}{P_{Etot}} \tag{4.30}$$

Extraction efficiency. The pumping power P_{excit} into the upper laser level cannot be extracted totally by the radiation field. To maintain the threshold population n_s equation (4.21), i.e. the necessary gain, a certain amount of pumping power is necessary, the threshold power, equations (4.16, 24):

$$P_{excit.th} = \eta_{excit} P_{th} = \left| \ln\left(V\sqrt{R}\right) \right| J_s F$$

This power is totally converted into fluorescence and can become rather large. The extraction efficiency is defined by:

$$\eta_{extr} = \frac{P_L}{P_{excit}} \tag{4.31}$$

The extraction efficiency depends on the losses and the reflectivity. Low reflectivity means high threshold gain factor and low extraction efficiency. The maximum extraction efficiency is obtained for the optimum reflectivity and can be taken from the diagram in Fig. 4.14.

Example

A Nd:YAG-rod with 5 mm diameter, $F = 19.6$ mm^2, a reflectivity of $R = 0.7$, a loss factor of $V = 0.95$ produces radiation losses by spontaneous emission of $P_{spon} = P_{excit,th} = 72$ W.

Determination of losses and excitation efficiencies

Findlay and Clay (1966) and Hollinger, Niedrig and Unruh (1992) developed a method to determine the losses and the excitation efficiency. The relation between reflectivity and threshold power reads according to equation (4.24)

$$P_{th} = \frac{\left| \ln \sqrt{R} + \ln V \right|}{\eta_{excit}} J_s F$$

P_{th} is measured for different reflectivities and plotted versus $-\ln\sqrt{R}$. The intersection with the x axis delivers the loss factor $-\ln V$, and from the slope (see Fig.4.16):

$$\gamma = -\frac{\partial P_{th}}{\partial \ln \sqrt{R}} = \frac{J_s F}{\eta_{excit}}$$

the excitation efficiency can be calculated. Errors may occur due to the cross section F, which is not known exactly in each case. Another difficulty is the assumption, that the excitation power is proportional to the electrical power of the lamp. To control this the fluorescence intensity which is proportional to the upper level population and proportional to P_{excit} has to be measured versus the electrical pumping power as shown in Fig. 4.16. There is a good linearity, but an offset exists, which has to be taken into account. An example for the determination of the loss factor V and the excitation efficiency η_{excit} for two different pumping devices is given in Fig. 4.17.

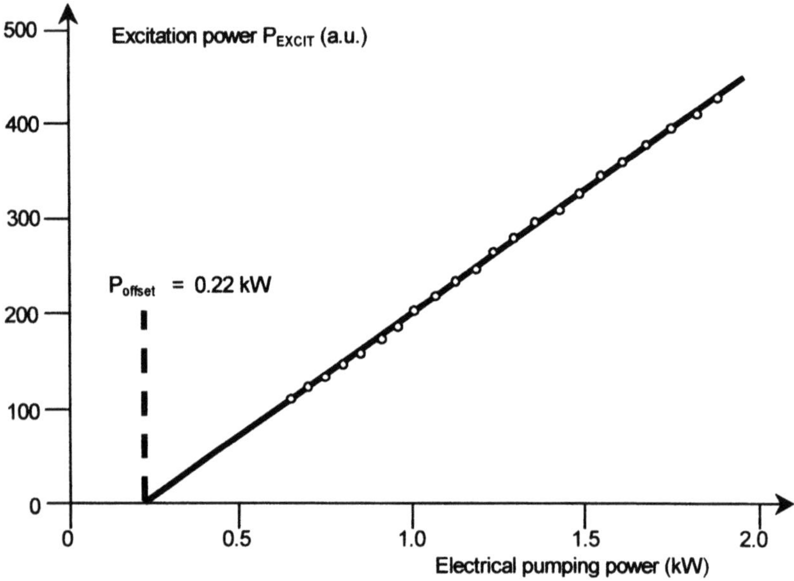

Fig. 4.16 Excitation power vs electrical power of the lamp. The offset of that particular device is 0.22 kW.

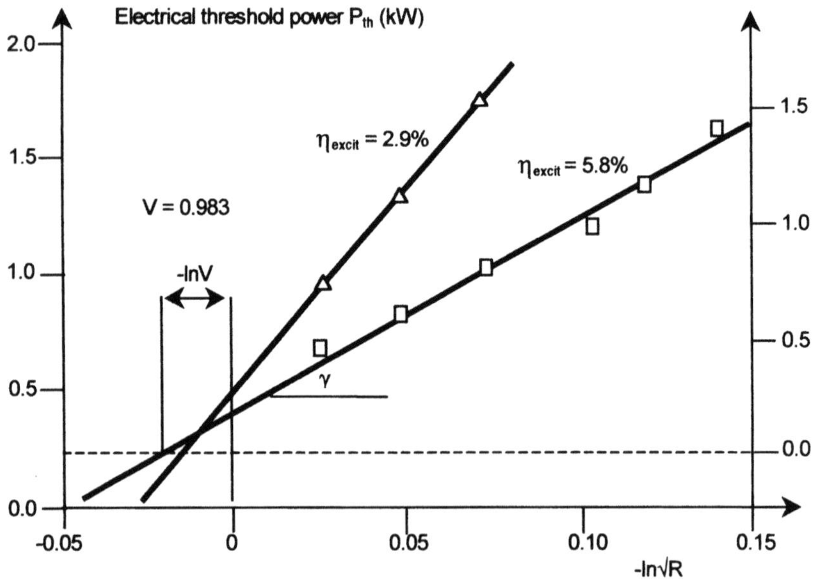

Fig. 4.17 Electrical threshold power for two different pumping devices, but the same crystal. The loss factor $V = 0.983$ is the same for both systems. The slope γ is different. The offset for the lamp is taken into account.

For fundamental mode operation errors of 10% can be obtained, but for higher order modes with unknown filling factors and inhomogeneous pumping light distribution the accuracy is much lower.

4.2.4 Thermal Effects

A considerable amount of the electrical pumping power P_E is converted into heat as shown in Fig. 4.15 and has to be removed by cooling. In particular the heat produced in the active medium is difficult to handle. The heat production inside the crystal is characterised by the heat efficiency

$$\eta_H = \frac{P_H}{P_E} \tag{4.32}$$

where P_H is the heat power generated in the active medium. To compare different pumping schemes and different lamps it is useful to refer the heat power to the output power

$$P_H = \frac{\eta_H}{\eta_{tot}} P_L$$

The active medium is cooled normally by water, and due to the heat conductivity κ of the active medium a temperature gradient occurs. In the following sections three geometries will be discussed as shown in Fig. 4.18:

1. cylindrical rod;
2. rectangular slab;
3. annular (tube) laser.

Rod geometry

The rod is cooled outside and heated more or less homogeneously inside. Then a radial, parabolic temperature profile occurs. The temperature difference inside the rod is

$$\Delta T = \frac{P_H}{4\,\kappa \pi l} \tag{4.33}$$

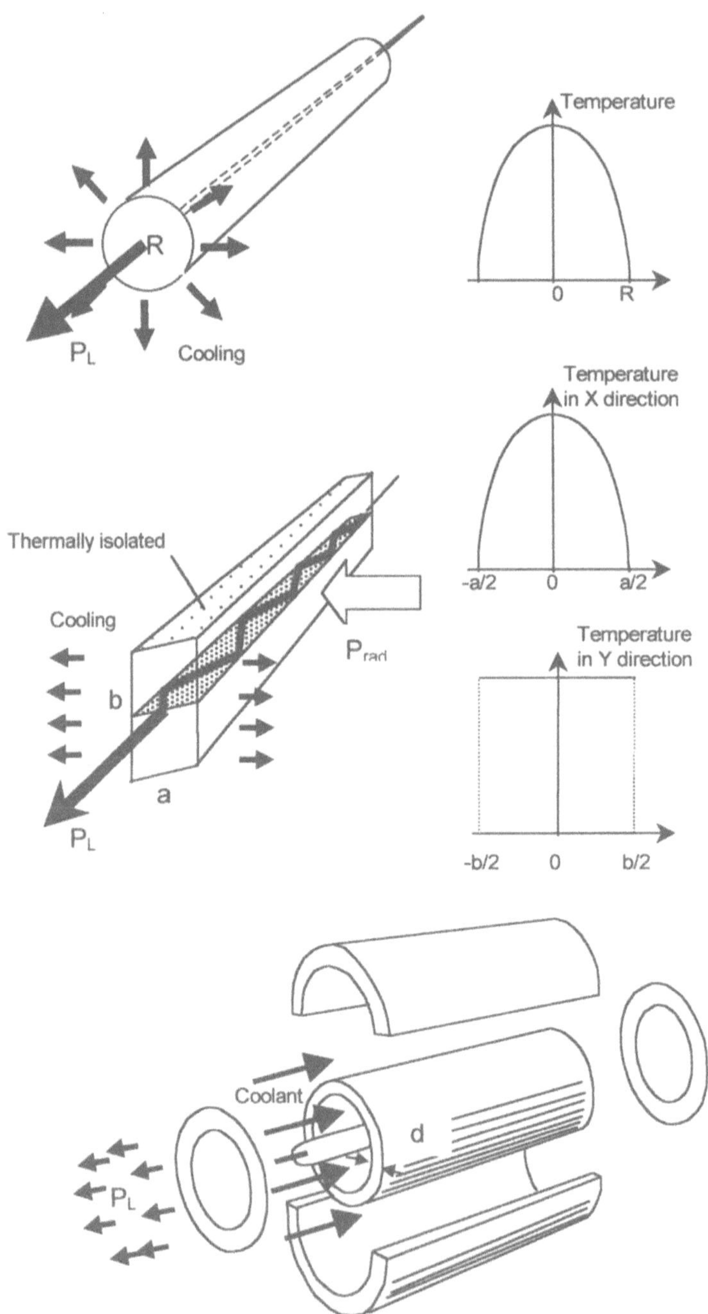

Fig. 4.18 Three geometries for high output power: rod, slab and tube.

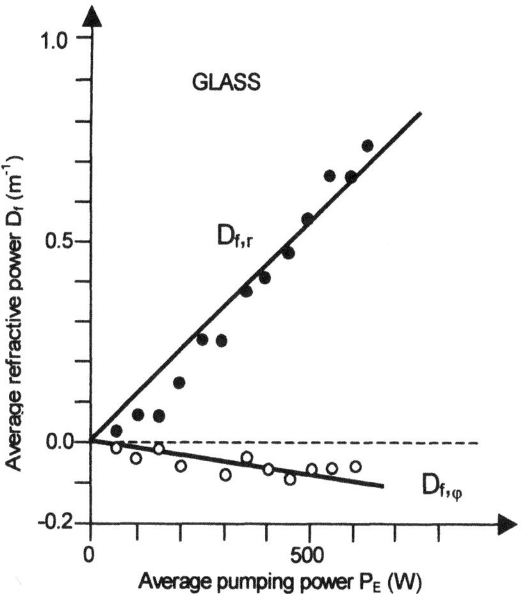

Fig. 4.19 Average refractive power D_f for a Nd:Glass rod (8 mm diameter, LG 760) vs average electrical pumping power P_E for the two polarizations r, φ.

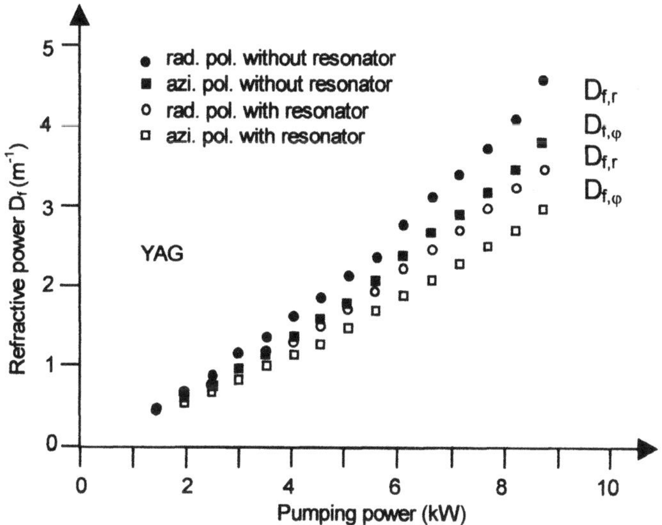

Fig. 4.20 Refractive powers $D_{f,r}$ and $D_{f,\varphi}$ for radial and azimuthal polarisation vs electrical pumping power P_E. Nd:YAG rod, 8 mm diameter.

The temperature gradient produces mechanical stresses and a parabolic refractive index gradient. The refractive index profile is caused ~ 85% by the temperature gradient and ~ 15% by stress. This part leads to circular birefringence, the refractive power is different for r (radial)- and φ (azimuthal) polarisation (Figs. 4.19,20).

The parabolic refractive index profile acts on a plane wave like a lens, and the rod can be characterised by a refractive power D_f. It is a thick lens, with the principal planes pp at a distance $h = l/2n_r$ from the surfaces as plotted in Fig. 4.21. n_r is the refractive index and l the rod length.

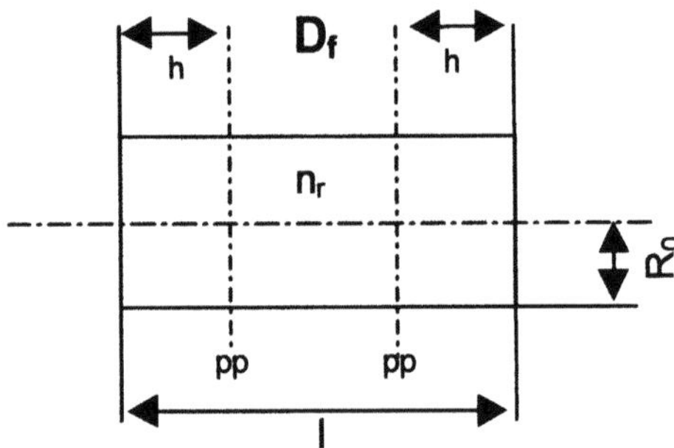

Fig. 4.21 The thermal lensing of a rod produces a thick lens with refractive power D_f and two principal planes pp.

The refractive power D_f is approximately proportional to the absorbed heat power P_H, and therefore proportional to the electrical power P_E of the lamp. The refractive power for unpolarised light is given by:

$$D_f = \frac{\beta P_E}{\pi R_0^2} \tag{4.34}$$

With β the thermal lensing coefficient and R_0 the rod radius. Some numbers are given in Table 4.6. This thermal lensing effect has a major influence on the beam quality, as will be shown in section 4.3. The stress σ caused by the temperature gradient has also a parabolic profile with the maximum at the rod surface. It does not depend on the rod radius, but only on the heating power per length.

The maximum stress at the rod surface $r = R_0$ is:

$$\sigma_{R0} = \frac{P_H}{l} \frac{\alpha E}{8\pi\kappa(1-v)}$$

(4.35)

where α = thermal expansion coefficient
   E = Young's modulus
   v = Poisson's ratio

The stress must be lower than the rupture stress σ_{max}. This implies an upper limit for the heating power P_H per length:

$$\frac{P_{H,max}}{l} < 8\pi R_t$$

with R_t a characteristic parameter of the material, the thermal shock parameter (Koechner, 1990)

$$R_t = \frac{\kappa(1-v)\sigma_{max}}{\alpha E}$$

(4.36)

The rupture stress σ_{max} depends on the particular crystal but also on the surface polishing. Therefore the maximum heating power must be lower by a safety factor of $s = 2...5$ than the above value.

Table 4.6 Thermal parameters of Nd:YAG and Nd:Glass

	Nd:YAG	Nd:Glass.	Nd:YLF
	*(*Experimental values, ** σ/π polarisation)*		
thermal lensing coefficient β (mm/kW)	0.023	0.16	-0.003/0.0014**
refractive power D_f (m^{-1})for $P_E = 1$ kW and $R_O = 5$ mm	0.3	2.1	-0.048/0.018**
thermal shock parameter R_t(W/cm)	7.9	0.6	2.2
maximum average output power per length $P_{L,max}$ /l (W/cm)*	35	1.5	10

With the efficiencies of section 4.2.3 the maximum output power per crystal length is obtained.

$$\frac{P_{L,\max}}{l} = \frac{8\pi R_t}{s}\frac{\eta_{tot}}{\eta_H} \tag{4.37}$$

Example

A glass rod with D = 8 mm is pumped with an average electrical power of $P_E = 1$ kW. The average refractive power according to Eq.(4.34) is $D_f = 12.5\ m^{-1}$, which corresponds to a focal length of f = 80 mm

---

Slab geometry

The slab is pumped and cooled from the two side surfaces, the upper and lower small surfaces are thermally isolated. Then the temperature in the y direction is constant and in the x direction a parabolic relation applies again, but with another temperature difference inside the slab.

$$\Delta T = \frac{a}{b}\frac{P_H}{8\kappa d} \tag{4.38}$$

The temperature difference can become smaller than for the rod, equation (4.33) if a small aspect ratio is chosen: $a/b < 2/\pi$. The same pumping power produces in that case a lower temperature difference, the stress is lower and the pumping power can be increased. Slabs with a small aspect ratio a/b can produce high output power per length. The aspect-ratio is limited by the mechanical stability, but also by the beam quality, as will be discussed in section 4.3. There is no thermal lensing in y- direction, because no heat flow appears in this direction. In the x direction thermal lensing remains, if the laser beam runs parallel to the z axis. If the beam crosses the slab in zigzag as shown in Fig. 4.18, the optical distortions are compensated to first order (see section 4.4.4). That is why a slab is of high interest. But the compensation of the distortions requires careful design of cavity and cooling arrangement.

Tube geometry

The tube is an annular rod, pumped inside (or outside) and cooled in and outside. Normally the thickness of the tube is small compared with the tube diameter, which allows efficient cooling. Then the slab formulas can be applied. The advantage of the tube is the high mechanical stability even for very small wall thickness, and the high excitation efficiency due to the close coupling of lamp and medium. The details will be discussed in section 4.4.

Thermally induced birefringence

Thermally induced birefringence means that an isotropic medium (glass, cubic crystals) becomes optically anisotropic due to the photoelastic effect, produced by the thermal load. In this case the propagation of light depends on the direction of the electric field vector with respect to a reference frame, which is related to the anisotropy.

As one example let us discuss the Nd:YAG rod, an isotropic (cubic) crystal. A homogeneously distributed heat source causes a parabolic temperature and stress profile as discussed above. Because the system is of circular symmetry, the three principal values of the stress tensor are σ_r, σ_φ, σ_z. If the light passes the rod parallel to the z axis, only σ_r and σ_φ are acting on the electric field vector $E = (E_r, E_\varphi, 0)$ in such a way, that the refractive index is different for both components as shown in Fig. 4.22.

This kind of polarisation is called radial/azimuthal and differs principally from the well-known linear, circular or elliptical polarisation state. The orientation of the E vector varies in the observation plane and cannot be transformed into normal polarisation by simple optical elements (retardation plates). In the centre of the field appears a singularity, which requires zero for the intensity, because of div $E = 0$. Such a mode is called a doughnut or donut. The refractive power of the rod now reads:

$$D_{r,\varphi} = \frac{\beta_{r,\varphi} P_E}{\pi R_0^2} \tag{4.39}$$

with $\beta_r = 0.25$ mm/kW and $\beta_\varphi = 0.22$ mm/kW for YAG. The birefringence even in non polarising resonators can produce radiation fields with purely radial/azimuthal polarisation or a mixture of both states. This means that the direction of the electric field vector is not uniform across the beam.

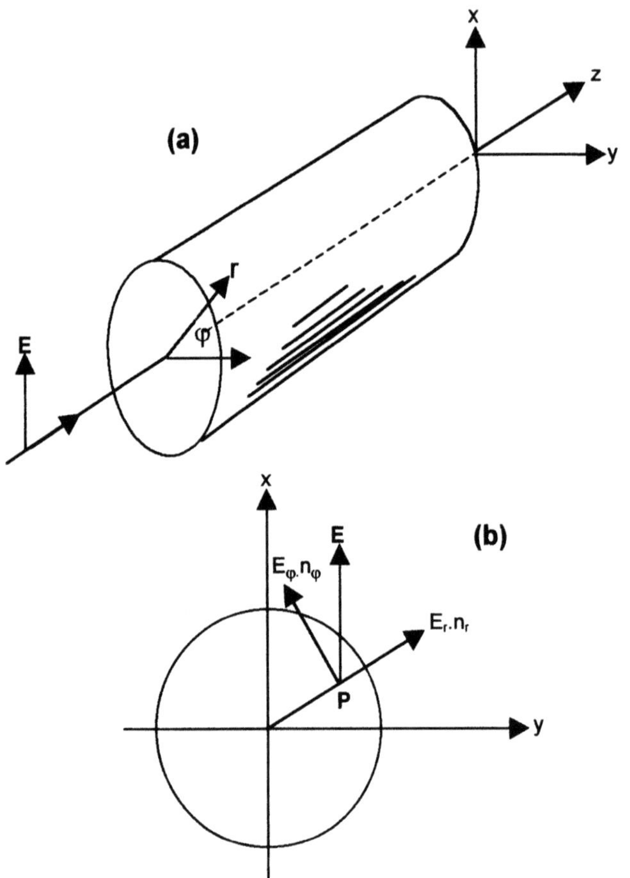

Fig 4.22 (a) A linearly polarised light wave becomes elliptically polarised after crossing a birefringent rod, the elliptical polarisation depends on the radial coordinate r. (b) The refractive index is different for the r and φ polarisation of the electrical field vector E.

A beam delivery system with polarizing elements (mirrors, Brewster plates) will affect the beam structure in an unpredictable manner. Moreover, fundamental mode oscillation is principally impossible in such a system and the beam quality becomes worse. If such a birefringent rod is observed between crossed polarizers the well known Geneva stop can be observed as shown in Fig. 4.23. In resonators with polarising elements the birefringence causes considerable losses. Compensation of these distortions at high power levels is difficult and no satisfying technical solution exists.

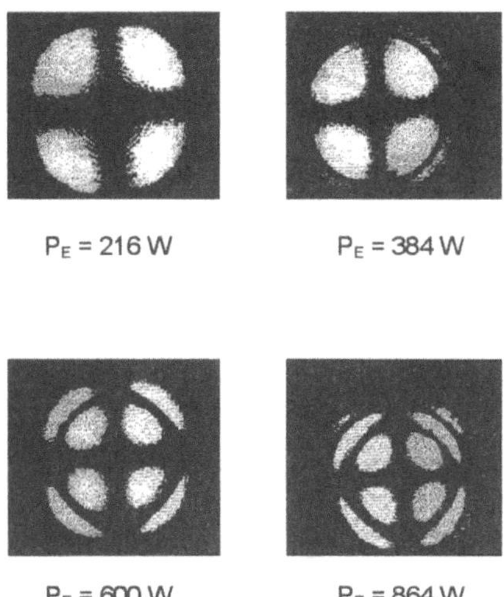

Fig. 4.23 A laser beam crossing a birefringent rod between crossed polarisers delivers the above intensity distribution (Geneva stop), depending on the pumping power P_E. Nd:Glass rod, EV4,3x1/4 inch.
upper left: 216 W; upper right: 384 W;
lower left: 600 W; lower right 864 W.

4.2.5 The pulsed mode of operation

The results of cw operation hold approximately for the pulsed mode operation as long as the pumping pulse duration Δt is long compared with the upper laser level lifetime τ_L. The situation is quite different, when the pumping pulses are short, or if the loss factor V of the resonator is modulated. Several pulsed modes of operation are possible as compiled in Table 4.7.

For material processing, the spiking mode is of major interest (i.e. drilling, spot welding) and will be discussed in more detail. Q-switch operation is important for precise material removal (trimming) and mode locking may become useful for ablation.

Table 4.11   Modes of pulsed operation with typical data of Nd:YAG systems

mode of operation	pulse repetition frequency f/Hz	pulse duration δt	peak power $P_{L,max}$
Spiking Pumping pulses with $\Delta t = 100\text{-}500$ µs	20-100 kHz	100-500 ns	10-100 kW
Q-switch,single pulse Optical switch with steep risetime (Pockels cell)	$> 1/\tau_L$	5-50 ns	10 MW
Q-switch, periodically periodical loss modulation with acoustooptical systems	1-100 kHz	10-100 ns	10-100 kW
Mode locking periodic loss modulation	100-200 MHz	10-50 ps	0.1-1 GW

Spiking

A typical spiking behaviour of the laser output is shown in Fig. 4.24. The laser oscillation starts a bit delayed after the pumping pulse, and consists of a sequence of sharp intensity peaks (spikes) which are damped and approach steady state. This transient behaviour is useful for special applications in material processing. The first spikes of high peak power interact with the matter nonlinearly and enhance the absorption, the following lower intensity is then much better absorbed. By adapting the spike repetition rate to the dynamics of material removal and plasma expansion, a much better quality in material processing can be achieved. The transient behaviour of the laser system is easy to understand and in principal equivalent to the transient behaviour of other oscillating systems, with one exception: the laser system is highly nonlinear.

After switching on the pumping light, the upper laser level is very fast populated and the ion density n exceeds the steady state value n_S.

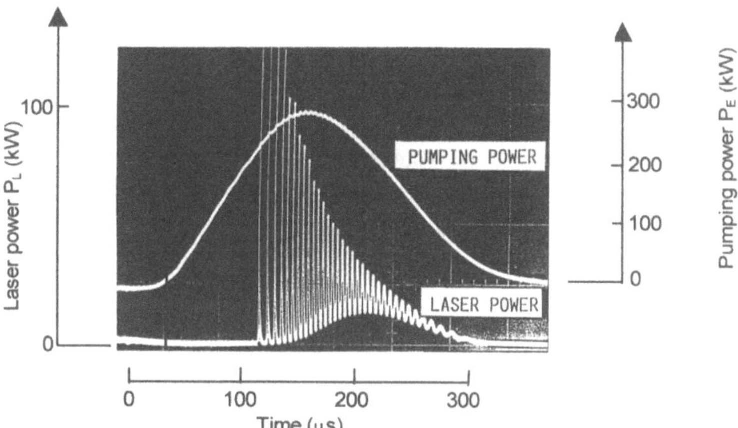

Fig. 4.24 Typical spiking emission of a fundamental mode Nd:YAG laser.

The intensity in the resonator is in the beginning very low, because it starts from noise. But once $n > n_S$ the total gain factor per bounce is larger than one and the noise intensity is amplified to a value $P_{L,max}$ higher than the steady state value P_L. Now, by induced emission, the population n is depleted below steady state, the total gain factor is below one and the laser power breaks down. This process can occur several times with a frequency f. Whether these oscillations are damped out or not depend on the parameters of the system (pumping power, mode structure) and shall not be discussed here. The dynamics of the oscillations can be only calculated numerically, and it is especially difficult to predict the peak power of the spikes. It depends on the mode volume and the spontaneous emission rate. For small deviations from steady state and fundamental mode operation a sinusoidal oscillation appears. Frequency f and damping time τ_D can be calculated (Siegmann, 1986; Iffländer, 1990):

$$2\pi f = \left[\frac{P_E / P_{th} - 1}{\tau_L \tau_R} \right]^{\frac{1}{2}} \qquad \text{spiking frequency} \qquad (4.40)$$

$$\tau_D = 2\tau_L \frac{P_{th}}{P_E} \qquad \text{damping time} \qquad (4.41)$$

$$\tau_R = \frac{\tau_c}{\left| \ln V \sqrt{R} \right|} \qquad \text{resonator decay time} \qquad (4.42)$$

τ_R is the decay time of the intensity inside the resonator, $\tau_c = L/c$ the cavity transit time. The above formulas hold within 50% accuracy. Larger deviations from experiment occurs for strong oscillations. The peak power of the first spike can exceed steady state by a factor of 10 to 100. Pulse widths down to 100 ns can be obtained. The normally damped spiking can be stabilised by modulating the losses V of the resonator with the spike eigenfrequency f. In that case undamped spikes appear. It is a typical resonance phenomenon. Very small modulation amplitudes are sufficient to produce strong spiking. This also means, that the laser system is very sensitive to mechanical vibrations in the frequency range of f. Typical values of f are 20–100 kHz.

Single pulse Q-switch

Due to the relative long lifetime of the upper laser level of $\tau_L = 200$ -300 μs, energy can be stored and then converted into a powerful laser output pulse. The stored energy is given by:

$$E_{st} = n_i h \nu_L Fl = \eta_{p,excit} E_P \tag{4.43}$$

n_i is the initial population in the upper laser level, produced by an electrical pumping pulse energy E_E without induced emission. The excitation efficiency $\eta_{p,excit}$ in that case depends on the shape, peak power and duration of the pumping pulse. To prevent laser oscillation, the feedback of the mirrors, at least of one, has to be interrupted. This can be done by optical shutters, mainly electrooptical devices (Pockels cells). The Pockels cell (PC) is a crystal which by a suitable voltage applied perpendicular to the z axis, works as a phase retarder. Linearily polarised light in the x direction, as shown in Fig. 4.25, is transformed into left circularly polarised light. After reflection at the mirror it becomes right circularly polarised and is by the PC again transformed into linear polarisation, but perpendicular to the x axis, and blocked by the polariser. Such a device works like an optical shutter and interrupts the feedback of the mirror, so no laser oscillation can be built up. Once the voltage at the PC is switched off, the laser oscillation can start. Due to the high gain, a pulse with a very fast rise time is built up and depletes the upper level. A great part of the stored energy is converted into electromagnetic energy and a powerful pulse is emitted. The principal temporal behaviour is outlined in Fig. 4.26, an example is given in Fig. 4.27.

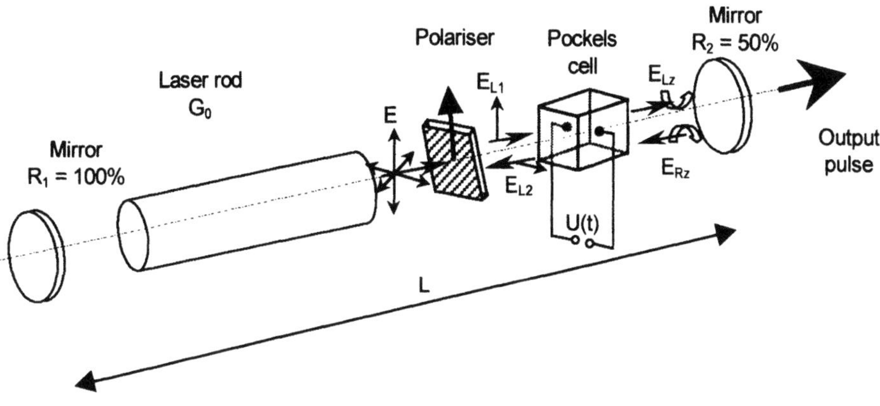

Fig. 4.25 Schematic setup for Q-switching.

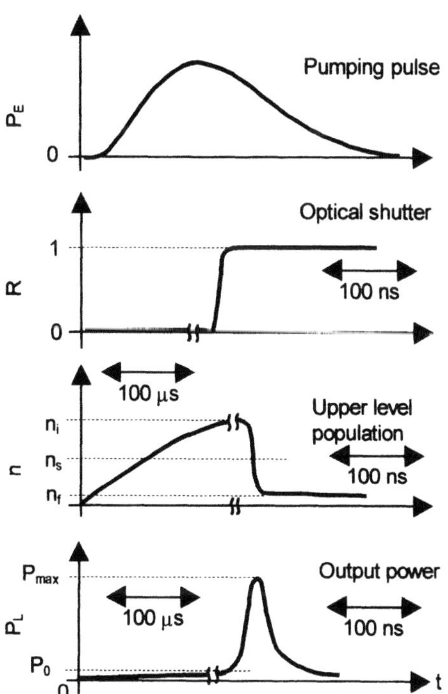

Fig. 4.26 Schematic time dependence of a laser system, operating in the single pulse Q-switch mode. The time scales of the plots are different.

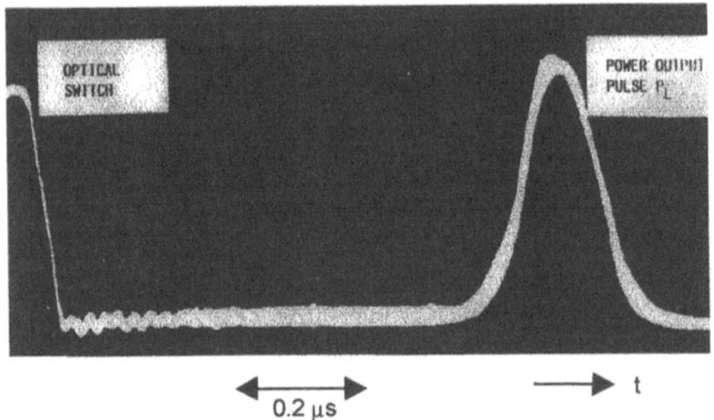

Fig. 4.27   Q-switch pulse generated with an electro-optical shutter. Width of display ≈ 1µs.

The whole process can be repeated, which finally leads to periodical Q-switching. But as long as the pulse repetition frequency f is small compared with the inverse upper level lifetime $1/\tau_L$ the periodic pulses have the same properties as the single pulse. Some simple relations can be found in literature (Yariv, 1970; Iffl), 1990). By storing a large amount of energy, short and high peak power pulses can be generated. An upper limit is given by parasitic oscillations. High initial population n_i means a high gain factor $G_0 = \exp(\sigma_L n_i l)$ and superradiance can occur, i.e. the population is depleted by amplified spontaneous emission or whisper modes oscillations. To prevent these parasitic oscillations, Nd:Glass can be employed, because for the same stored energy (same n_i), the gain is much lower due to the lower cross section σ_L for glass.

Periodical Q-switching

If the repetition frequency f of the Q-switching is much larger than the inverse upper level lifetime $1/\tau_L$, no energy can be stored on average. At frequencies of $f > 1$kHz, it becomes difficult to modulate the exciting lamp. In that case the excitation operates in steady state and the resonator losses are modulated by an acoustooptical element. An example is given in Fig. 4.28. The energy available per pulse is just the energy which can be stored between two pulses.

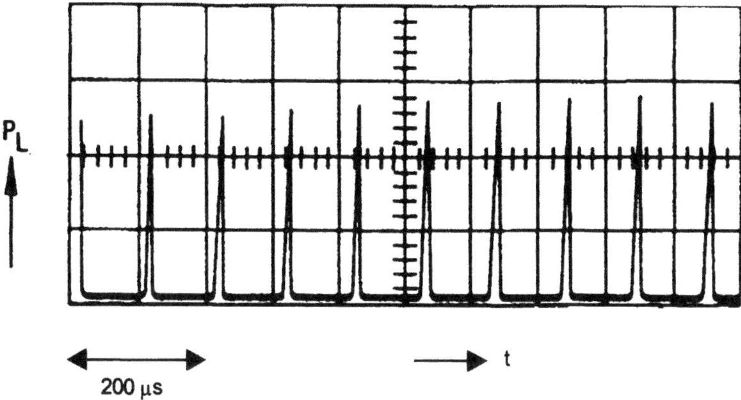

Fig. 4.28 Periodical Q-switch generated with an acousto-optical modulator.

If P_E is the electrical power and f is the pulse repetition frequency, the stored energy reads:

$$E_{st} = \eta_{excit}\,\frac{P_E}{f} \tag{4.44}$$

By some simple considerations the pulse parameters can be estimated (Ifflander, 1990). The dynamics of periodic Q-switching depend on a function $m(f)$ given by:

$$m(f) = \frac{1 - \exp(-1/f\tau_L)}{1 + \exp(-1/f\tau_L)} \tag{4.45}$$

Using this function output energy and peak power reads:

$$E_L = 2Fe_s m\big(\big|\ln\sqrt{R}\big|\big)\left(\frac{P_E}{P_{th}} - 1\right) \tag{4.46}$$

$$P_{l,max} \approx Fe_s\,\frac{\tau_c}{2\tau_R^2}\,\big|\ln\sqrt{R}\big|\left[m\left(\frac{P_E}{P_{th}} - 1\right)\right]^2 \tag{4.47}$$

The average power reads with equation (4.23) $P_{l,av} = fE_L = 2f\tau_L\,m\,P_L$ and is lower than in the unmodulated case, because during two pulses energy is lost by spontaneous emission. For high modulation frequency $f\tau_L \gg 1$ the average power approaches the cw power. The pulse duration is defined by the ratio of output energy to peak power.

$$\delta t = \frac{E_L}{P_{L,\max}} = 4\frac{\tau_R^2/\tau_c}{m\left[(P_E/P_{th})-1\right]} \tag{4.48}$$

Pulse peak power, duration and energy depend strongly on the frequency f. Increasing frequency at constant pumping power level means longer pulses, lower energy and reduced peak power as shown in Fig. 4.29.

Example

Nd rod with $D = 5mm$ and $l = 100$ mm is pumped four times above threshold $P_E/P_{th} = 4$. The outcoupling mirror has a reflectivity of $R = 0.7$. The resonator has a loss factor of $V = 0.85$ and a length of $L = 400$ mm with a transit time of 1.33 ns and a decay time $\tau_R = 3.92$ ns. The system is modulated at $f = 10$ kHz, $f\tau_L = 2.3$. The function m has the value $m = 0.21$. The output energy per pulse is $E_L = 16.5$ mJ, peak power $P_{L,\max} = 456$ kW. The average power is $P_{L,av} = 165$ W. The cw power without modulation (equation 4.23) is $P_L = 172$ W. The pulse duration is $\delta t = E_L/P_{L,\max} = 360$ ns.

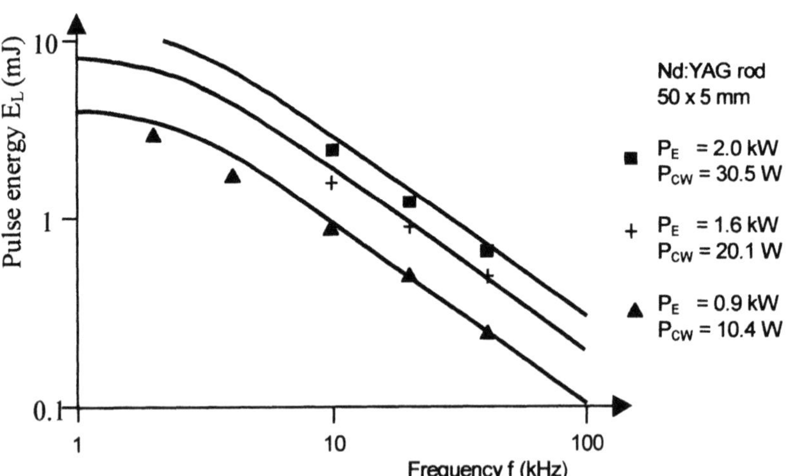

Fig. 4.29 Output energy of a periodically Q-switched Nd:YAG laser vs modulation frequency f for different pumping powers P_E (Iffländer, 1990). P_{cw} is the corresponding output power without modulation.

4.3 RESONATORS FOR HIGH POWER AND BEAM QUALITY

For high power solid state laser resonators with spherical mirrors are mainly used. They are characterised by their g parameters:

$$g_i = 1 - \frac{L}{\rho_i} \qquad i = 1,2$$

where L = optical distance between the mirrors
ρ_i = radius of curvature of the mirrors

Only for annular systems are nonspherical (toroidal) resonators under discussion (Wittrock, 1993). Two types of spherical resonators can be distinguished as shown in Fig. 4.30:

1. stable resonators with $0 < g_1 \cdot g_2 < 1$ and confined mode structures, which means that the mode diameter is determined essentially by the g_i parameters
2. unstable resonators with $g_1 \cdot g_2 < 0$, or $g_1 g_2 > 1$ and unconfined modes. The mode diameter depends on the internal apertures, mainly on the diameter of the active medium.

The resonator with two plane mirrors $g_1 = g_2 = 1$ is in between and widely used for rod lasers.

4.3.1 Lens resonators

The resonator with an internal lens as plotted in Fig. 4.31 can be replaced by an equivalent resonator without lens, characterised by new parameters $g_i{}^*$, L^*. For circular symmetry they are given by:

$$g_i^* = g_i - D_f d_j \left(1 - \frac{d_i}{\rho_i} \right) \qquad i,j = 1,2 \quad i \neq j$$

$$g_i = 1 - \frac{d_1 + d_2}{\rho_i} \tag{4.49}$$

$$L^* = d_1 + d_2 - D_f d_1 d_2$$

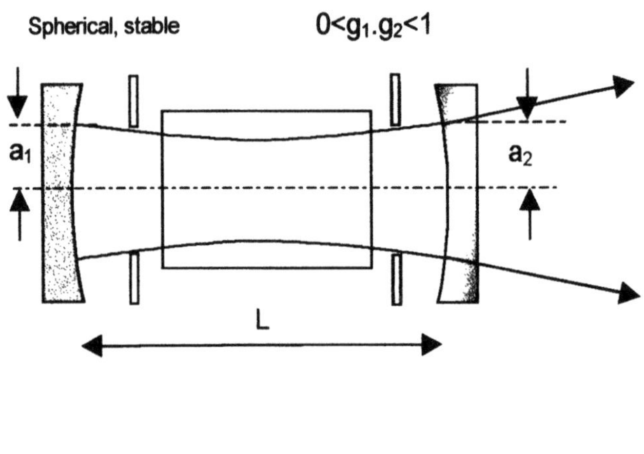

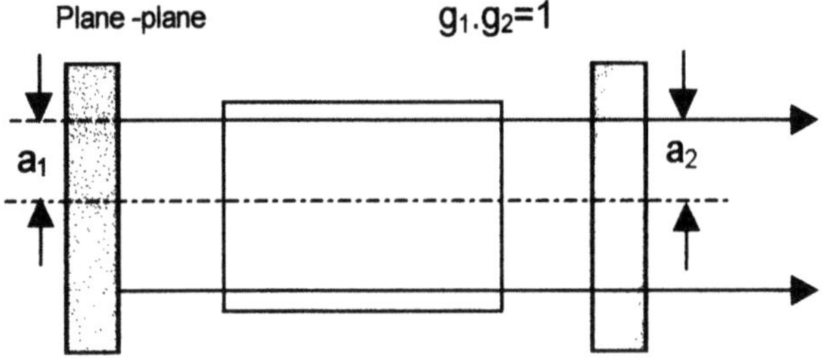

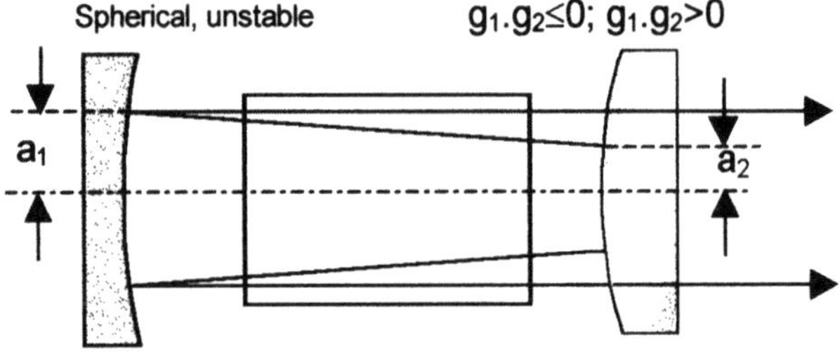

Fig. 4.30 The spherical stable and unstable resonator. The plane-plane resonator is just between stable and unstable.

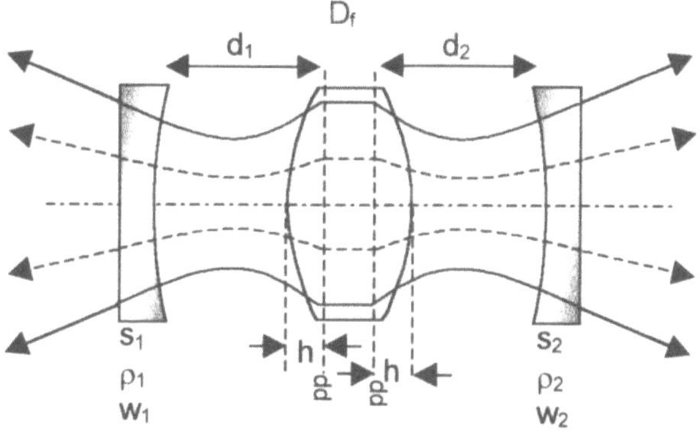

Fig. 4.31 The spherical resonator with an internal lens of refractive power D_f and the principal planes pp in a distance $h = l/2n_r$ from the rod surfaces,
l: geometrical length of the rod. The diameter $2R_0$ of the lens corresponds to the diameter of the active medium. The dotted lines indicate the slope of the fundamental mode. w_1, w_2: beam radius on the mirrors S_1, S_2.

where D_f = refractive power of the lens or rod, respectively.
  d_i = geometrical distance between the mirrors S_i and the
      corresponding principle plane $h = l/2n_r$
  l = geometrical length of the rod
  n_r = refractive index of the rod

Using the above parameters the formulas for normal empty resonators can be applied (Hodgson and Weber, 1997). A resonator can be represented in the g diagram (Fig. 4.32) by its coordinates $g_1{}^*$, $g_2{}^*$. Increasing refractive power D_f due to the thermal lensing of the rod means according to equation (4.49) that the resonator moves on a straight line in this diagram, starting with $D_f = 0$ at the point (g_1, g_2). With increasing pumping power P_E the refractive power normally grows linearly equation (4.34) and the resonator crosses stable and unstable regions in the g diagram.

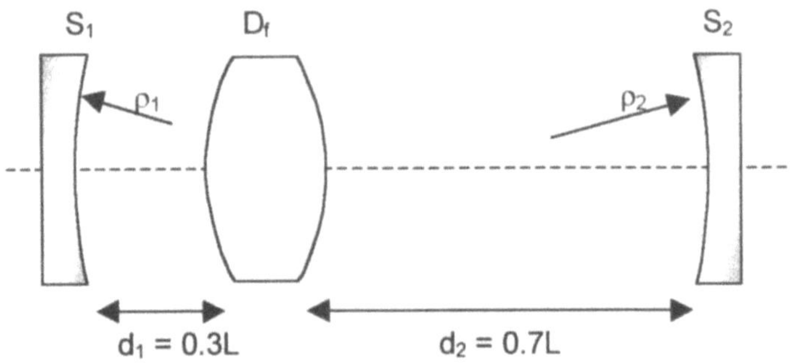

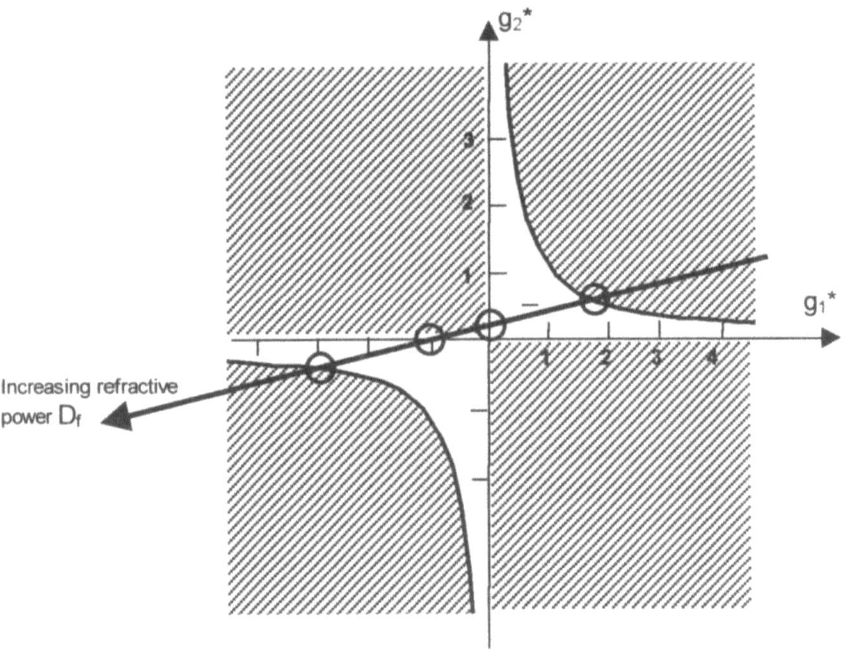

Fig. 4.32 A resonator of circular symmetry can be represented by a point $P = (g_1, g_2)$ in the g-diagram. With an internal lens, the resonator is characterised by new parameters g_1^* g_2^*. Increasing refractive power D_f means that the equivalent resonator moves in the diagram on a straight line. The plotted line represents a resonator with $L = 1$m, $\rho_1 = -1.5$ m, $\rho_2 = 2.5$ m, starting with $D_f = 0$ in the point $g_1 = 1.6$, $g_2 = 0.6$.

The transition stable $\rightarrow$ unstable occurs at the axes $g_1 = 0$ and $g_2 = 0$ and at the hyperbolae $g_1 g_2 = 1$. When the g parameters approach these limits the fundamental mode diameter increases very rapidly and approaches infinity; in reality the mode diameter is then limited by the internal apertures. A special point is $g_1 = g_2 = 0$, a singularity, where the value of the mode diameter is not unique. It depends on which way in the g diagram this point is approached. By means of a pinhole inside the resonator the mode structure can be defined.

4.3.2 Stable resonators ($0 < g_1 \cdot g_2 < 1$)

The slope of the radiation field inside the lens resonator is plotted in Fig. 4.31. Two waists w_{01}, w_{02} appear and the divergences of the left- and right-mirror output fields are different. It is straight forward, but tiring to calculate waists and waist positions applying the ABCD-law. The resulting formulas are lengthy and give not much insight into the physics. They can be taken elsewhere (Hodgson and Weber, 1997). The following discussion is restricted to the symmetric plane plane resonator as plotted in Fig. 4.33. Symmetric means that the rod is placed in the centre of the resonator. This type of resonator is widely used for high power systems. It is characterised by:

$$g_1 = g_2 = 1$$
$$d_1 = d_2 = d$$

Its equivalent parameters read:

$$g_1^* = g_2^* = 1 - D_f d \qquad (4.50)$$
$$L^* = d\left(2 - D_f d\right)$$

Equal beam waists appear on both mirrors. Using well known resonator formulas (Siegmann, 1986; Hodgson and Weber, 1997) the fundamental beam waist diameter D_o on the mirrors can be calculated

$$D_o^2 = \frac{4\lambda d}{\pi} \sqrt{\frac{2 - D_f d}{D_f d}} \qquad (4.51)$$

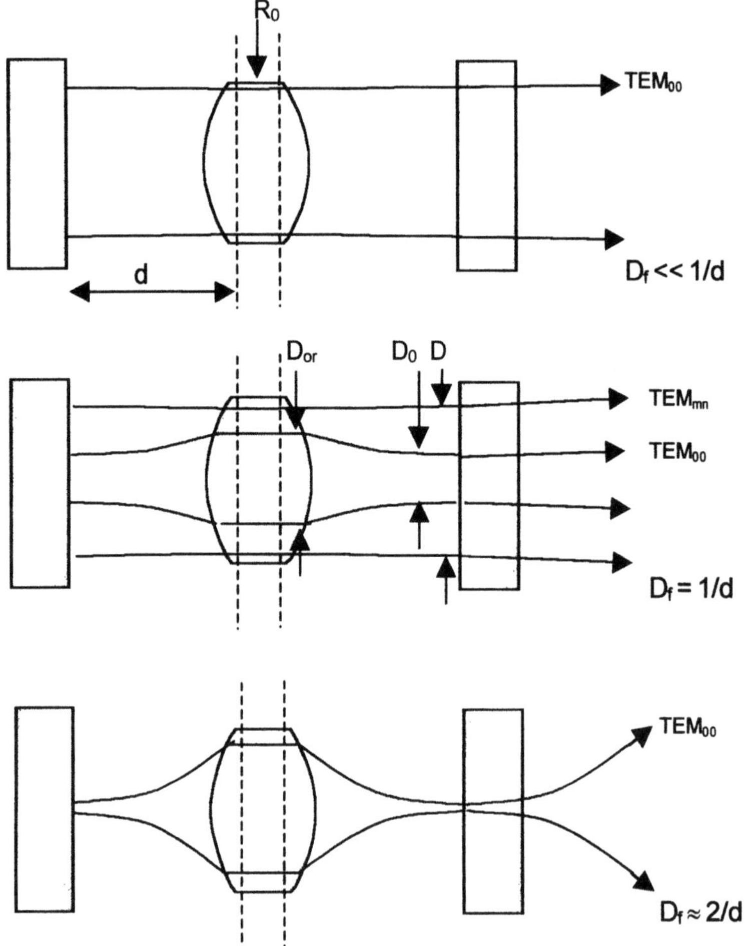

Fig. 4.33 The symmetric plane-plane- resonator with a thermal lens inside. With increasing refractive power D_f, the fundamental mode diameter D_0 on the mirrors varies from a very large value to nearly zero, whereas the mode diameter D_{0r} in the rod runs from the rod diameter $2R_0$ via a minimum to $2R_0$ again.

and the full angle of divergence is given by

$$\theta_0 = \frac{4\lambda}{\pi D_0} \tag{4.52}$$

Note: the unit of θ is radians (rad). Beam diameter and beam divergence depend on the thermally induced refractive power D_f of the rod and will vary with the pumping power P_E, equation (4.34).The fundamental beam diameter at the principal planes in the rod can be obtained by applying the ABCD law for beam propagation:

$$D_{0r}^2 = \frac{8\lambda d}{\pi}\, \frac{1}{\sqrt{D_f d(2 - D_f d)}} \tag{4.53}$$

Without thermal lensing, $D_f = 0$, the fundamental mode diameter D_{0r} is limited by the rod diameter $2R_0$, high diffraction losses occur, and the system will not oscillate. With increasing D_f the beam waist D_{0r} becomes smaller. The laser will start when $D_{0r} < 2R_0$ i.e. when the fundamental mode fits the active medium. This condition together with equation (4.53) delivers:

$$D_f d > D_{fc1} d \approx 2\left(\frac{\lambda d}{\pi R_0^2}\right)^2 \tag{4.54}$$

D_{fc1} is called the first critical refractive power. A second critical point occurs at

$$D_{fc2} d = 2 - D_{fc1} d \tag{4.55}$$

Example

A resonator with $d = 0.5$ m and a Glass-rod of $R_0 = 3$ mm has for $\lambda = 1$ µm a first critical refractive power, Equation (4.54) $D_{fc1} = 9.3 \times 10^3$ m^{-1}. This small refractive power is produced by an electrical pumping power of $P_E = 1.6$ W. This means that the system becomes immediately stable. The second instability occurs at $D_{fc2}\,d \approx 2$, $D_{fc2} \approx 4$ m^{-1}. The corresponding electrical pumping power is $P_E = 706$ W.

The plane-plane resonator is at threshold already a stable, spherical resonator, as is demonstrated by the example in the footnote. With increasing pumping power, i.e. increasing refractive power, both waists D_0 on the mirrors and D_{0r} in the rod decrease as schematically outlined in Fig. 4.33. When $D_f = 1/d$ the confocal point $g_1 = g_2 = 0$ is reached and D_0 has a minimum. Beyond this minimum D_0 increases again and for $D_f \approx 2/d - D_{fc2} \approx 2/d$ the resonator becomes unstable again. When the fundamental mode does not fill the rod completely, parts of the active medium remain unsaturated. The total gain is larger than one and the system becomes unstable. In that case higher order modes TEM_{mn} will appear with a larger beam diameter D_r in such a way that the rod is completely filled with the radiation field. At the principal planes the beam diameter is:

$$D_r = MD_{0r} = 2R_0 \tag{4.56}$$

where M is a factor, defined by equation (4.56), which characterises the mode order. The same holds on the mirror

$$D = MD_0 \tag{4.57}$$

From mode theory is known (Siegmann, 1986; Hodgson and Weber, 1997) that for higher order modes beam divergence and waist diameter increase equally, therefore

$$\theta = M\theta_0 \tag{4.58}$$

The characteristic parameter of the output radiation is the beam parameter product, which results from the above equations

$$\frac{D\theta}{4} = M^2 \frac{\lambda}{\pi} \tag{4.59}$$

M is given by equations (4.53, 56)

$$M^2 = \left(\frac{2R_0}{D_{0r}}\right)^2 = \frac{\pi R_0^2}{2\lambda d}\sqrt{D_f d(2 - D_f d)} \tag{4.60}$$

Some examples have already been given in Table 4.6. Equation.(4.60) holds only in the stable region $D_{fc1} \le D_f \le D_{fc2}$. At the critical points M is nearly equal to one and the resonator oscillates in the fundamental mode, if no other distortion occurs. Additionally in this region the system is very unstable. For refractive powers larger than D_{fc1} the parameter M increases and reaches a maximum at $D_f = 1/d$ with

$$M^2_{\text{max}} = \frac{\pi R_0^2}{2\lambda d} \tag{4.61}$$

and than decreases again. The maximum beam parameter product is immediately obtained by equation (4.4).

$$\left[\frac{D\theta}{4}\right]_{\text{max}} = \frac{R_0^2}{2d} \tag{4.62}$$

Example

A resonator with a total lenth of L = (about) 1 m, d = 0.5 m, a rod radius of R_0 = 3 mm, has for λ = 1μm a maximum beam propagation factor of M^2 = 28.3. The beam parameter product is $(D\theta/4)_{max}$ = 9.6 mm.mrad.

The relation between refractive power and beam parameter product according to equations (4.59, 4.60) is parabolic as shown in Fig. 4.34. The above discussion holds for the symmetric plane plane resonator only, but similar results are obtained for other stable resonator configurations. It can be proved (Hodgson and Weber, 1997) that the plane plane-resonator is that with the largest stable region. The stable range of refractive power is approximately:

$$\Delta D_f d \approx 2 \tag{4.63}$$

The refractive power is related to the pumping power by equation (4.44) which delivers for the corresponding pumping power range:

$$\Delta P_E = \frac{\Delta D_f \, \pi R_0^2}{\beta} = \frac{2\,\pi R_0^2}{\beta d} \tag{4.64}$$

equations (4.63, 64) deliver an interesting relation between the maximum beam parameter product and the range of stability:

$$\frac{\Delta P_E}{(D\theta/4)_{max}} = \frac{4\pi p}{\beta} \tag{4.65}$$

where $p = 1$ for symmetric resonators
$\quad\quad\quad p = \frac{1}{2}$ for all other resonators

This relation was derived for the symmetric plane plane resonator, but holds for all symmetric, stable resonators. For asymmetric resonators a factor of p = ½ appears, because the resonator does not cross the singularity ($g_1 = g_2 = 0$) and two stable regions of half the width exist.

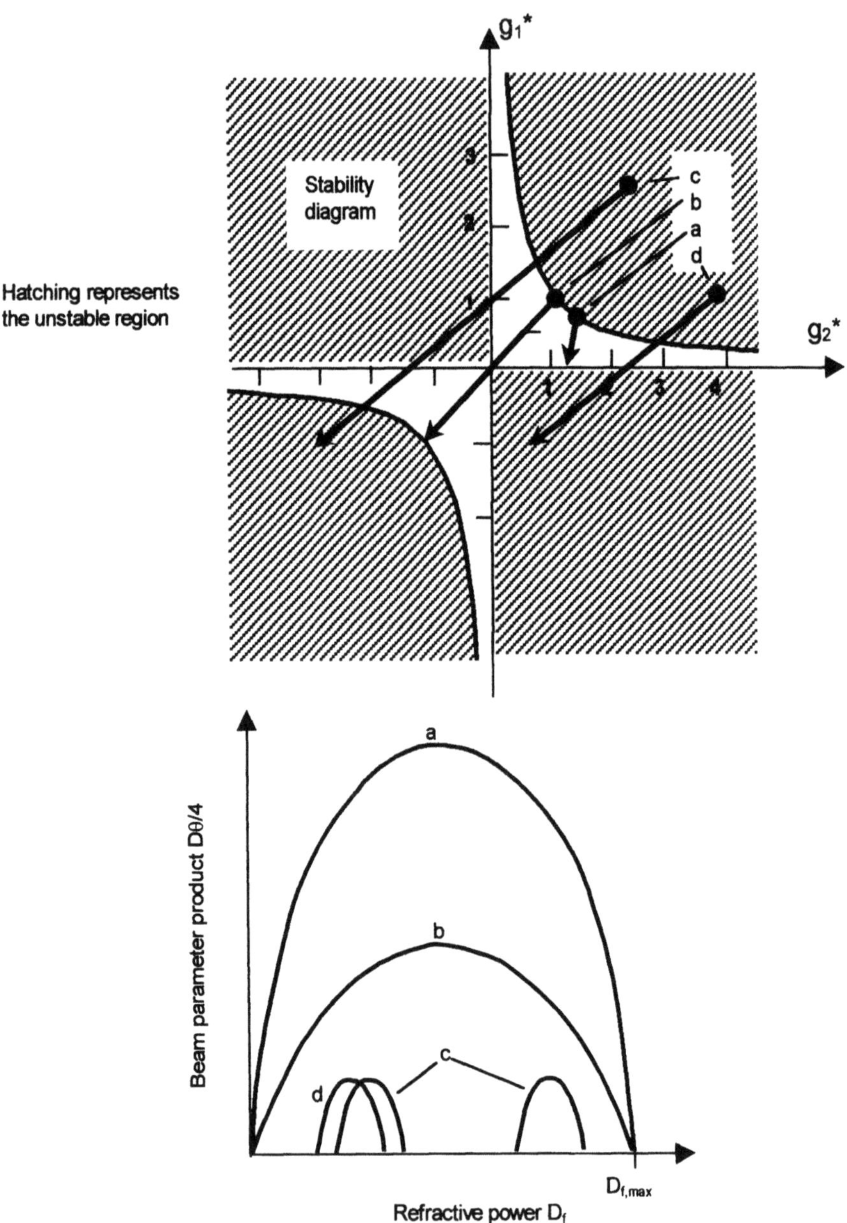

Fig. 4.34 The beam parameter product Dθ/4 vs refractive power D_f for different resonator configurations. The ratio of maximum beam parameter product to the range of stability is smaller by a factor of 2 for the symmetric plane-plane resonators than for all other configurations.

The above relation depends only on the thermal lensing coefficient β, not on the resonator configuration, nor on the rod dimensions. If a small beam parameter product is desired, according to equation.(4.62) a long resonator $L = 2d$ has to be used, but then the range of stability is reduced equation.(4.64). The same holds for the rod radius R_0. To improve the system, that means to have a low beam parameter product and a broad range of stability, the thermal lensing coefficient has to be reduced. This can be done by a well adapted pumping light spectrum (diode pumping).

In Fig.4.35 some experimental results are compiled, which confirm the above discussion.

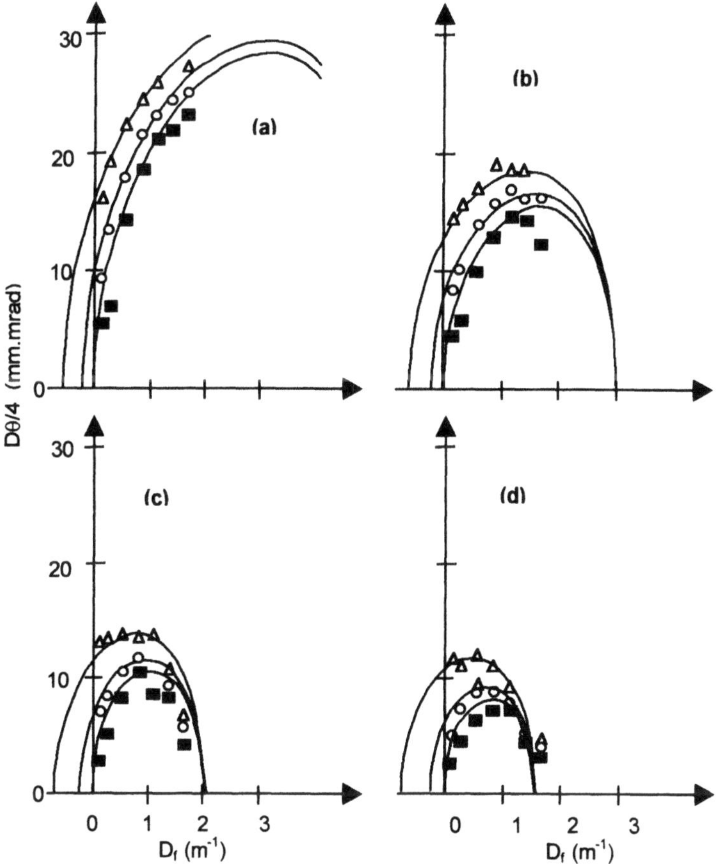

Fig. 4.35 Experimental results. Beam parameter product of Nd:Glass rod (LG 706, $\ell = 100$ mm, 6 mm diameter) vs refractive power of the rod for different resonators. $d_1 = 0.033$ m, $\rho_1 = \infty$, $\rho_2 = \infty$ (■), 5 m (o), 2 m (Δ). a) $d_2 = 0.15$ m, b) $d_2 = 0.3$ m, c) $d_2 = 0.45$ m, d) $d_2 = 0.6$ m.

If a laser with low beam parameter product has to be designed, this can be done by a suitable resonator configuration as shown in the figure, but then the system has to operate at a fixed power and only a small variation is allowed. At high pumping levels, when the refractive power approaches the upper critical value D_{fc2}, theory predicts again fundamental mode oscillation ($M = 1$). But now additional distortions occur:

- the thermal conductivity is temperature dependent and the refractive index profile becomes non-parabolic.
- the thermally induced lens is no longer a spherical one, which reduces beam quality.
- birefringence becomes important and reduces beam quality too.

An experimental example of the beam parameter product of a Nd:YAG laser is given in Fig. 4.36.

Conclusion

High power lasers in stable resonators produce a large beam parameter product (low beam quality). Typical values are given in Fig. 4.49. Moreover, the beam parameter product depends on the pumping power (output power), which means that the focus diameter and the focal length are not constant. This is a severe disadvantage for applications. The focus position is constant in the case of a plane plane resonator, if a telescope is used to image the output mirror waist on the fibre input or working piece.

4.3.3 Unstable resonators ($g_1g_2 > 1$; $g_1g_2 < 0$)

The geometrical approach

In unstable resonators the radiation field is not confined. The field diameter D increases from bounce to bounce until it is limited by the internal apertures. An example is outlined in Fig. 4.37. In the steady state the field can be approximated (geometrical optics) by two spherical waves starting from the virtual or real foci Q_1, Q_2.

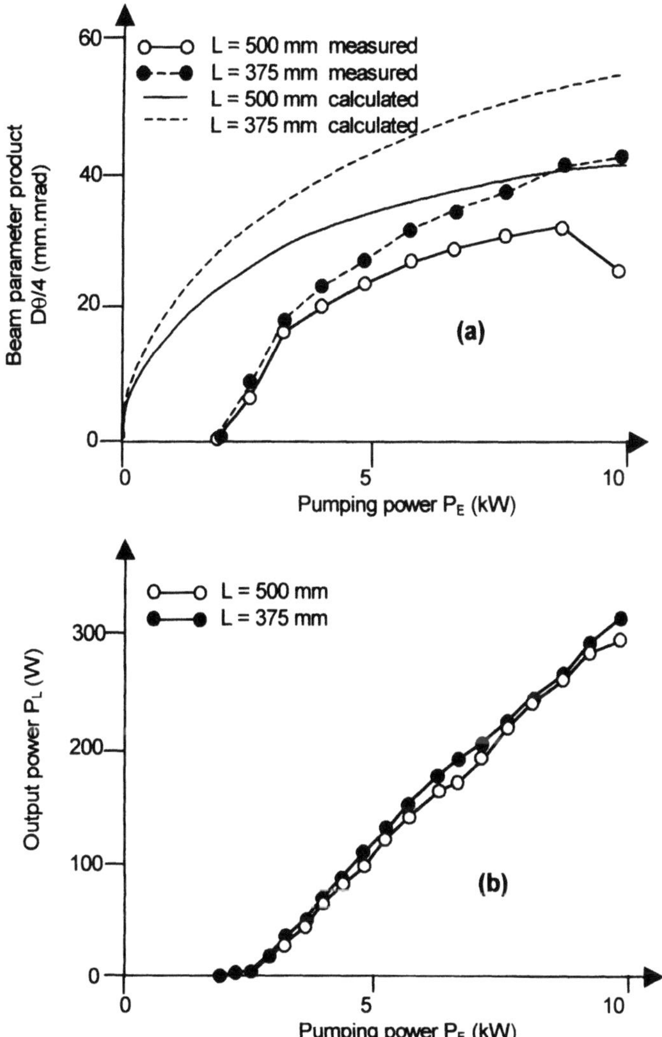

Fig. 4.36 Beam parameter product (a) and output power P_L (b) vs pumping power P_E of an 8mmx152mm Nd:YAG rod for two resonator lengths L.

Two branches are distinguished:

1. $(g_1^* g_2^* > 1)$ positive branch, virtual foci;
2. $\{g_1^* g_2^* < 0)$ negative branch, real foci inside the resonator.

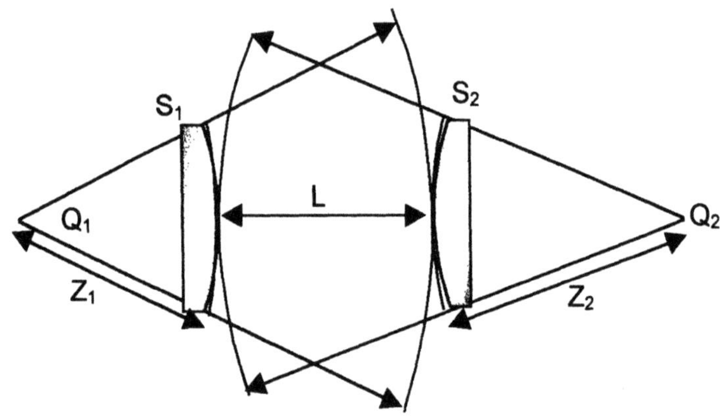

Fig. 4.37 Schematic setup of an unstable resonator.

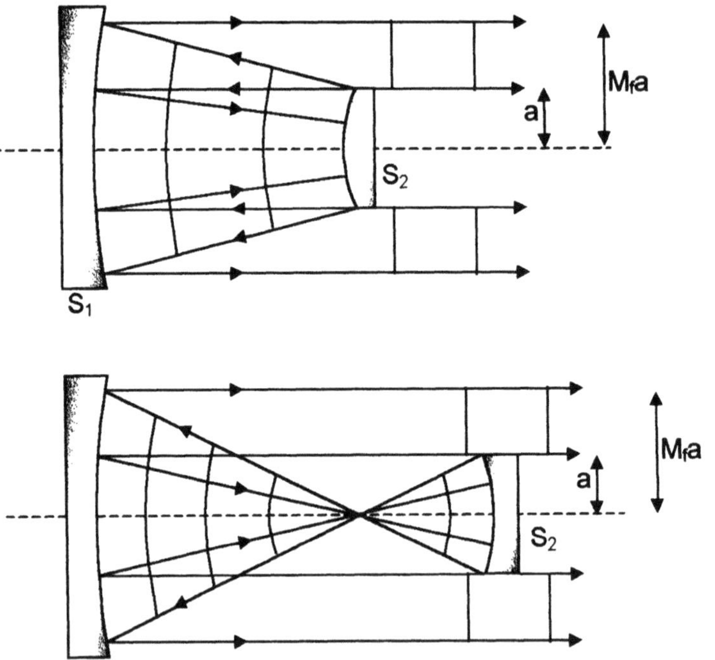

Fig. 4.38 The confocal, unstable resonator.
Top: positive branch: Bottom: negative branch.

$g_1{}^* g_2{}^*$ are the g parameters for a resonator with an internal lens, as defined by equation.(4.49). For most systems the positive branch is preferred in order to avoid hot spots in the active medium. Moreover a special configuration, the confocal resonator, is used as shown in Fig. 4.38. Confocal means that the foci of the two mirrors coincide. This has the advantage, that the system delivers a collimated output beam. The condition for confocal resonators is:

$$g_1^* g_2^* = g_1^* + g_2^*$$

(4.66)

The magnification M_f per bounce can be calculated (Hodgson and Weber, 1997) and reads:

$$M_f = |G| + \left(G^2 - 1\right)^{1/2}$$

(4.67)

$$G = 2g_1^* g_2^* - 1$$

(4.68)

The magnification depends on the g parameters only. M_f = const are the lines of constant $g_1{}^* g_2{}^*$ as shown in Fig. 4.39.

In the case of thermal lensing the resonator moves in the g diagram on a straight line as outlined in Fig. 4.32. Lines of constant magnification are crossed. That means with increasing pumping power for YAG systems operating in the positive branch, the magnification starting at a point with g_i > 0 becomes smaller, and approaches 1 at the limit to stability. The resonator enters the stable region. For confocal systems the magnification simplifies to:

$$M_f = \frac{g_2^*}{g_1^*}$$

(4.69)

In the case of unstable resonators the mirrors are highly reflective with $R_1 = R_2 = 1$. One mirror (S_1 in Fig. 4.38) is large compared with the field diameter, the other one has a radius a, adapted to the active medium radius R_0. The field starting on the mirror S_2 with radius a is reflected from S_1 and magnified by M_f. It bounces again on the mirror S_2, but only a part is reflected back, another part leaves the resonator and is the outcoupled beam, now with a hole in the centre. The inner radius is a and the outer radius $M_f \cdot a$. The loss in the geometrical approach is equal to the relative ring area

$$\Delta V = \frac{M_f^2 - 1}{M_f^2}$$

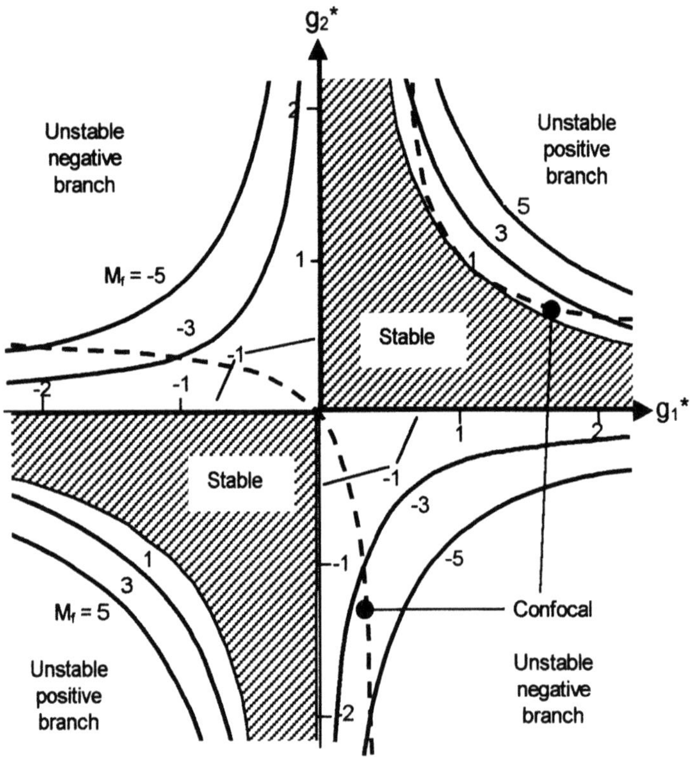

Fig. 4.39 The g-diagram with lines of constant magnification M_f.

from which the loss factor V per bounce becomes:

$$V = \frac{1}{M_f^2} \tag{4.70}$$

This hollow beam becomes compact and gaussian shaped in the far field $z \gg z_f$ or in the focal plane of a lens. An example is given in Fig. 4.40. Thermal lensing, i.e. varying g_i^*, means that the magnification, the loss factor and the outcoupled power of the laser field vary. It also means that the resonator is confocal only for one special refractive power D_f. A problem difficult to handle.

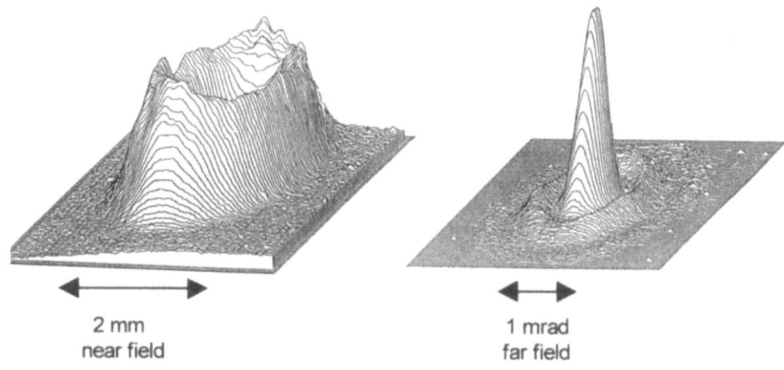

Fig. 4.40 Near and far field of an unstable resonator with $g = 1.1$.

So far, the resonator has been discussed in the geometrical approach. If diffraction is taken into account, field structure and loss factor are slightly different (10–50%). Therefore the above relations are only a rough estimation. The influence of diffraction is discussed in detail in the textbooks by Yariv (1970), and the influence of the active medium on field structure and loss factor by Hodgson and Weber (1997). It is difficult to optimise an unstable resonator, which can be done only numerically. To obtain maximum output power the following parameters have to be considered:

1. The outcoupling mirror diameter $2a$ must be adapted to the rod diameter $2R_0$. But a is not equal the rod radius, because the field can be divergent or convergent inside the resonator, depending on the G parameter of equation (4.68).
2. The outcoupling factor $T = 1 - R_{opt}$. R_{opt} can be taken from the diagram in Fig. 4.14, if the other losses are known (scattering losses, losses of the highly reflective mirror).
3. If G and $V = R_{opt}$ are known, the magnification M_f can be evaluated (Hodgson and Weber, 1997).
4. Finally thermal lensing D_f must be taken into account, which depends on the output power. The system can be optimised for one pumping power level only. But by choosing a suitable set of parameters this dependence can be minimised.

The whole procedure is difficult and not very reliable. Therefore it is easier to use the approach of geometrical optics and to determine the correct values by trial and error.

Beam quality of unstable resonators

The beam parameter product is more difficult to discuss compared with the stable resonator, because no analytic solutions exist, and numerical calculations have to be performed. But again, geometrical optics are a good approach. In this approximation the output field of the confocal system is a collimated plane wave. The divergence θ is given by diffraction at the mirror S_2. The far field structure of an annular plane wave is shown in Fig. 4.41.

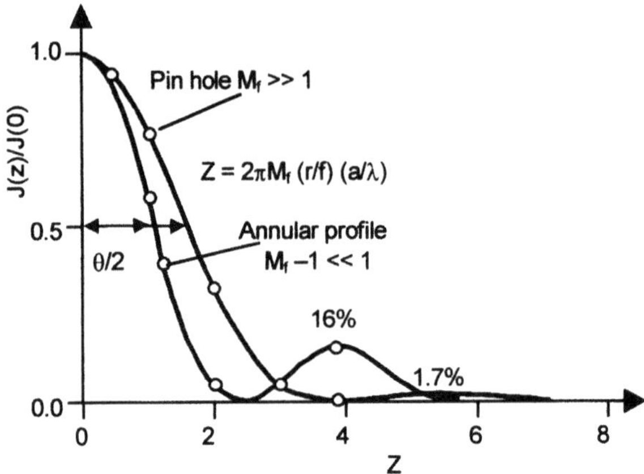

Fig. 4.41 Far field structure of an annular field and a plane wave limited by a pinhole of radius *a*. *Z* is a normalised coordinate in the far field (focal plane of a lens with focal length *f*), *r* is the radial coordinate in this plane.

It depends on the width of the ring $\Delta a = (M_f - 1)a$. For values of M_f near 1 the divergence: $\theta \approx 0.35 \ \lambda/a$ and the beam parameter product is

$$\frac{\theta D}{4} \approx 0.7\lambda \qquad M_F - 1 \ll 1 \tag{4.71}$$

this is near the diffraction limited value of λ/π. But a lot of power is contained in the side lobes . For large values of the magnification $M_f >> 1$ the near field is approximately a pinhole of radius $M_f a$ with the divergence $\theta \approx 0.5 \lambda / M_f a$ which results in a beam parameter product of:

$$\frac{\theta D}{4} \approx \lambda \qquad M_F >> 1 \tag{4.72}$$

The side lobe intensity is strongly reduced, but still remains. In both cases the beam parameter product is nearly diffraction limited, that means the K value is near one. An example of the field structure is given in Fig. 4.42.

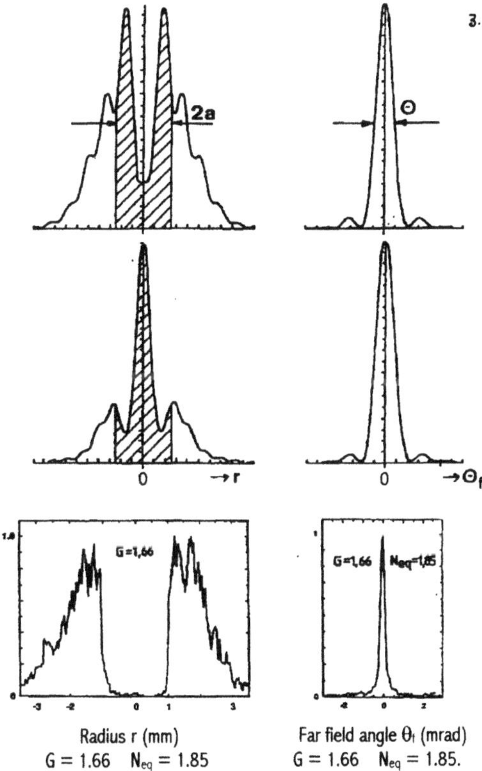

Radius r (mm)
G = 1.66 N_{eq} = 1.85

Far field angle θ_f (mrad)
G = 1.66 N_{eq} = 1.85.

Fig. 4.42 Near and far field distribution of different modes of a Nd:YAG laser with an unstable resonator. a = mirror radius, θ_f = beam angle coordinate in the far field, θ = beam divergence.
Upper picture: numerical results, lower picture: experimental results.
N_{eq}: equivalent Fresnel number = $(a^2/\lambda L).(G^2 - 1)^{1/2}/2g_2^*$ a normalised number, characterising the unstable resonator.

High beam quality with the main power in the centre maximum is obtained for large magnification, but this requires high pumping power to overcome the high losses according equation (4.70). Unstable resonators are not useful for low gain media. The thermal lensing does not affect the beam parameter product very much as shown in Fig. 4.43, and in any case , the unstable resonator delivers better beam quality, but lower output power because of the smaller filling factor.

There is one severe disadvantage. Due to the thermal lensing the resonator is confocal only for one specific pumping power level. In the non confocal case, a spherical wave will be emitted with the origin depending on the pumping power. If the output field is focused by an optical system, the focus will move with varying thermal lensing, i.e. with input/output power. This cannot be avoided, only minimised. An example is given in Fig. 4.44.

Table 4.12 Comparison of stable and unstable resonators

	stable resonator	*unstable resonator*
	(Varying refers to the pumping/output power)	
beam parameter product	varying with power, high	constant, low
focus position	constant	varying
focal length z_f	varying	constant
output power	-	10–20% lower
side lobes in the far field	no	yes
output coupling	constant	varying

Another problem of unstable resonators is the sensitivity against laser light reflected from the target back into the resonator. It produces a converging wave with spots inside the resonator. Optical components can be damaged and the mode structure is distorted.

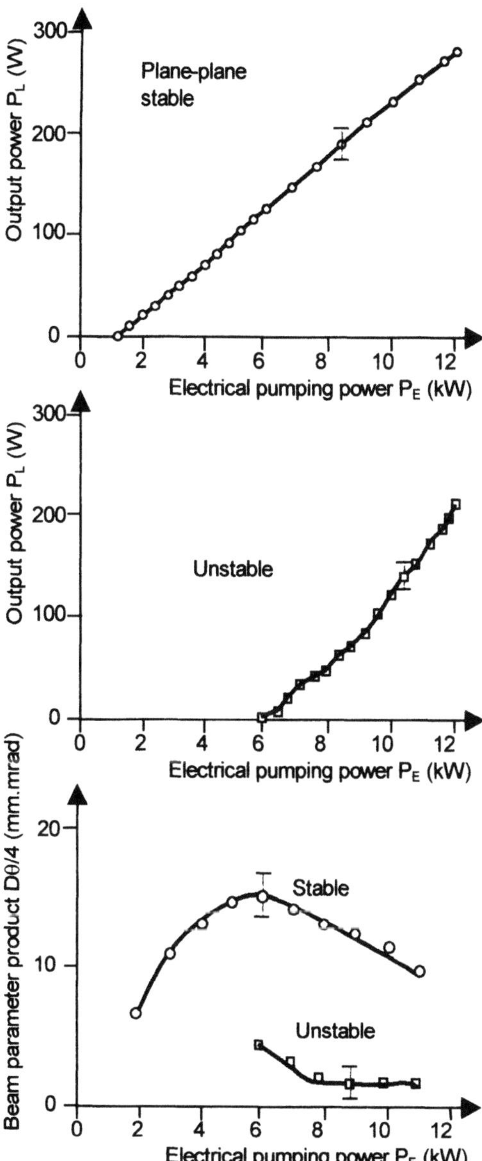

Fig. 4.43 Output power and beam divergence vs pumping power for a stable and unstable laser resonator. Note the higher threshold power for the unstable system.

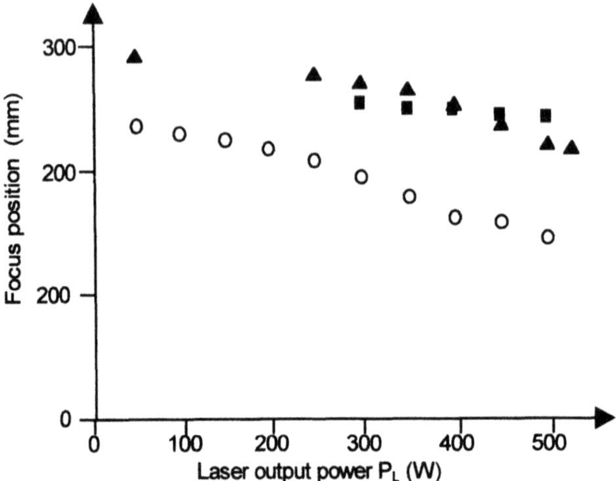

Fig. 4.44. Focus position vs output power for an unstable Nd:YAG system. Focusing lens 25 mm behind the output mirror with $f = 0.2$ m for different resonators (Hodgson and Weber, 1997):

a = outcoupling mirror diameter; $d_1 = 0.033$ m.

O $\quad$ $L = 0.6$ m; $\rho_1 = -0.5$ m; $\rho_2 = -0.36$ m; $d_2 = 0.20$ m; $a = 2$ mm

■ $\quad$ $L = 0.7$ m; $\rho_1 = 0.3$ m; $\rho_2 = -0.30$ m; $d_2 = 0.30$ m; $a = 4$ mm

▲ $\quad$ $L = 0.5$ m; $\rho_1 = \infty$ $\quad$ $\rho_2 = -0.36$ m; $d_2 = 0.25$ m; $a = 2$ mm

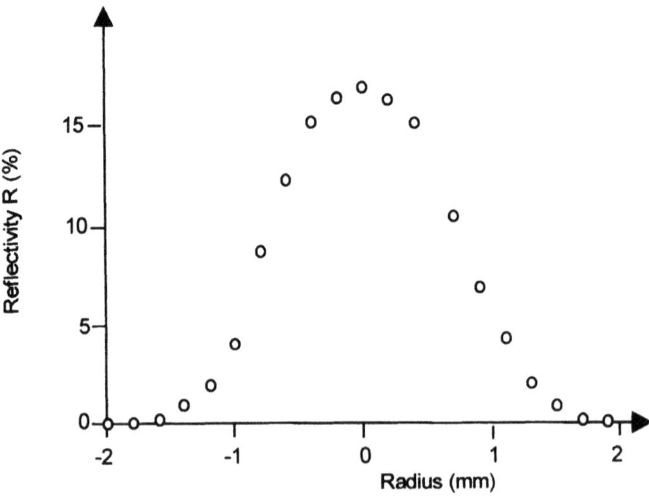

Fig. 4.45 Reflectivity profile of a graded mirror.

Graded mirrors

One disadvantage of unstable resonators is the amount of power in the side lobes. Although the far field intensity peaks in Fig. 4.43 look rather low, the integrated power can become considerably and often broad shoulders besides the main peak appear. The reason for the high side maxima is diffraction at the hard apertures of the output mirror. If a smooth aperture (graded mirror) is used, these diffraction effects can be reduced. The typical reflectivity profile of such a mirror is shown in Fig. 4.45. It can be approximated by

$$R(r) = R_{max} \exp\left[-\left(\frac{r}{r_0}\right)^{2n} \right] \qquad n = 1,2,3...$$

where R_{max} is the centre reflectivity. $n = 1$ is a mirror with a gaussian reflectivity profile. In that case the Gauss-Laguerre polynomials of the stable resonators are eigensolutions, but with complex g parameters, which makes an analytical description rather difficult. A profile with $n = 1$ is not very useful, because the modes do not fit the active medium very well, the filling factor is low. Therefore super gaussian mirrors with $n = 2...3$ are preferred. An experimental example is given in Fig. 4.46. Comparison with Fig. 4.42 demonstrates that the shoulders disappear. A detailed review on unstable resonators with graded mirrors is given by Morin (1997).

4.3.4 Prism resonators and folded resonators

One mirror can be replaced by a Porro prism. This resonator is in one direction less sensitive against misalignment of the prism. An example is given in Fig. 4.47. Instead of the Porro prism a triple-prism can be taken. Then the system is less sensitive against misalignment in both directions. The disadvantage of these prisms is the imperfectly polished roof of the prism. Moreover, the field becomes polarised, which is often not desired. For these reasons prisms are rarely used in high power systems. But they are used for slab lasers to reduce the height b of the slab, and in order to obtain a more square-like profile. Some examples are given in Fig. 4.48 and discussed in detail in literature (Dong, 1991).

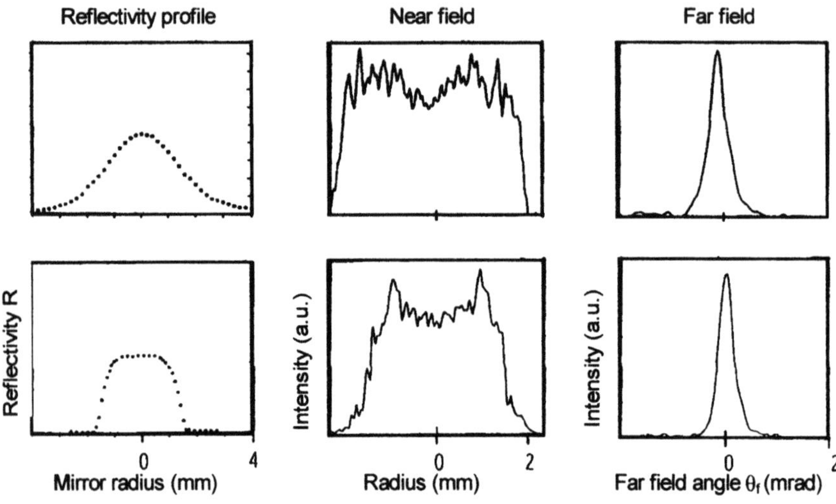

Fig. 4.46. Left: different graded mirror reflectivity profiles. Middle: near field. Right: far field of an unstable system with $M_f = 1.5$ (Bostanjoglo, 1994).

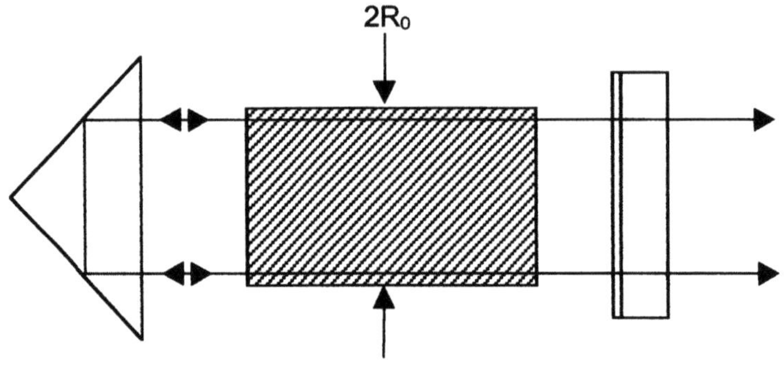

Fig. 4.47 A prism resonator.

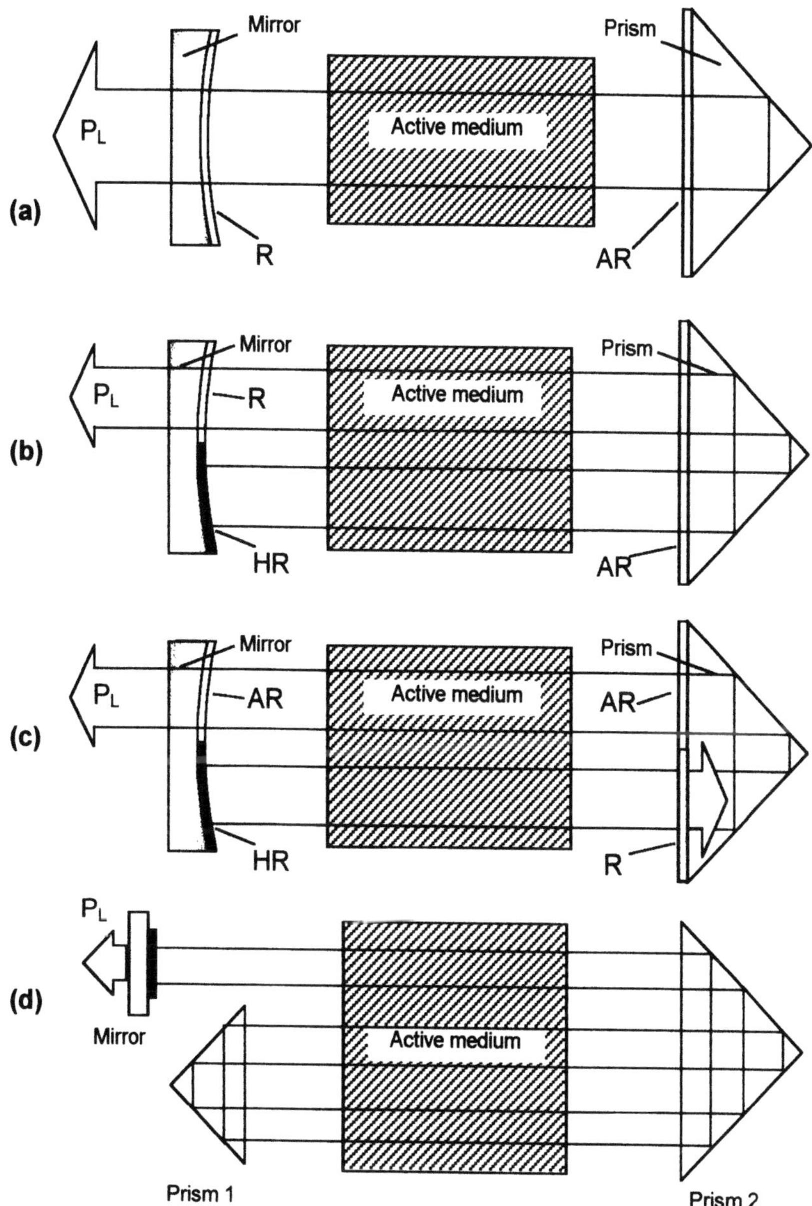

Fig. 4.48 Different possibilities for obtaining a square output field from slabs (Hodgson and Weber, 1997).

4.4. REALISATION OF HIGH POWER SYSTEMS

In this section the technical data of high power lasers are briefly discussed. Rod and multi rod systems up to 6 kW are on the market. Slab lasers with about 1kW output power are available. Other systems such as diode pumped lasers, fibre lasers and disk lasers are under investigation. The relation between output power and the beam parameter product is plotted in Fig. 4.49, and compiled in Table 4.13. A special paragraph is dedicated to the diode pumped system, although high power systems are not yet on the market.

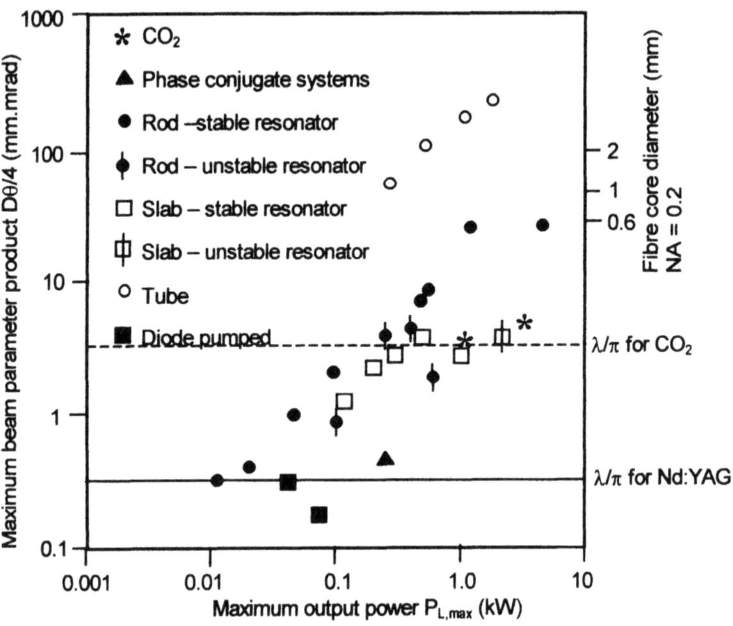

Fig. 4.49 Beam parameter product vs average output power for different systems. The right hand scale indicates the minimum core diameter of an $NA = 0.2$ fibre for safe beam delivery.

4.1 The Single rod laser

A typical system is shown in Fig. 4.50. It is called close coupled, because lamp and rod are as close as possible together, surrounded by a diffuse reflecting ceramic cavity. The maximum output power is limited by the rupture stress and the length of the crystal. Nd:YAG-rods up to 180 mm are available and a lamp pumped system delivers about 600 W average power with an total efficiency of 4–5%. The laser can operate in cw and in the pulsed mode with pulse length of 0.1–10 ms and corresponding repetition rates, limited by the maximum average power. The beam parameter product can be taken from Fig. 4.49.

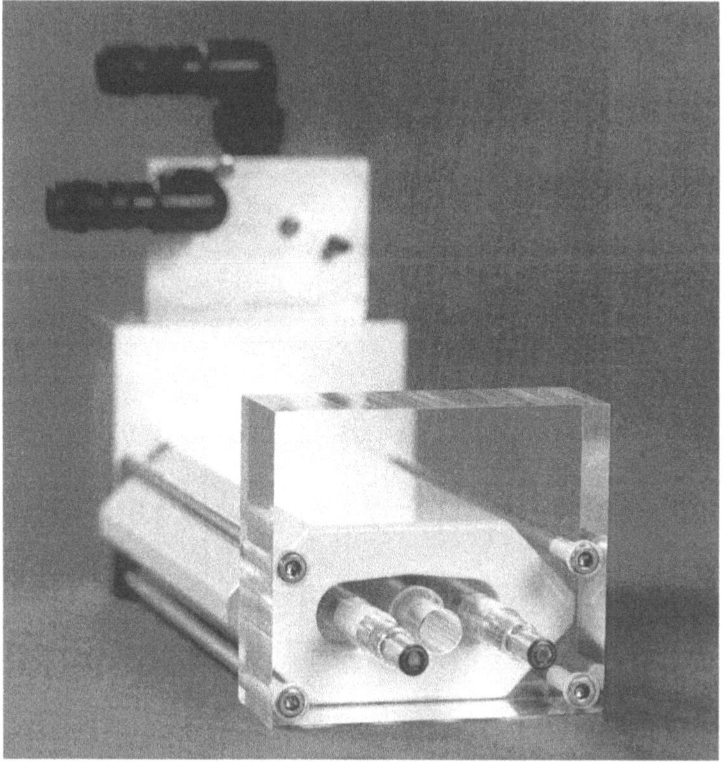

Fig. 4.50 A single rod cavity. (Courtesy Lumonics UK).

4.4.2 Multirod systems, stable resonators

The output power can be increased by using several rods in series This can be done inside or outside the resonators. If the beam intensity is comparable with the saturation intensity J_s (section 4.2.1) the efficiency in both cases is approximately the same. The beam quality of the system is nearly independent of the number of rods, as long as the setup is symmetric. The distances between rods and mirrors must be chosen as plotted in Fig. 4.51.

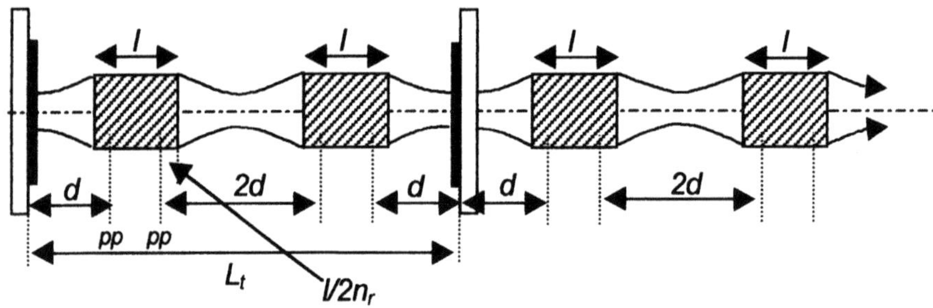

Fig. 4.51 A multirod resonator with amplifier stages, pp: principal planes.

In that case the rods are filled completely by the radiation field (Fig. 4.52) and a high efficiency comparable with that of the single rod can be obtained.

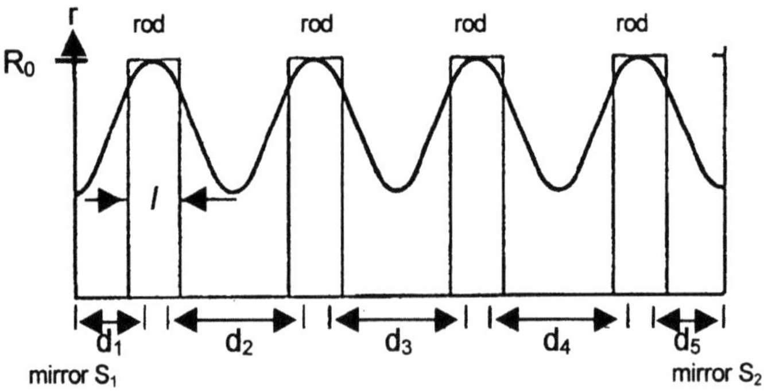

Fig. 4.52 Slope of the beam diameter inside a four rod resonator. High efficiency (high filling factor of the active medium) and conservation of the beam parameter product requires $d_1 = d_5 = L/2$, $d_2 = d_3 = d_4 = L$.

This type of resonator crosses the singularity $g_1^* = g_2^* = 0$, which requires careful alignment and equal refractive power of all rods. Otherwise a dip will appear at the singularity. Fig. 4.53 gives an example of that dip for a single rod system. This can be used to determine the refractive power of the rod, because at that point $D_f = 1/d_1$ (if $d_1 < d_2$).

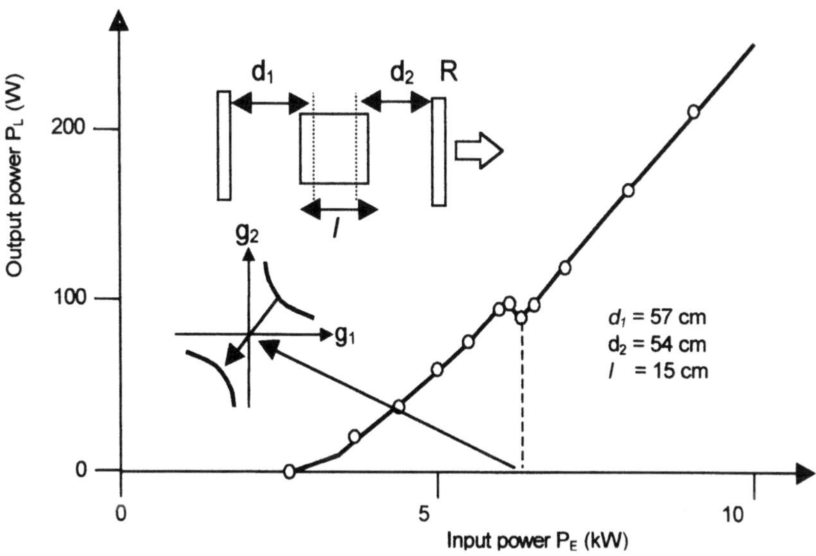

Fig. 4.53 Output power P_L of a single rod cavity vs pumping power P_E.
At 6.3 kW input power, the singularity (confocal point) $g_1^* = g_2^* = 0$ is crossed.

Output powers up to $P_L = 2.7$ kW and beam parameter products of $(D\theta/4) = 25$ mm·mrad are achievable with 4 rods (Fig. 4.54). More rods (about 8) have been realized, but the system becomes more and more sensitive to misalignment and differences in the thermal lensing of the rods. The above multi rod systems operate with plane mirrors. Spherical mirrors are also possible, but this reduces the stable range of the system as discussed in section 4.3. An experimental setup is shown in Fig. 4.55.

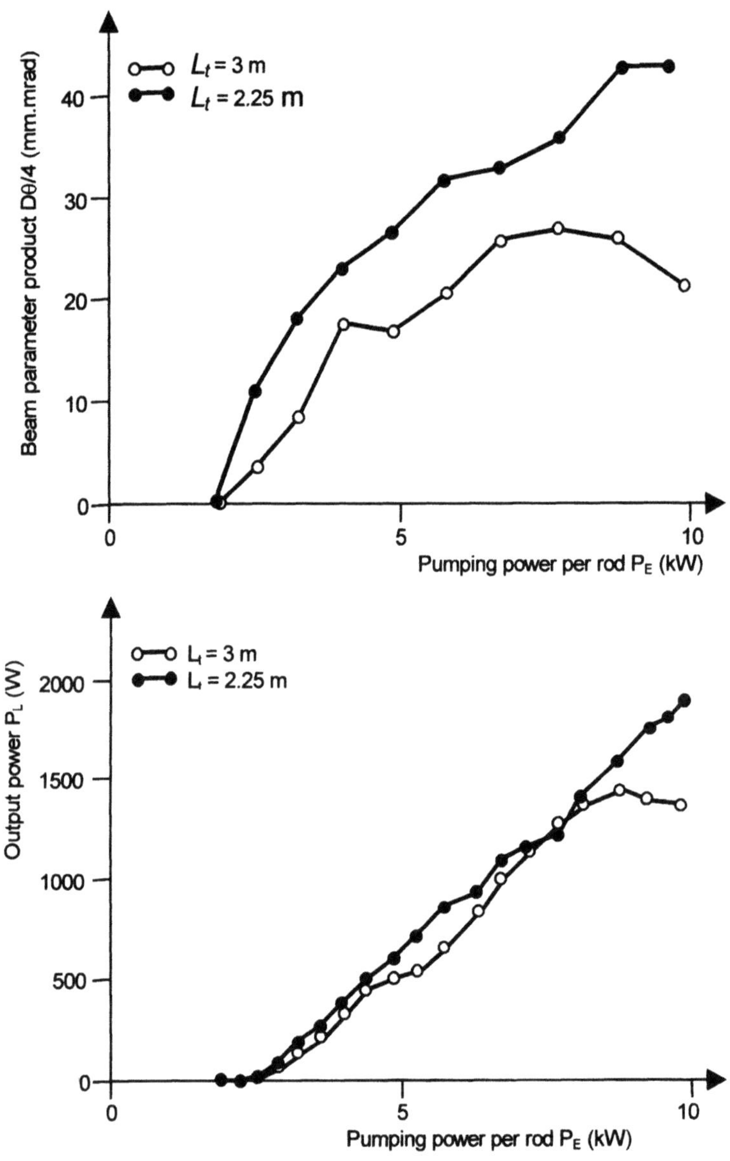

Fig. 4.54 Beam parameter product ($D\theta/4$) and output power P_L vs pumping power P_E per rod of a multirod system for different total resonator lengths L_t (Kumkar et al, 1992).

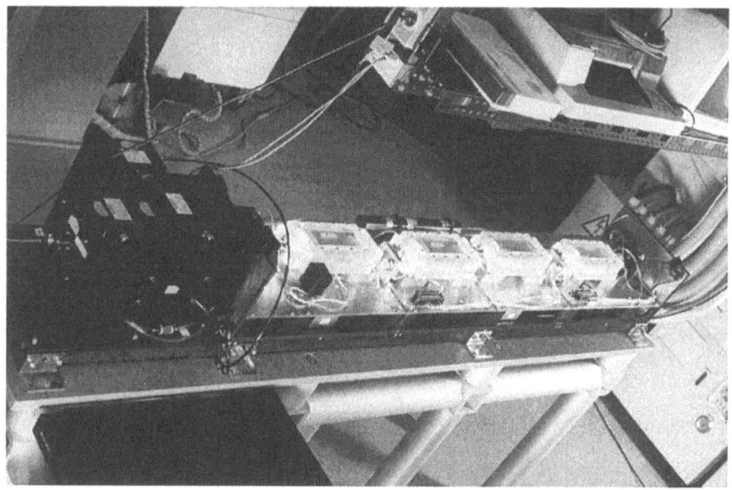

Fig. 4.55 A Multirod setup (Courtesy FLI Berlin).

4.4.3 Unstable multirod systems

Unstable resonators can be used with multiple rods to improve the beam quality. Up to three rods have been used, with internal lenses to adapt the field to the rods properly (Fig. 4.56). Some results are shown in Fig. 4.57. The difficulty is to obtain a high filling factor for all rods. Therefore the output powers of the 2 and 3 rod system are nearly the same. A detailed discussion is given in Hodgson, Bostanjoglo and Weber (1993). Up to now unstable systems are only used in the one cavity version.

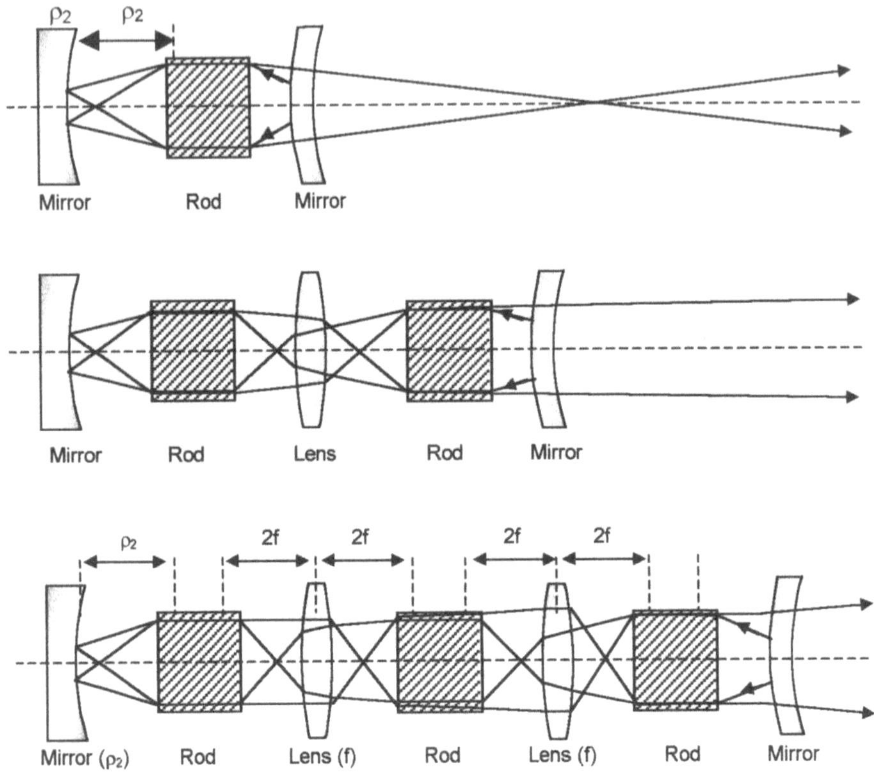

Fig. 4.56 Multirod unstable configurations (Hodgson et al., 1993)

4.4.4 Slab and multi-slab-systems

As discussed in section 4.2 the slab configuration has the advantage of

- higher rupture limit, which means higher pumping power;
- lower thermal lensing.

The beam passes the slab in zigzag fashion, crossing the parabolic profile of temperature gradient and refractive index as outlined in Fig. 4.58. The thermal distortions are therefore compensated to a first order as shown in Fig. 4.59.

Table 4.13   Performance of high power lamp pumped Nd:YAG lasers (Wittrock, 1993). (c-commercially available, d-demonstrated, p-possible)

| | | Output power P_L (kW) | | | | | |
| | | Multi-Rod | | Slab | | Tube | |
		stable	unst	stable	unst	stable	unst
Beam parameter product							
> 40 mm.mrad	c			0.5			
	d			1.2		2	
	p			1.4		4	
5–40 mm.mrad	c	5					
	d	7		0.6	0.6	0.4	0.5
	p	10		1	0.8	0.5	
< 5 mm.mrad	c	0.2			0.35		
	d	1			0.8		
Efficiency at 30 mm.mrad							
η_{tot} (%)	c	5		3.5			
	d	5.5		6		10	5
	p					10	

The upper left interferogram is taken for straight through transmission of the light and no pumping. About one fringe appears, which characterises the optical quality of the cold slab. In the interferogram of straight through transmission of the pumped rod ~ 10 fringes appear, different in x and y directions, from which the strong thermal lenses in both directions can be evaluated. The interferogram of the zigzag transmission at the same pumping level demonstrates the compensation of the thermal effects in the x direction, but distortions in y direction remain. They are caused by inhomogeneous cooling and effects due to the ends of the slab. To reduce losses, the slab at the entrance surfaces is cut at the Brewster angle.

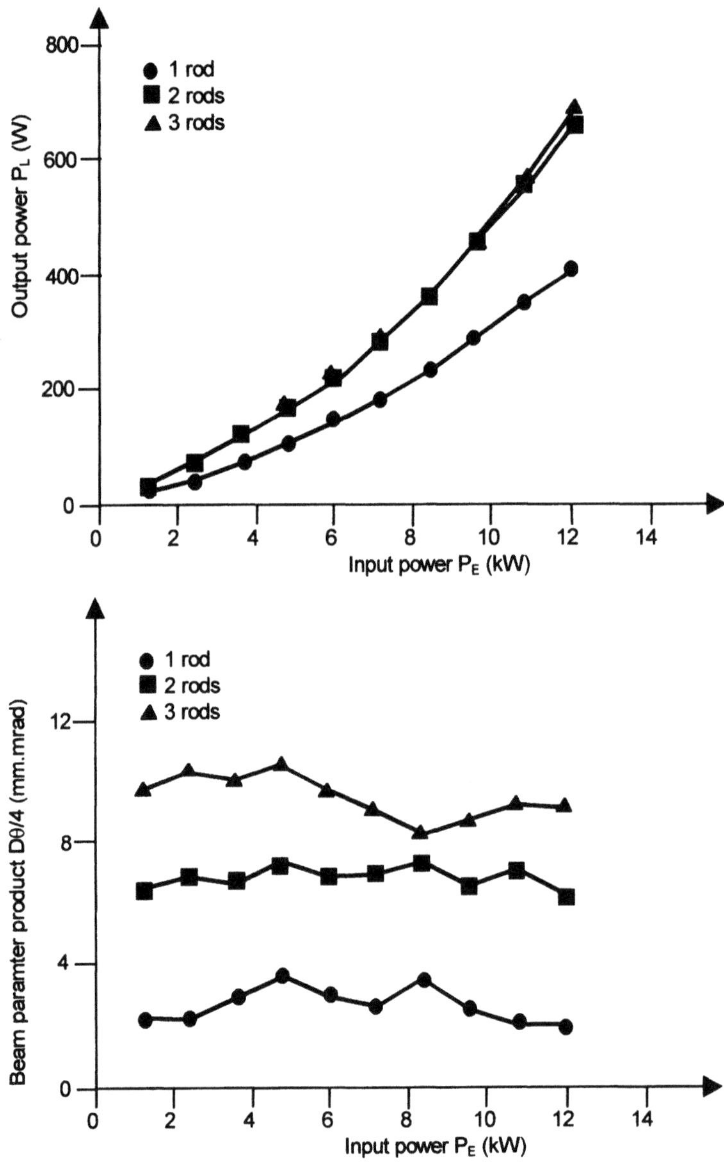

Fig. 4.57 Output power P_L and beam parameter product $D\theta/4$ of unstable multirod systems with 1, 2 and 3 rods (Hodgson et al., 1993).

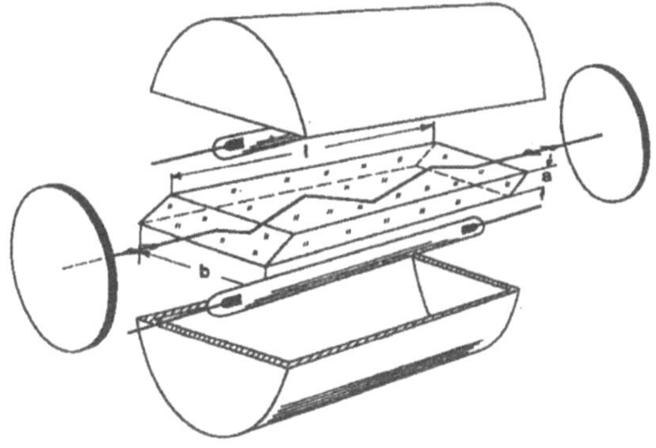

Fig. 4.58 The zig-zag-slab.

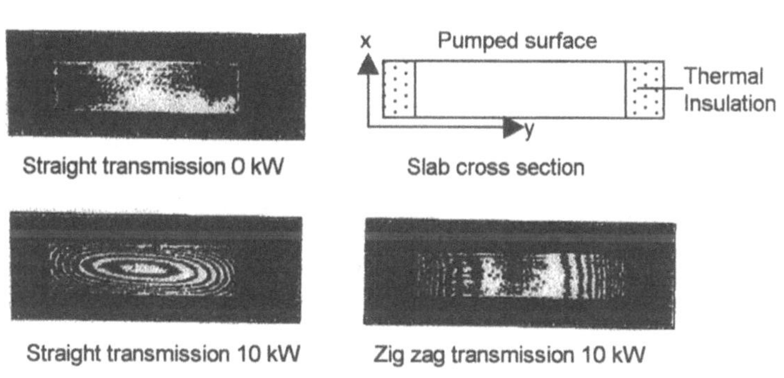

Fig. 4.59 Interferograms showing optical distortions in unpumped and pumped slabs.

However severe disadvantages appear:

- rectangular output beam
- very homogeneously pumping and cooling required
- sophisticated and expensive construction

Nevertheless, these problems can be overcome by careful design. An example is given in Figs. 4.60 and 4.61. Output powers of more than 1kW per unit (slab dimensions $a = 7$ mm, $b = 2$ 6mm, $l = 130$ mm) were obtained with 2 or 3 times diffraction limited beam quality (Hodgson, Bostanjoglo and Weber, 1993).

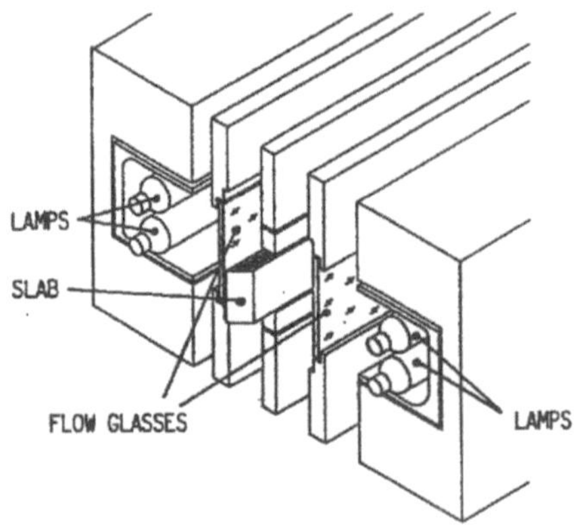

Fig. 4.60 Schematic of end of a slab cavity.

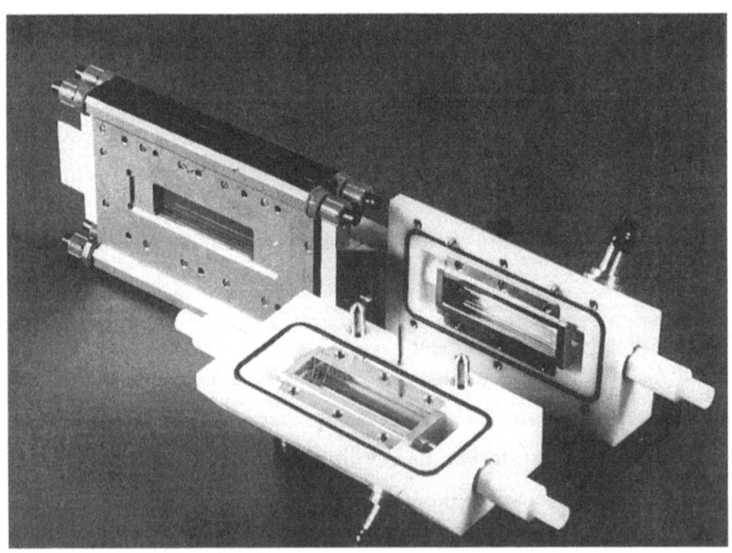

Fig. 4.61 Section of a slab head (Courtesy FLI Berlin).

To overcome the unfavourable rectangular beam profile various resonators can be used such as folded (section 4.3) or unstable systems. Some of these setups are shown in Fig. 4.62. Results of an unstable configuration (plane-unstable-off-axis, folded) are given in Fig. 4.63. Compared with normal stable resonators, the output power is reduced, but the beam quality is very high and power independent. Higher output power can be obtained by multislab systems (Fig. 4.64) in resonator or resonator/amplifier configuration. The results are compiled in Fig. 4.65. It can be seen that steady state values are below the values of short time operation (< 10 s), which indicates thermal distortions.

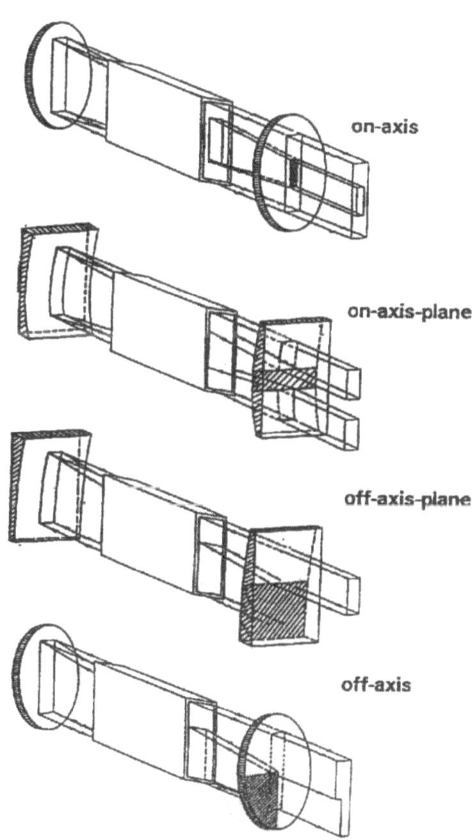

Fig. 4.62 Unstable resonator configurations for slab lasers.

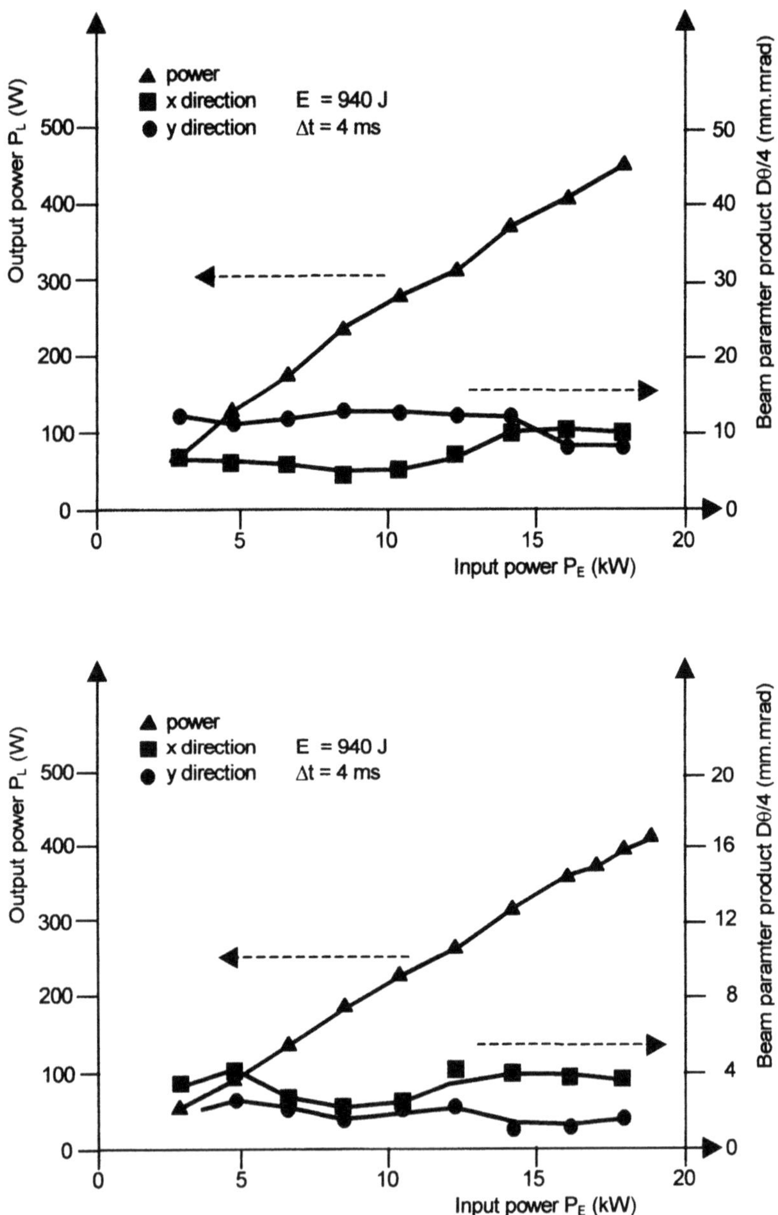

Fig 4.63 Beam parameter product $D\theta/4$ and output power P_L of a single slab (top) in a folded resonator, and (bottom) a plane unstable off-axis configuration. The beam parameter product is different in x, y-directions.

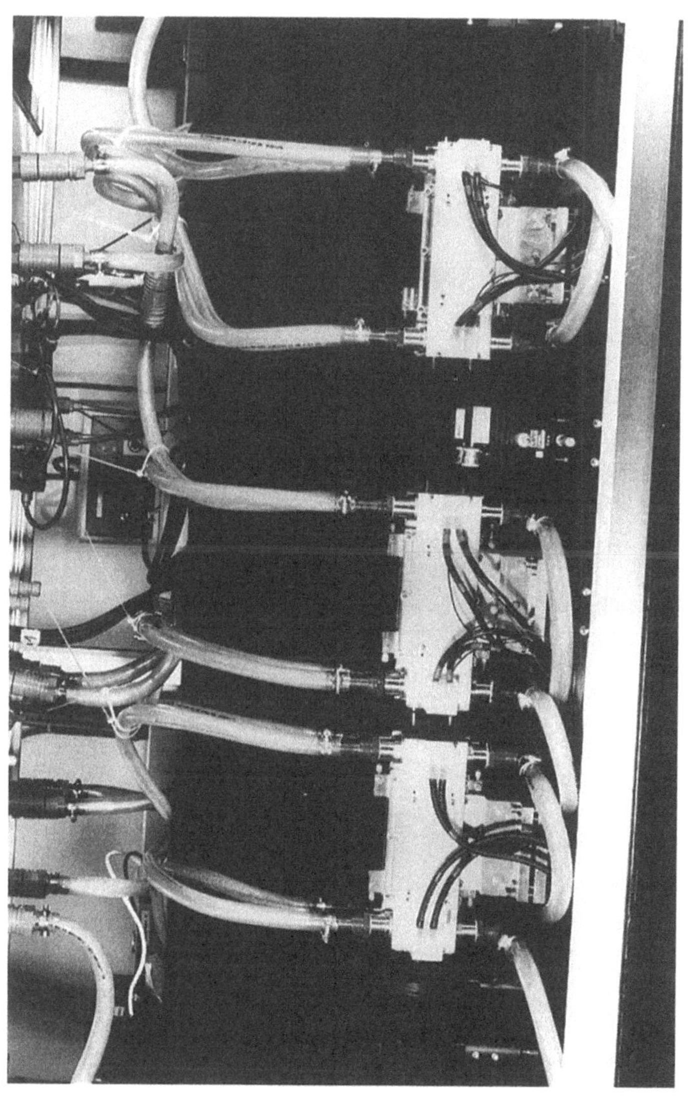

Fig. 4.64 Multi slab setup /Courtesy FLI Berlin).

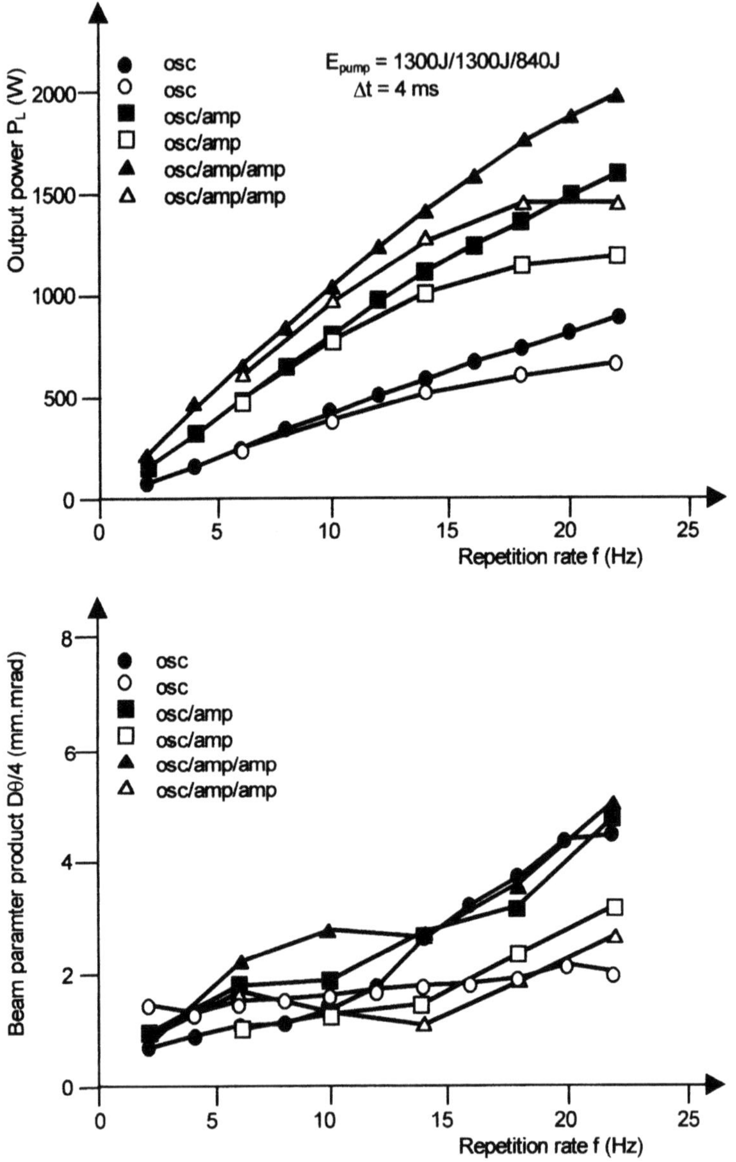

Fig. 4.65 Top: Output power P_L vs repetition rate f of the pumping light for different multislab configurations. Full symbols: start time operation; open symbols: steady state operation. Bottom: Beam parameter product vs repetition rate for different configurations. Full symbols: x direction; open symbols: y direction..

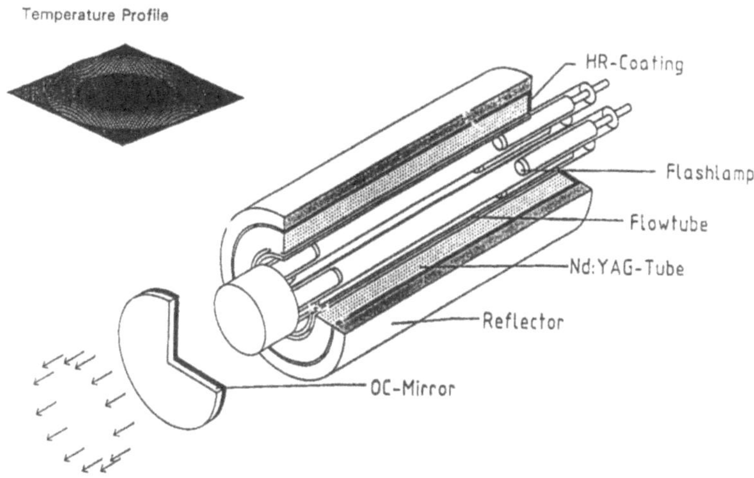

Fig. 4.66 The inside pumped tube laser.

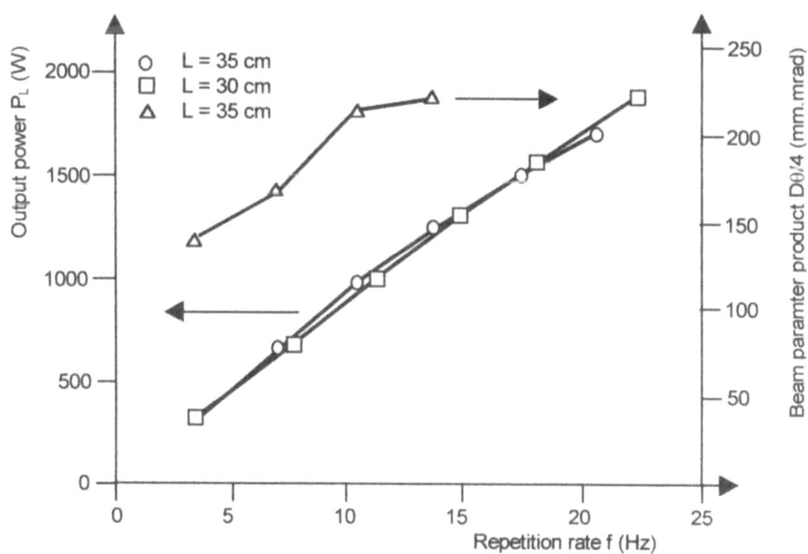

Fig. 4.67 Beam parameter product $D\theta/4$ and output power P_L vs pumping power P_E for a Nd:YAG tube laser.

4.4.5 Tube configurations

The tube laser is the system with the highest efficiency of about 10%. This is obtained by the efficient absorption of the pumping light. The lamps are inside the hollow crystal and most of the pumping light is absorbed in the first transit. A schematic drawing is given in Fig. 4.66. Output powers up to 1.7 kW with a Nd:YAG tube (35 mm inner diameter, 53 mm outer diameter, 180 mm long) were realised (Wittrock, 1992) but with low beam quality as Fig. 4.67 demonstrates. Up to now fibre transmission of tube laser radiation requires fibres with about 2 mm core diameter, which is for most applications too large. An experimental setup is shown in Fig. 4.68. Tube lasers might be useful for applications where a high beam quality is not required (surface treatment). It might be possible to improve the beam quality by suitable resonators (toroidal type, multipass systems).

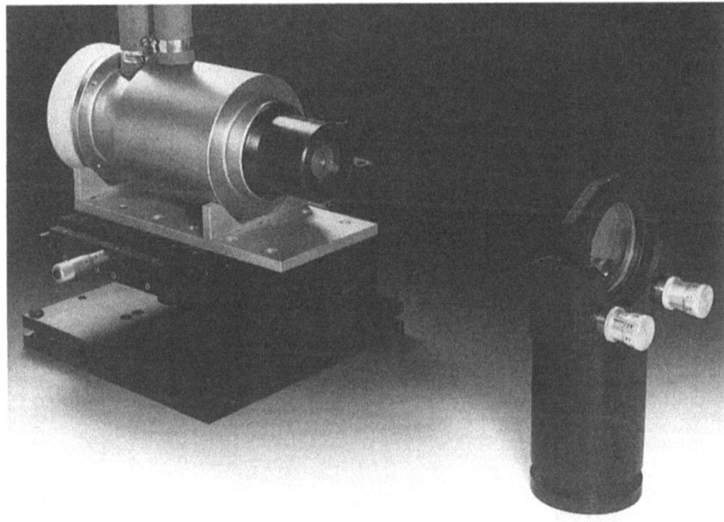

Fig. 4.68 A tube laser in the laboratory (Courtesy FLI Berlin).

4.4.6 Diode pumped Nd:YAG laser

Diode pumping looks very promising and has several advantages compared with lamp pumped systems :

- higher efficiencies (up to 20 % longitudinal pumped, 10% transversely pumped)
- lower thermal load (30–50%)
- higher lifetime

Today, the high diode price is an obstacle which prevents economic systems. Up to now, only lower power systems (10 W) are on the market. But high power lasers of 1-2 kW have been realised in the lab and will be available in some years.

All relations concerning output power, beam quality and resonators of the previous sections hold. Only the lamps are replaced by diodes. The efficiency η_{excit} increases by a factor of 2–3, and the heat production is reduced by a factor of 2 in pulsed systems, and by 30 % in cw systems.

Three principal configurations are under investigation. They are shown schematically in Fig.4.69. The longitudinal pumping, (upper picture) delivers output powers of some tens of watts and is not suitable for high power. For powers up to some hundred watts, side pumped rods will be used. Higher output powers will be obtained with side pumped slabs. One example is given in Fig.4.70. In Figs.4.71 and 4.72, experimental systems are shown. The diodes have to be cooled, which can be done at low powers with Peltier elements. At higher power levels water cooled systems have to be used, because the efficiency of Peltier elements is too low. Some results are compiled in Table 4.14. A survey is given in Comaskey (1992).

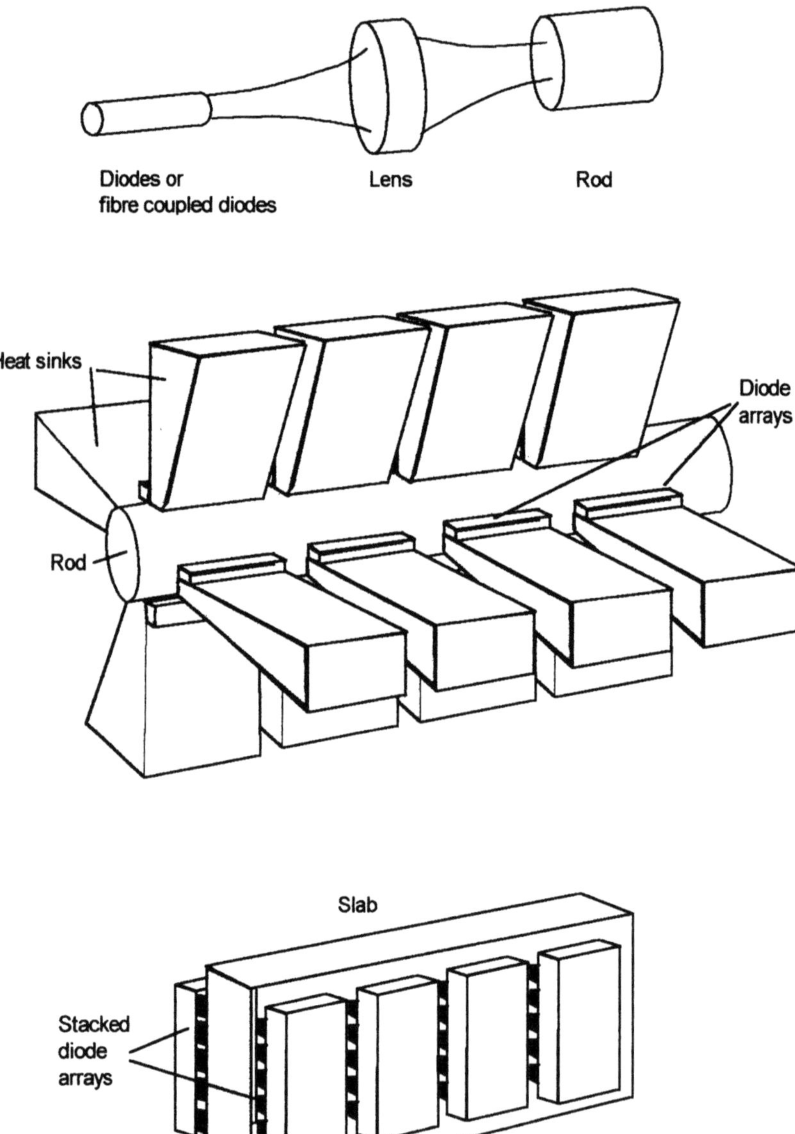

Fig. 4.69 Diode pumped solid state lasers. Top: longitudinally pumped rod. Middle: transversely pumped rod. Bottom: transversely pumped slab.

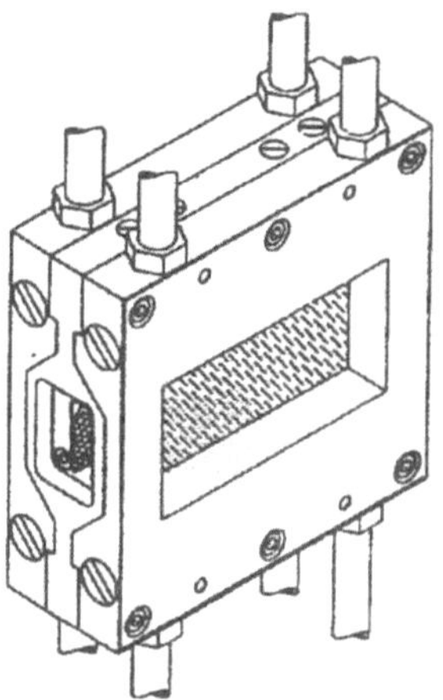

Fig. 4.70 Diode pumped slab head for an output of appr. 1 kW.
Dimensions 110mm x 220mm x 40mm. (Courtesy FLI Berlin).

Fig. 4.71 Experimental setup of a transversely diode pumped rod, output power
100 W (Courtesy LMTB Berlin).

Table 4.14 Some results of diode pumped Nd:YAG systems (June 1997) (c-commercially available, L-laboratory results, BP-beam parameter product)

cw systems

Institution	Power P_L (W)	BP (mm.mrad)	remarks
Various companies (Nd:YAG rod)	10	0.3	c
Opt. Inst. T.U. Berlin	70	0.3	L
LMTB Berlin (Nd:YAG rod)	1100	30	L
LZH Hannover (Nd fibre)	32	0.33	L
IFSW Stuttgart (Yb disk)	255	~5	L

pulsed systems

Institution	Average Power $P_{L,av}$ (W)	Pulse energy E_L (mJ)	Repetition rate (Hz)	Remarks	
Various	0.3	3	100	c	
companies	0.75	0.5	500	c	
	3	3	1000	c	
	100	1250	40	free running	c
	30	750	40	Q switch	c
	550(BP=30mm.mrad)	275	200	long pulses	
	300(BP= 2mm.mrad)				
Lawrence Livermore	1000	400	2500	L	
FLI Berlin	1–150	500–5	$20–10^4$	L	
McDonnel Douglas	32	1000	32	L	

4.4.7 The MOPA concept

Nd:YAG-lasers with fundamental mode operation are available with powers up to some tens of watts. Using Nd:YAG amplifiers this moderate power can be increased considerable. Such a system is called Master Oscillator Power Amplifier. Of course the thermal distortions of the amplifier remain and will destroy the high beam quality of the oscillator. It is possible to compensate these distortions for pulsed systems by using a phase conjugating mirror and crossing the amplifier twice. The phase conjugating mirror is a nonlinear optical device, which has the ability to reverse the phase front of a light beam. Any distortion of the phase is reversed by the mirror and cancelled at the second transit. Unfortunately the conjugation does not work for the polarisation and the birefringence of the rod is not compensated. Therefore the Nd:YAG crystal is replaced by a Nd:YALO crystal, which has less thermally induced birefringence due to its strong natural birefringence. This crystal has a lower cross section σ_L for laser emission and reduced efficiency. Another possibility is to use two identical rods with a quartz-rotator and an optical system, which compensates polarisation distortions. An example is given in Fig. 4.72. Up to now these phase conjugating systems operate in the pulsed mode (periodically Q-switched) with pulse durations in the 10–100 ns range. 500 W output nearly fundamental mode has been obtained (Eichler, 1994, 1997) An interesting application may be the sublimation cutting of metals and precise drilling.

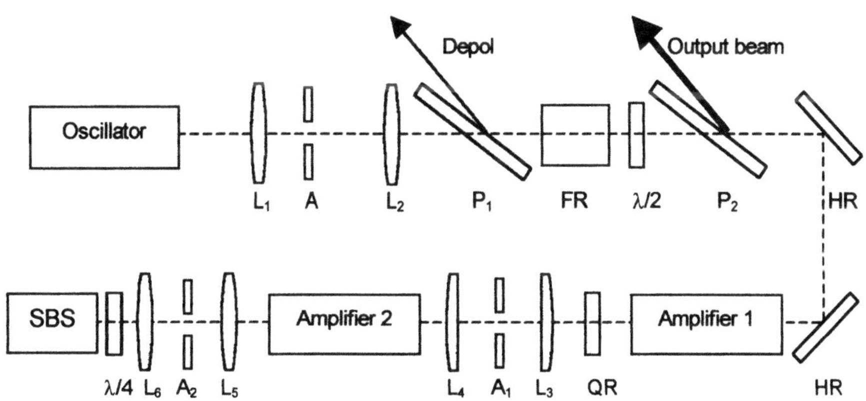

Fig. 4.72 Experimental setup of a oscillator-amplifier system with phase conjugation (Seidel and Mann, 1996). L1-L6: Lenses; P1, P2: thin film polarisers; FR: 45^0 Faraday rotator; $\lambda/2$, $\lambda/4$: retardation plates; HR: high reflective mirrors; Amp.1,Amp.2: Nd:YAG amplifiers; QR: Quartz rotator; A1, A2, A3: apertures; SBS: Stimulated Brillouin scattering cell; Depol.: Radiation with distorted polarisation ($\sim 5\%$).

4.5 BEAM DELIVERY BY FIBRES

One main advantage of Nd:YAG lasers is the possibility to deliver the output beam by fibres, even at high output levels in the kW range (Table 4.15). At the wavelength of $\lambda = 1.06$ µm the internal losses of silica fibres are extremely low. Most losses are caused by the in- and out-coupling optics due to Fresnel reflection and mismatching. Transmission efficiencies of 93% have been achieved with transmission lines of up to 200 m length (Ishide, 1990; Reng and Beck, 1993). Fibres are easy to handle. They are flexible and can be used like coaxial cables with comparable connectors as shown in Fig. 4.73.

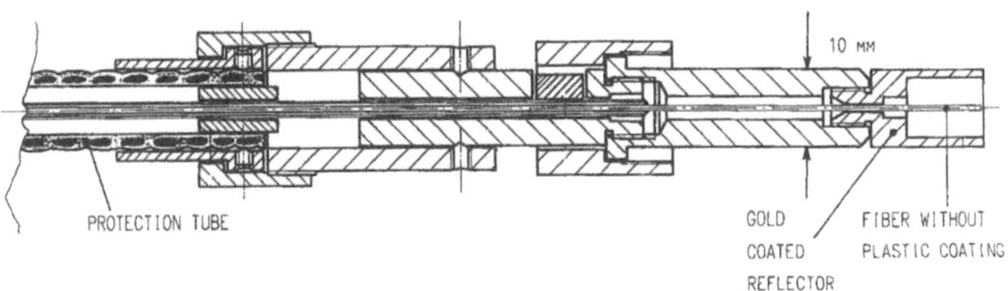

Fig. 4.73 Fibre connector.

4.5.1 Fibre types

For high average power only silica fibres are used with core diameters in the range of $d_{core} = 0.4–1$ mm. Core diameters much below 0.4 mm are not useful, because of the high power density on the fibre surface, and the difficulty to couple the laser beam into the fibre. Core diameters above 1 mm impede the desired flexibility. As a rule of thumb: bending radius > 200 core diameters. Another problem is the focusing of the fibre output onto the workpiece. For cutting, small foci of about 0.2–0.4 mm are required. To reduce the 1 mm diameter of the fibre to a 0.2 mm spot needs a sophisticated and heavy optical system. Because d_{core} is large compared with the wavelength, geometrical optics can be used to describe the beam propagation (Okoshi, 1982; Das, 1990)

The laser beam in the fibre is guided by special refractive index profiles, produced by doping the quartz glass with Ge or F. The two mostly used fibre types are shown in Fig. 4.74.

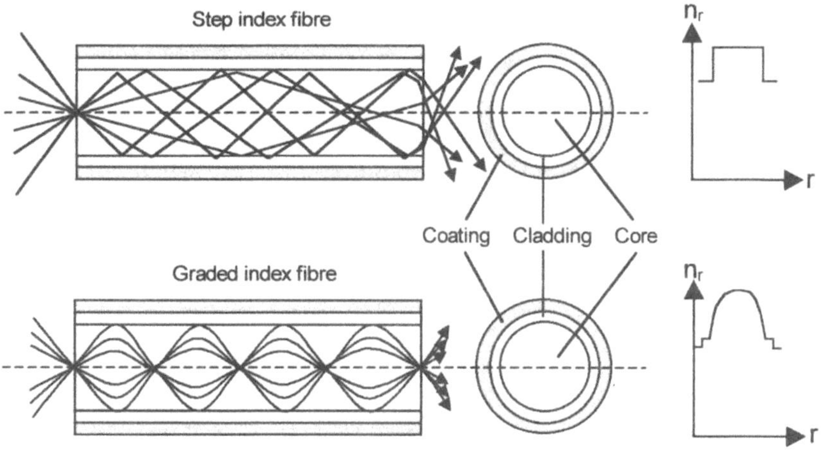

Fig. 4.74 The step index and graded index fibre.
Left: schematic representation of the rays.
Right: refractive index profile $n_r(r)$ (Beck et al, 1993).

Step-index fibre (SI)

The step index fibre has a core of about 0.4–1 mm diameter with a refractive index n_{co} = 1.45 surrounded by a cladding with a slightly lower index n_{cl}. The refractive index difference is:

$$\Delta n_r = n_{co} - n_{cl} = 0.01...0.03 \tag{4.73}$$

The fibre cladding is covered with a plastic coating, which protects the fibre against fracturing and environmental influences. The beam inside the fibre is guided by internal total reflection at the cladding. The angle of incidence has to be lower than θ_{max} with:

$$\sin\left(\frac{\theta_{max}}{2}\right) = \sqrt{n_{co}^2 - n_{cl}^2} = NA \tag{4.74}$$

which is called the numerical aperture NA of the fibre (Fig. 4.75). Rays with $\theta_{in} > \theta_{max}$ are not guided, but will enter the cladding and the coating. This causes losses and destroys finally the fibre. Rays with $\theta_{in} < \theta_{max}$ are totally reflected by the cladding, but due to the wave character of light it will penetrate some tenths of a µm. Therefore the cladding has to be lossless too.

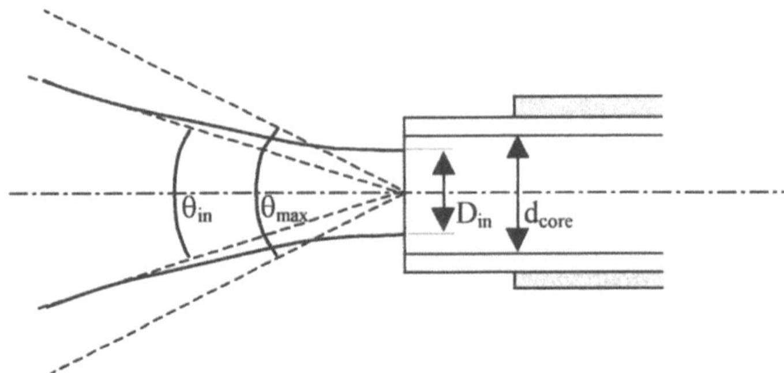

Fig. 4.75 Coupling a laser beam into a fibre requires $D_{in} < d_{core}$; $\theta_{in} < \theta_{max}$.

The rays propagate in a zigzag path with different path lengths, depending on the entrance angle. The shortest zigzag (pitch length) is given by:

$$l_p = \frac{d_{core}}{NA} \tag{4.75}$$

Example

A 600 µm fibre with with NA = 0.2 (θ_{max} = 11.5°) has a pitch length of l_p = 3.0 mm. The maximum beam parameter product accepted by this fibre is $d_{core}NA/2 \approx 60$ mm.mrad. For high power transmission, the beam parameter product has to be lower, otherwise the intensity on the claddding would be too high.

After some ten centimetres the core diameter is homogeneously filled with the radiation field:

$$D_{out} \approx d_{core} \tag{4.76}$$

The input angle θ_{in} of the laser beam is conserved if the fibre is not bent too much.

$$\theta_{out} \geq \theta_{in} \tag{4.77}$$

Anyway, the beam parameter product increases by fibre transmission. The characteristics of a step fibre transmission are shown in Fig. 4.76

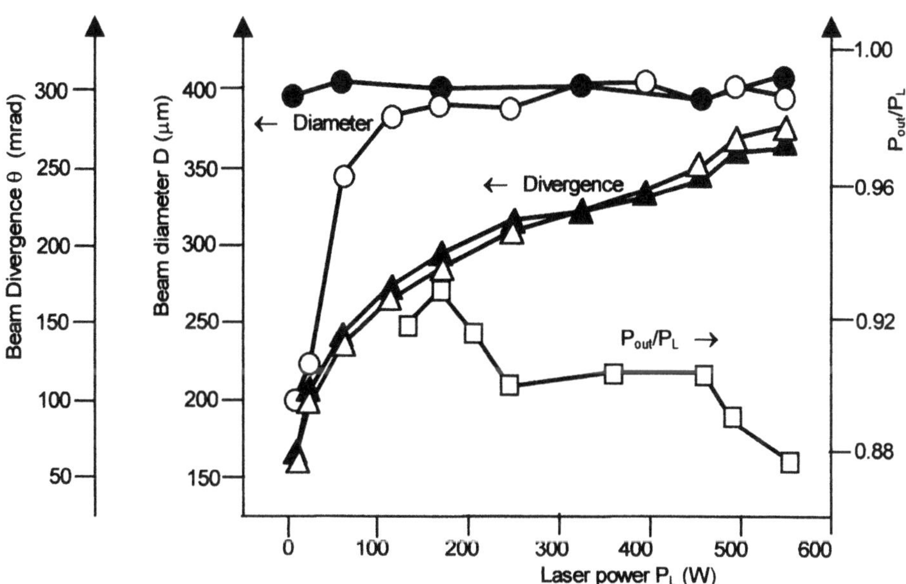

Fig. 4.76 Transmission properties of a 400 μm step index fibre.

○:   Beam diameter input   ●:   Beam diameter output

Δ :   Beam divergence input   ▲:   Beam divergence output

□ :   Power transmission

All parameters are plotted vs laser input power P_L.

Table 4.15 Transmission properties of fibres (Beck et al, 1993)

	step index fibre		*graded index fibre*	
d_{core} (mm)	0.4	0.6	0.4	0.6
d_{clad} (mm)	0.48	0.66	0.55	0.85
coating	acrylate	silicone	silicone	silicone
$P_{in, max}$ (W)	549	795	1620	2070
$P_{out, max}$ (W)	482	677	1420	1890
Transmission (%)	88	85	90	91
Intensity (MW/cm^2)	0.44	0.28	1.29	0.73
$D_{in, max}$ (mm)	0.40	0.57	0.40	0.61
$D_{out, max}$ (mm)	0.41	0.66	0.50	0.67
$\theta_{in, max}$ (mrad)	277	256	406	301
$\theta_{out, max}$ (mrad)	264	310	444	410

Graded index fibre (GI)

The graded index fibre has a parabolic shaped refractive index profile with:

$$n_{co} = n_r\left(1 - \delta.r^2\right) \tag{4.78}$$

This profile works lenslike, i.e. a periodic focusing of the beam occurs with the periodicity

$$l_p = \frac{\pi}{\sqrt{2\delta}} \tag{4.79}$$

l_p is again in the range of some millimetres. In principle the beam parameter product in a graded index fibre should be constant. But aberrations and deviations from the ideal profile (which is never really parabolic) make the beam quality worse. Beam diameter and beam divergence increase slightly but less than for the step index fibre. An example is given in Fig. 4.77.

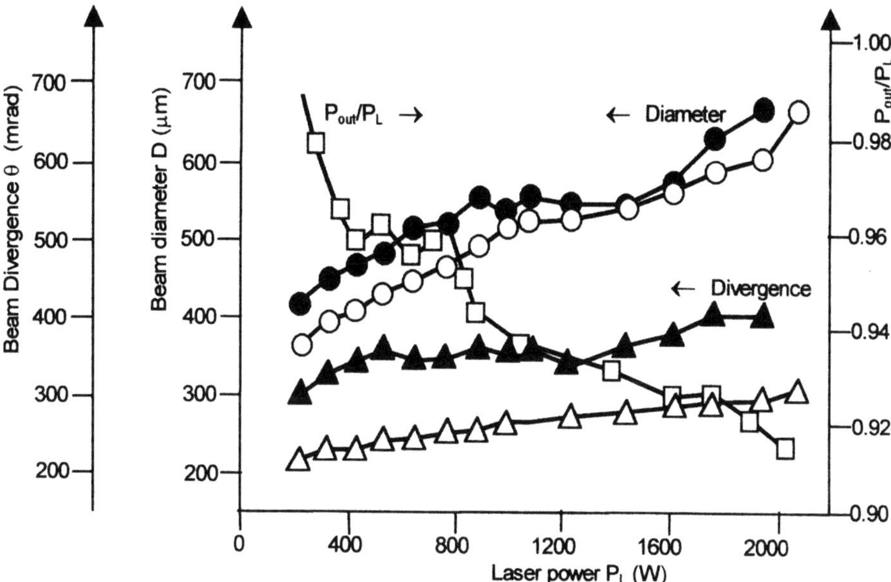

Fig. 4.77 Transmission properties of graded index fibres.

○:   Beam diameter input   ●:   Beam diameter output

Δ :   Beam divergence input   ▲:   Beam divergence output

□ :   Power transmission

All parameters are plotted vs laser input power P_L.

The transmission is a bit higher. There are graded index fibres with a double cladding. In this case, a part of the laser beam can be guided by the cladding without affecting the plastic coating. This leads to a spot diameter at the output which can be larger than d_{core}

Laser beam diameter and divergence have to be adapted to the fibre dimensions, core diameter and numerical aperture. This can be done by a suitable optical system. Anyway, the beam parameter product has to be smaller than that of the fibre. To avoid damage the condition to be fulfilled, reads:

$$[D_{in}\theta_{in}]_{laser} < 2s[d_{core}NA]_{fibre} \tag{4.80}$$

which for high power lasers is not possibly to achieve in any case. s is a safety factor and depends on the beam shape and power. It is in the range of 0.5–0.7.

4.5.2 Three in one

It is possible to couple several single fibres into one fibre with larger product $d_{core}\,NA$. This can be done by using an optical system or by direct fibre coupling as shown in Fig. 4.78.

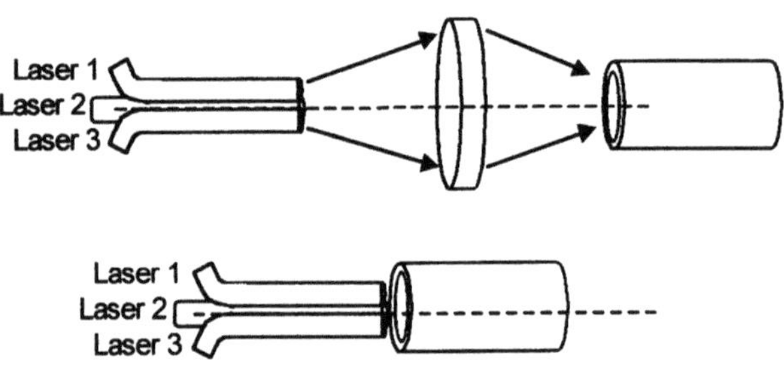

Fig. 4.78 Coupling three fibres into one.
Top: By an optical system: Bottom: End on.

In both cases the output power can be increased, but the beam parameter product increases too. In principle the resulting beam parameter product is proportional to the square root of the number of coupled fibres (incoherent coupling), in reality it will be larger. Anyway the beam quality becomes worse. But there are applications, where low beam quality is not a disadvantage. One example is the welding of sheet metals in automotive industry, where due to the low tolerances broad gaps occur.

4.5.3 Limitations and problems

For cutting and drilling fibre diameters ($D_{core} \leq 0.4$ mm) with powers in the 3–5 kW range would be useful. The limitation is the laser beam quality, which requires fibres with high core diameters or large numerical apertures NA. To increase the NA is possible, but this can only be done by higher doping of the fibre and produces losses. Moreover, high NA means large beam parameter product $d_{core}NA/2$ at the fibre output, which is difficult to handle by the optics, and after focusing produces a short focal length z_f, as discussed in section 4.1. Other problems are the variations of the focus diameter and the focus position at the incoupling optics. Both depend on the laser pumping power. Careful alignment and special optical systems are necessary to overcome these problems. The positioning must be very precise; 5% misalignment means that 100 W of a 2 kW beam hits the cladding and destroys the fibre. Safe transmission of up to 6 kW by fibres of 0.6 mm core diameter is possible

The fibre surface treatment is a very delicate problem. Average power densities of more than 10^6 Wcm^{-2} occur on the surface. To avoid damage, a very clean surface is necessary. The surface can be AR coated to reduce reflection losses, but the better way seems to be the use of cleaved surfaces. The cleaving is limited to core diameters about of 0.6 mm. Fibres of larger core diameters are difficult to cleave. The surface is not plane and has to be polished very carefully.

So far cw power transmission has been discussed. The above relations hold for pulsed radiation too. The limitation is given by the damage threshold of the fibre surface, which for a clean surface is in the range of some 10 GWcm^{-2}. That means for a Q-switch pulse of 10 ns pulse duration transmitted by a 0.6 mm fibre, an upper limit for the pulse energy is about 250 mJ. For safety reasons the limit is lower.

4.6 REFERENCES

Fundamentals of laser physics and solid state lasers

Hodgson, N. and Weber, H., (1997) *Optical Resonators*, Springer, NY, London

Iffländer,R., (1990) *Festkörperlaser zur Materialbearbeitung, Laser in Technik und Forschung*, Springer Verlag, Berlin

Koechner, W., (1990*) Solid State Laser Engineering*, Springer Verlag, Berlin

Siegmann, A.E., (1986) *An Introduction to Lasers and Masers*, Mc Graw Hill, N.Y

Yariv, A., (1970) *Quantum Electronics*, J. Wiley and Sons, N.Y.

Specific References

Beck, T., et al, (1993) Transmission properties of all-silica for high power cw Nd:YAG lasers, *Laser Focus World,.* **29**, (10), p11

Bostanjoglo, G., (1994) Unstable multirod Nd:YAG lasers with variable reflectivity mirrors, *SPIE*, **2206**, p459

Budgor, A.B., Esterowitz,L. and De Shazer L.G., eds (1981*) Tunable Solid State Lasers*, **47**

Chernoch, J., (1990*)* Characteristics of a 1kW Nd:YAG Face Pumped Laser, *ICALEO*

Comaskey, B.J., et al, (1992) *IEEE J.QE,* **28**, p992

Das, P., (1990) *Lasers and Optical Engineering*, Springer Verlag, NY

Dong, S., et al, (1991) *Opt.Comm.* **82**, p514

Eichler, H.J., et al, (1994) 100 W Average output power 1.2-diffraction limited beam from pulsed Nd- single road amplifier with SBS phase conjugation, *IEEE J.QE.* **30**

Eichler, H.J., et al, (1997) *Proceedings of the Munich Laser Conference 1997* (to be published)

Findlay, D., Clay, R.A., (1966) *Phys. Lett.* **20**, p277

Hammerling, P., Budgor, A.B. and Pinto A., eds (1986) *Tunable Solid State Lasers II*, **52**

Hodgson, N., Bostanjoglo, G. and Weber, H.,(1993) *Appl.Optics*, **32**, p5209

Hollinger, F., Niedrig, R., Unruh, I., (19920 Excitation Efficiency Measurement in Multimode Solid State Lasers *Optics and Laser Technology*, **24**, *p353*

Ishide, T., et al; (1990) *Proc. SPIE* **1277**, p188

Kaminskii, A.A., (1981) *Laser Crystals*, Springer Series in Optical Sciences, **19**, Springer Verlag, Berlin

Kortz, H.P., Weber,H., *Appl.Opt*, **20**,1936

Kumkar,M., Wedel,B. and Richter, 4., ,(1992) *Optics & Laser Technology*, **24**, no.2, p67

Morin, M. (1997) *Opt. Quant. Electr.* **28** (to be published)

Okoshi, T., (1982) *Optical fibers*, Academic Press

Reng,N. and Beck,T., (1993) Transmission Properties of all-silica fibres for materials processing, *SPIE*, **1983**, p747

Rigrod, W.W., (1978) *IEEE J.QE* **14**, p377

Schindler, Q.M., (1980) *IEEE J.QE* **16**, p546

Seidel, S., and Mann, G., (1996) *SPIE* **2788** p183

Siegman, A., (1990) *Proc. SPIE*, **1224** p2

Weber, M.J., (1986) *Handbook of Laser Science and Technology* **I**, CRC Press, Boca Raton Florida

Wittrock, U., (1993) *High Power Rod, Slab, and Tube Lasers, Solid State Lasers: New Developments and Applications*, NATO ASI Series, Plenum Press

Wittrock,U., (1992) *Advanced Materials*, **4**, No.4

Wolf, E., (1978), *J.Opt.Soc.Am*,.**68**, p6

Wright, D. et al (1992) *Opt. Quant. Electr.* **24**, S993

DIN EN ISO 11146 Optics and optical instruments, test methods for laser beam parameters.

4.7 SYMBOLS

a	m	slab dimension, aperture radius
α	$°C^{-1}$	thermal expansion coefficient
$\beta, \beta_r, \beta_\varphi$	$mm \cdot kW^{-1}$	thermal lensing coefficient
b	m	slab dimension
d_i	m	distance mirror $S_i \rightarrow$ principal plane of rod
D_{core}	mm	fiber core diameter
D_{foc}	mm	focus diameter
D	mm	laser beam waist diameter
D_O	mm	fundamental mode waist diameter
D_{0r}	mm	fundamental mode diameter in the rod (pp)
D_f	m^{-1}	refractive power of rod, unpolarized light
$D_{f,r}$	m^{-1}	refractive power of rod, radial polarization
$D_{f,\varphi}$	m^{-1}	refractive power, azimuthal polarization
$D_{fc,1,2}$	m^{-1}	critical refractive power
D_{mn}	mm	multimode waist diameter
δt	s	laser pulse duration
Δt	s	pumping pulse duration
E	GPa	Young's modul
ΔE	Ws	energy difference lower laser level -ground state
E_{st}	Ws	stored energy
E_L	Ws	laser output energy per pulse
E_{th}	Ws	electrical pumping pulse threshold energy
E_E	Ws	electrical pumping pulse energy
e	Ws/cm²	energy density of laser pulse
e_O	Ws/cm²	energy density of incident laser pulse
e_S	Ws/cm²	saturation energy density
F	cm²	cross section of active medium

f	Hz	pumping pulse frequency
G	-	generalized g-parameter
G_0	-	small signal gain factor
G_s	-	saturated gain factor
g_i	-	resonator parameter
g_i^*	-	lens resonator parameter
$h\nu_L$	Ws	photon-energy of laser transition
$h\nu_P$	Ws	photon-energy of pumping light
h	cm	principal plane position
J	W/cm²	intensity of laser beam
J_0	W/cm²	incident beam intensity
J_s	W/cm²	saturation intensity
J^+, J^-	W/cm²	intensity inside the resonator
κ	W/m°C	thermal conductivity
K	-	beam quality number
kT	Ws	thermal energy
L_R	W/ster	m² radiance
L	cm	optical distance of the mirrors
ℓ	cm	length of active medium
ℓ_p	mm	pitch length in fibre
λ	μm	wavelength
λ_L	μm	laser wavelength
λ_p	μm	pumping wavelength
M	-	beam propagation factor
M_f	-	magnification
m	-	modulation function
NA	rad	numerical aperture of fiber
n	cm⁻³	ion density in upper level
n_i	cm⁻³	initial ion density
n_1	cm⁻³	ion density in lower laser level
n_s	cm⁻³	steady state ion density
n_r	-	refractive index of the laser crystal

n_{co}	-	fibre core refractive index
n_{cl}	-	fibre cladding refractive index
η	-	efficiencies, see table 2.2
ν	-	Poisson's ratio
ν_p, ν_L	- Hz	pumping light/ laser light/ frequency
Ω	ster	solid angle
P_{excit}	W	pumping power into upper laser level
P_L	W	laser output power
P_E	W	electrical pumping power
P_{th}	W	electrical threshold pumping power
P_P	W	pumping power into absorption band
P_H	W	heating power
$P_{L,max}$	W	laser output peak power
P_{min}	W	minimum electrical threshold power for $R_1=R_2=1$
R	-	reflectivity of output mirror
R_0	mm	rod radius
r	cm	radial coordinate
ρ_i	m	radius of mirror curvature
s	-	safety factor
σ_L	cm^2	cross section of laser transition
σ_{max}	MPa	rupture stress of crystal
$\sigma_{r,\varphi,z}$	MPa	stress tensor components
σ_{R0}	Mpa	maximum stress at rod surface
τ_L	s	upper laser level lifetime
τ_c	s	cavity transit time
τ_R	s	cavity decay time
T	°C	temperature
ΔT	°C	temperature difference in the active medium
θ	mrad	beam divergence
θ_f	mrad	angle coordinate in the farfield
θ_{max}	mrad	maximum acceptance of a fibre

V	-	loss factor per transit
x,y	cm	transverse coordinates
z	cm	longitudinal coordinate, beam propagation
z_f	cm	focal length
Z_i	cm	distance of real or virtual foci to mirror S_i

5

Excimer lasers

C. Fotakis, C. Kalpouzos and T. Papazoglou

5.1 INTRODUCTION

Excimer lasers, celebrating over two decades of development, have reached a level of maturity which makes them attractive for a variety of applications. Currently, excimer lasers play a key role in many specific applications that utilise their unique features, for diverse applications in material processing, remote sensing, biomedicine and research in basic physics. The interplay between the available excimer laser technology and the quality of applications is critical. On the one hand there are new demanding applications which require short wavelengths and tunability, and higher standards of excimer laser technology. On the other hand there is an increasing competition from convenient solid state laser systems emitting in the near infrared (IR), whose output can be extended with high efficiency at shorter wavelengths.

Excimer lasers may produce radiation at several ultraviolet (UV) wavelengths, typically 193, 248 and 308 nm. Depending on the type of excimer, average powers that range from a few milliwatts in waveguides to 1 KW in an X ray preionised XeCl laser have been obtained. The typical pulse duration of excimer lasers which is 15–25 ns can now be extended beyond 100 ns for beam delivery through optical fibers as required in several medical applications to lower than 100 fs in ultrashort pulse, high peak power systems used to study novel aspects of fundamental physics. It should be pointed out that excimer lasers are one of the few sources of high UV output suitable for pumping tunable dye lasers used in spectroscopic research. The recent availability of tunable Optical Parametric Oscillators

(OPOs) has opened new prospects for the use of excimer lasers as pumping sources.

Recent advances in the technology of excimer lasers include higher average powers, more compact lasers, extended gas lifetime, easier gas handling, safe and user friendly operation. Finally, improved beam quality, which is an important aspect in several processing application has also been achieved. This chapter addresses several issues related to the above subjects and discusses existing tendencies in current excimer laser technology especially in regard to industrial and medical applications.

5.2 THE PHYSICS OF THE ACTIVE MEDIUM

5.2.1 Excimer transitions

The interaction of two ground state closed shell systems, for example atoms of s^2p^6 configuration, is repulsive except for a weak long range Van der Waals electrostatic interaction. In contrast, the interaction of an excited state with the ground state of the same fragments may be attractive and lead to the formation of a strong chemical bond. In the case of two interacting atoms, the diatomic molecule which corresponds to the bound excited state is called an **excimer**, as an abbreviation for excited dimer. The bound free transition from the bound excited states to the ground repulsive state defines the excimer emission system.

Typical potential surfaces for an excimer system are shown for the KrF excimer laser in Fig. 5.1. In this case, the fluorine atom exists in an open shell, degenerate state (p state) and more than one molecular state may correlate with the degenerate asymptote.

Rare gas dimers and primarily rare gas halides are the most common active media of excimer lasers. Triatomic classes of excimers have also been examined as laser candidates. Finally, excimer transitions may be obtained between excited states, involving bound and lower repulsive states of stable molecules such as the molecular halogens F_2 and I_2.

The spectroscopy of rare gas halide excimers, which are the most efficient excimer laser media, became the issue of extensive studies during the early seventies (Velazco and Setser, 1975; Golde and Thrush, 1974).

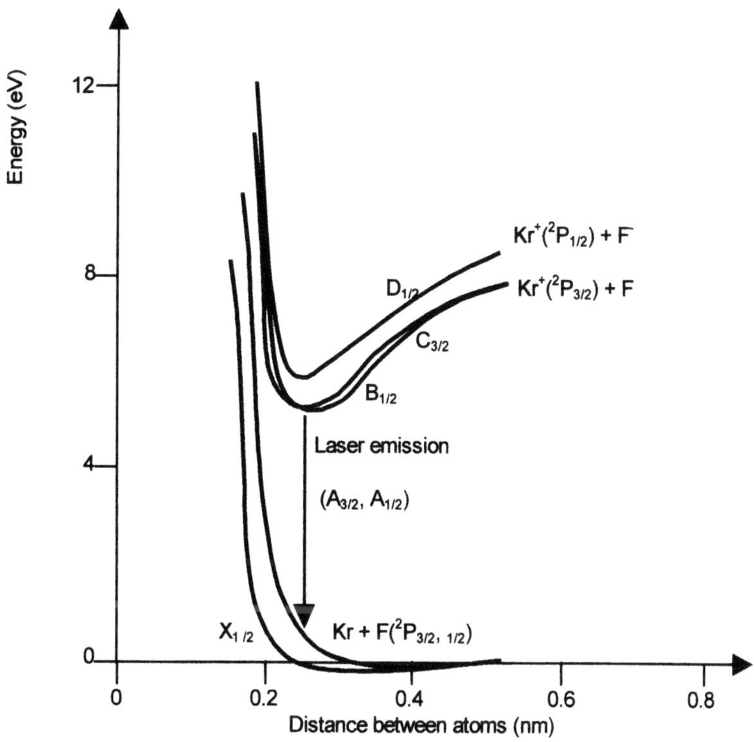

Fig. 5.1   Potential surfaces for the KrF excimer laser.

Lasers based on these principles were first realised in 1975 (Searles and Hart, 1975). In these lasers, the active medium is composed of an inert gas atom and a halide atom. As mentioned earlier, the rare gas halide molecule is bound only in electronically excited states, while its electronic ground states are purely repulsive or weakly bound. Emission takes place primarily from the lowest vibrational levels of the upper bound state. The small gain signal cross section of the laser transitions will be proportional to:

$$\sigma \approx \lambda^2 a_{21} g(\varepsilon)$$

$$(5.1)$$

where a_{21} is the Einstein coefficient for spontaneous emission from the upper state to the lower state, λ the wavelength of the transition and $g(\varepsilon)$ the continuum line shape factor, determined largely by the Franck Condon factors for the bound free transition. The potential energy curve parameters of the excimer molecule i.e. the degree of repulsive and attractive character, the relative equilibrium internuclear separation and vibrational spacing will largely determine the magnitude of Franck Condon factors and in turn the magnitude of σ.

Rare gas halide excimer lasers emit photons of high quantum energy in typical pulses of 15–20 ns duration. These are appropriate for the efficient excitation of atoms or molecules at highly excited states through the absorption of one or more photons and induce photochemical processes in polymeric materials or biological tissue. Table 5.1 includes spectroscopic data for the most commonly used rare gas halide excimers. Typical linewidths for excimer laser systems are of the order of 3 cm^{-1}.

Table 5.1 Spectroscopic data for common rare gas halide excimers

Excimer	$\lambda(nm)$	$r(\text{Å})$	$\omega_c(cm^{-1})$	$\sigma(\text{Å}^2 ns)$	$\tau(a_{21}^{-1})(ns)$
ArF	193	2.2		12	4.2
KrF	248	2.3	310	17	6.7
KrCl	222	2.8	210		
XeF	351	2.4	309	64	12
XeCl	308	2.9	194	50	11

where: λ = transition wavelength,
r = equilibrium internuclear separation,
ω = fundamental vibrational frequency of the excited state,
σ = stimulated emission cross section,
τ = radiative lifetime.

5.2.2 Rare gas halide kinetics

The excited rare gas halides states are of Rydberg or ion pair character and have become the issue of extensive experimental (Golde, 1975) and theoretical (Kraus and Mies, 1979) studies. The Rydberg states are characterised by smaller internuclear separations and deeper potential wells than the ion pair states, which correlate with the positive ion of the noble gas and the negative ion of the halide. Many of the excited ion pair states are strong emitters and give rise to lasing transitions.

In a mixture of noble gas atoms and halogen molecules, the mechanism for selectively populating the ion pair states relies in a sequence of collisional energy transfer processes with energetic electrons originating from the pumping mechanisms (electric discharge or electron beam pumping). A simplified scheme describing the excimer formation is as follows. Excited state decay mechanisms are ignored in this scheme. The high energy electrons ionise or excite the noble atom gas :

$$e^- + Kr \rightarrow Kr^+ + 2e \tag{5.2}$$

$$e^- + Kr \rightarrow Kr^* + e \tag{5.3}$$

where Kr* denotes an excited Kr Atom.

In addition, due to their high electron affinity fluorine gas atoms may attach an electron :

$$e^- + F_2 \rightarrow F^- + F \tag{5.4}$$

Several energy and charge transfer processes may take place in this complex mixture. In the final stage, the ion pair state may be formed as follows :

$$Kr^+ + F^- + M \rightarrow KrF^* + M \tag{5.5}$$

where KrF* denotes the excited excimer and M a third body, which may be He or Ne, taking up the excess energy and thus stabilising the KrF*.

5.3   PREIONISATION AND PUMPING CONSIDERATIONS

The threshold values for population inversion in excimer lasers may be very high due to the short wavelength and the considerable line widths of the relevant transitions. The gain expression (equation 5.1), when no population exists in the lower lasing level may be written as :

$$\sigma = \frac{\lambda a_{21}}{8\pi\Delta v} N_u \tag{5.6}$$

where N_u is the concentration of molecules in the upper state, Δv is the width of the gain profile, a_{21} is the spontaneous emission probability and λ is the wavelength. Typical values for excimer lasers are $a_{21} \sim 10^7$–10^8 sec^{-1}, $\lambda \sim (2$–3)x 10^{-5} cm, and for a threshold gain of 10^{-2} cm^{-1}, the concentration of active species is $N_u \sim 10^{14}$–10^{15} cm^{-3}. Such concentrations can only be obtained by means of very high pump energy densities ($\sim 10^{-2}$ J/cm^3) supplied in a short time interval (10^{-8} –10^{-7} s), achieved by passing a high -intensity beam of fast electrons or a high power discharge pulse through a high density gas.

Excimer lasers are invariably pumped by a power supply that discharges the stored electrical energy directly into the active medium. It is also well accepted that pumping schemes, i.e. power supply and laser cavity, involve the switching of this stored electrical energy to the cavity in a very short time and with a well defined and controlled profile. Discharge uniformity is very closely related to spatial uniformity of the laser beam. This places severe requirements to the pumping scheme in terms of peak currents and voltage risetimes.

The power supplies comprise, besides the charging section, a pulse forming network (PFN), in most instances constructed by a network of inductors and capacitors, placed in close proximity to the laser cavity. This network is charged to the required high voltage and the discharge is held off by a high -voltage switch which is triggered at a suitable time. The PFN section aims to match closely the dynamic complex resistance of a gas discharge, where the ohmic resistance is rapidly changing, and feed the stored electrical energy in a controlled time interval. The usual high power switches, such as thyratrons and spark gaps (single arc or in a rail geometry), are used as discharge initiators, but are currently at the limits of their operative parameter range. Since these switches must withstand the high voltage and current during a discharge cycle (voltages up to 40 kV, peak currents in the 100 kA range, *dI/dt* about 10^{11} A/sec), it is clear that they are the most stressed component in the system.

In addition they are always supported by auxiliary sections that prepare and preionise the active media for the main discharge pulse that follows. This glow type preionisation effect is critical in producing even and uniform discharge profiles with minimum filamentation in the main discharge. The critical parameters that affect preionisation, such as threshold preionisation electron density and preionisation uniformity, are heavily dependent on cavity considerations such electrode profile, type of electrode (screen or solid), gas pressure, duration of preionisation, preionisation electron loss processes, time delay between preionisation and the application of the main pulse, main pulse voltage risetime, and on the overall cavity geometry (Taylor, 1986). Two well known preionisation methods are

1. Preionisation electrodes situated next to the high voltage main electrode, in small and medium size excimer lasers;
2. X ray preionising sources feeding the cavity volume, in larger systems.

A novel concept for the preionisation in excimers attempts to eliminate the use of a thyratron or spark gap, thereby minimising high stress components in the system (Stehle, 1993). The power supply itself slowly charges the storage capacitors, at such a rate that no discharge can occur within the cavity. The laser cavity itself must be specially designed to hold off the operating high voltage. This charging phase is followed by a fast corona discharge through the perforated cathode electrode when the laser is triggered. The main discharge pulse follows this corona discharge. The scheme eliminates all stressed components and improves overall lifetime.

5.3.1 Thyratron based designs

Early power supply designs, were constructed around a triggered C-L-C network, with the following typical arrangement shown in Fig. 5.2. By closing the switch the energy stored in capacitor C1 is transferred in typically 100–150 ns to the peaking capacitors C2. Depending on the ratio of capacitance, a voltage enhancement can be obtained.

Ideally, the cavity design is such that the gas breaks down when the maximum energy is transferred.

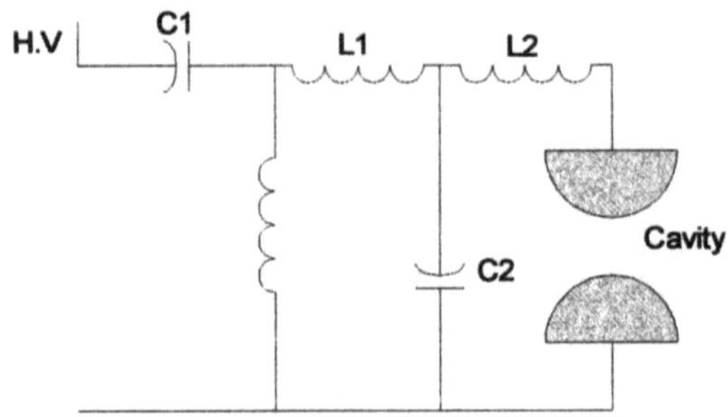

Fig. 5.2 Typical C-L-C transfer circuit in excimer lasers.

The key element in such a setup is the trigger switch. The spark gaps used in the beginning quickly gave way to thyratons which have seen considerable improvement since, in their reliability and lifetime characteristics. They must withstand voltages of about 30–40 kV, current surges of 5–15 kA and current rise times of 30–40 ns. The initial standard thyratron shown in Fig. 5.3(a) has seen numerous developments to that shown in Fig. 5.3(b).

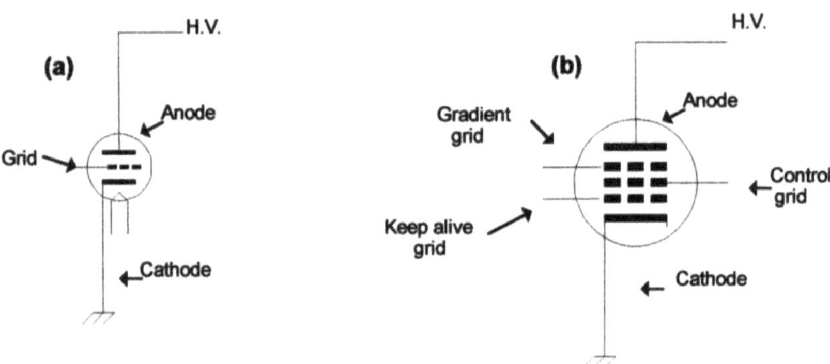

Fig. 5.3 Thyratrons used in excimer lasers. (a) Simple arrangement. (b) Recent designs allowing better control of the firing process.

A thyratron tube is filled with H_2, by heating a hydrogen metal reservoir (palladium). The Hydrogen pressure determines the hold off voltage of the tube, which decreases with pressure. The cathode is covered as in normal electron tubes, so that a supply of free electrons is provided for to ensure a homogeneous discharge. The free electrons are released by the heater behind the cathode. The grid is negative biased with respect to the cathode, so that the free electrons are kept close to the cathode. Applying a positive trigger pulse, the electrons can penetrate the grid baffles and the thyratron switches to the conductive mode. The lifetime, reliability and trigger accuracy of the thyratron is largely based on this conductive interval to give a glow discharge covering much of the electrode active areas. Allowable current can be scaled by increasing the area of the cathode. The current risetime in a thyratron is limited by the penetration time of the electric field into the cathode electron cloud of the cathode. It depends on the cathode geometry and should be less than 10^{11} A/sec. The tube's reverse voltage value is related to the reverse current maintained by the ions. A thyratron acts like a diode, and if the reverse current exceeds 10% of the maximum forward current, an arc will occur which will damage the cathode. The maximum reverse voltage should not exceed 500–1000 V.

A series of improvements have since occurred. The cathode area has been increased to 5 inch (125 mm) diameter. To achieve higher hold off voltages, a gradient grid is added to lower the field of the grid. For more reliable triggering a keep alive grid is added that maintains a continuous current from the cathode. Another grid can be added for prepulsing the device and achieve a more homogenous discharge. The hollow anode design has the anode perforated with holes connecting it to a chamber. During conduction the chamber is ionised and acts as a plasma source for reverse current. The design allows a higher reverse current.

5.3.2 Magnetic switching

A typical magnetic switch is shown in Fig. 5.4. Capacitor C1 is discharged via the thyratron and the inductance L1 to Capacitor C2. The additional inductance LMS allows the thyratron to operate well below its risetime limits. When the energy is on C2, the magnetic switch closes rapidly, transferring the energy to C3. This happens when the inductance LMS << L1. Physically the switch is an inductance that can be saturated in a short time. During the charging cycle of C2, LMS >> L1, and a small leakage current goes through the switch. This saturates the material whereby LMS becomes much less than L1; the time interval for this delay depends on the amount of magnetic material present.

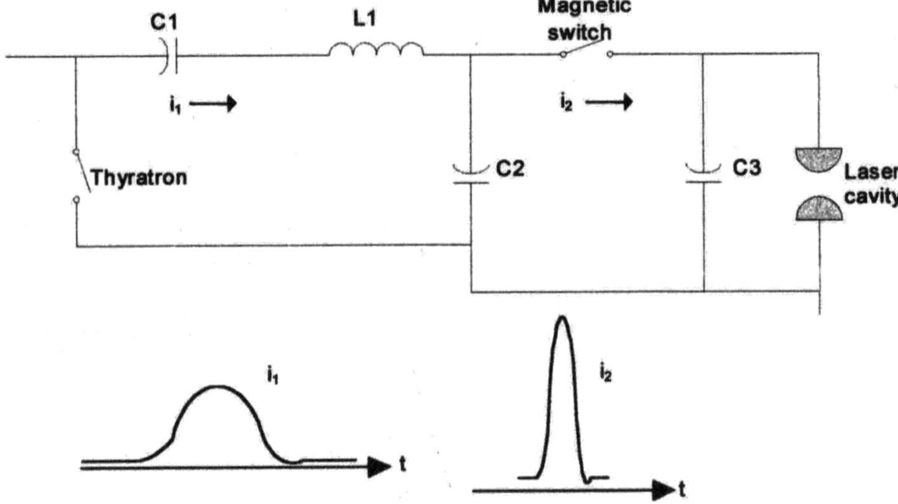

Fig. 5.4 Schematic of a single stage magnetic switch circuit, indicating the compression achieved in the current pulse delivered to the laser cavity.

Multistage compression schemes have been used, which result in acceptable compression factors but have 10–20% energy loss per stage. Further designs have incorporated a voltage multiplication stage (typically by a factor of 2) to lower the stresses on the thyratron.

5.3.3 Preionisation

The large cross section high voltage discharges needed in excimer lasers need a start up electron density of 10^7–10^8 cm^{-3}, in order to break down homogeneously (glow discharge). This preionisation can be achieved from UV light from sparks generated in the active laser volume, before the main discharge cycle. The arrangement is usually a large number of pins situated in close proximity to the H.V. laser electrode on either side. UV light is rather strongly absorbed by the gases, and only a limited depth can be preionised in this manner. This is a critical parameter in scaling discharge volumes. This can be partially compensated by electrode design, but then electrode shape becomes a rather critical factor. Typical preionised discharge volume widths are 10–25 mm, with XeCl lasers allowing the largest widths. X ray absorption based preionisation can occur to larger depths allowing a more uniform glow in larger volumes, and is used mainly in large aperture cavities.

5.4 COOLING SYSTEMS AND CORROSION PROBLEMS

Excimer lasers, operating at about 2% total conversion efficiency between input electrical power and output optical power, are subject to the usual cooling requirements of a laser system. The excess energy must be removed as excess heat efficiently, in order to obtain reasonable optical powers and repetition rates. The major problem with a gas laser cooling system, is the poor heat transfer that occurs between the gas phase of the active medium and a heat exchanger. Usually the active medium is contained in a pressure vessel of sufficient volume to allow good mixing of the components via forced circulation, and usually made of aluminum. An internal fan keeps the active gases well mixed and renewed in the lasing region, while at the same time forces the gas to flow through the heat exchanger. The heat exchanger must have sufficient contacting area, to allow reasonable temperature stability of the gases at all operating conditions. The cooling medium in the heat exchanger is usually water, either in a closed loop or an open loop system. An overall cavity schematic is shown in Fig. 5.5.

Such a cooling system suffers mainly from corrosion problems, as the surfaces in contact with water are composed of dissimilar metals, which are subject to galvanic corrosion. Galvanic corrosion is due to the different electronegativities of the metals and their alloys. The most commonly used metals in excimer cavities are aluminum, copper and nickel. Aluminum and copper are quite dissimilar in their galvanic properties, and in an aqueous environment an enhanced electron flow from aluminum to copper is absorbed, causing increased corrosion activity.

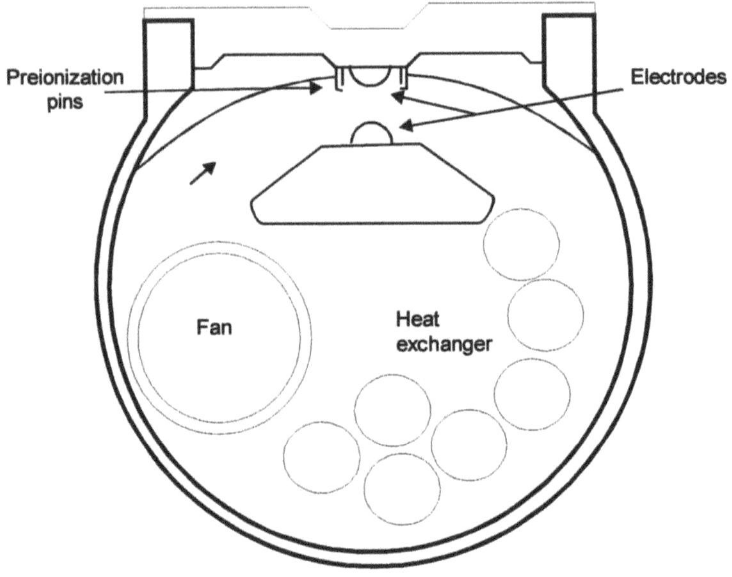

Fig. 5.5 Cavity detail of an excimer laser, depicting the heat exchanger and gas circulation in cross section. The units extend the whole length of the vessel.

This type of corrosion can be aggravated by the geometric ratio of such metals in the aqueous environment, when the area of the aluminum anode is smaller than that of the copper cathode. The surface area and the potential differences are important in a water cooling system that is left unprotected. In addition, the external walls of the system are also subject to the corrosive attack of the halogen gases.

The gas circulation system is of major concern in the design of the laser. For medium energy levels, internal fan circulators are sufficient to keep the gases well mixed, flow new mixture through the active lasing volume, cool them through the heat exchanger lines, and by passing them through scrubbers clean them of any particles that may result from ablation from electrodes, walls, etc. This is not sufficient in larger aperture cavities or for high repetition systems (Fig. 5.6). In such cases the blower system encompasses the laser cavity.

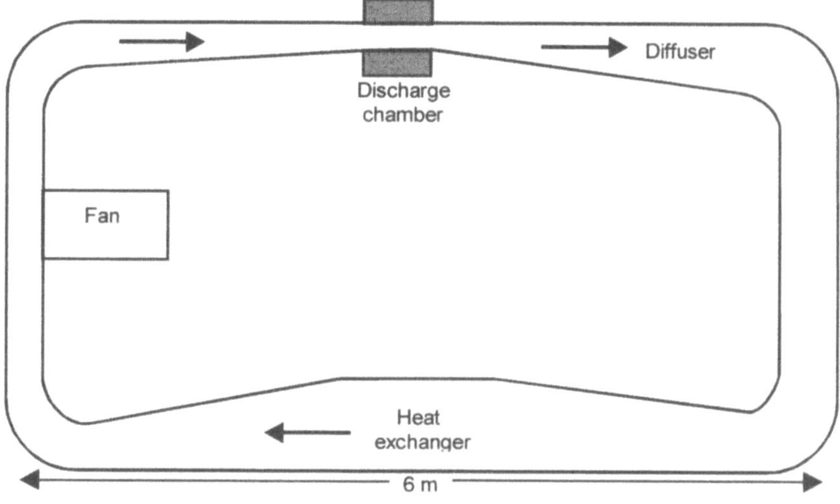

Fig. 5.6 Gas flow loop for the 1 kW excimer laser, with a 150 m/sec rate.

Corrosion problems in an excimer cavity are not really limited to the cooling system. The circulation of corrosive gases such as F_2 and HCl in a cavity, gives rise to strong interaction with the cavity walls, electrodes and optical elements. Gas lifetime is a very crucial question in an excimer laser as a result. Photodegradation of any halogen resistant polymers by intense UV light in the cavity is a well known source of organic contaminants of the laser gas. Wall material, electrode material and any other foreign matter will strongly interact with the halogen gases and degrade performance. Gas circulation helps to extend gas lifetime, as the gases are continuously cleaned through special filters.

Electrical discharges between the electrodes also create additional problems, as material sputtering to the cavity windows is not conducive to good laser operation. Usually mirrors are outside the cavity body, but the necessary windows attached to the gas vessel are prone to the same problems. They are usually kept free of deposits, by geometrically offsetting them form the discharge region and/or maintaining gas flow over their surface to carry away any material. In addition sputtering of material from electrodes implies electrode erosion which may be more critical, as the resulting discharge homogeneity is compromised. The plated electrodes lose their dynamic ohmic character when the plating material is lost, and surface irregularities can act as initiators of filamented discharges which eventually destroy the glow discharge required for proper laser operation.

New generation ceramics are making their way into the design of new cavities, for the critical plasma region of the laser, replacing the polymer based materials used up to now. The metal surfaces are made of special carbon and silicon free alloys, in order to avoid the detrimental effect on lasing of SiF_4 and CF_4 that may be produced, even on the ppm level. All electrical discharge components are made of corrosion resistant materials to reduce sputtering problems. Sealed cavities are developed that isolate the system from the environment, and coupled with a sealed off gas delivery system, extend laser and gas lifetime (Rehban, 1995).

5.5 BEAM MANIPULATION SYSTEMS

The demand for ultraviolet optical systems has increased in recent years, because of advancing applications in microlithography, optical inspection, medicine and industrial micro machining. The use of ultraviolet radiation for these applications creates special challenges for optical design, materials and methods.

The UV spectral region considered here extends from 180 to 400 nm, bounded by the vacuum ultraviolet on the short wavelength end and the region of visible wavelengths on the long wavelength end. In turn, the UV range comprises the so called UV-A, UV-B and UV-C bands; this constitutes widely accepted nomenclature for smaller wavelength intervals within the UV region. Light sources for this region include excimer, argon, helium cadmium and nitrogen lasers, as well as harmonically generated wavelengths from longer wavelength lasers. Mercury, xenon and deuterium lamps also offer useful emission in this region.

To maximise the UV processing power of industrial excimer lasers in microlithography, micromachining and marking applications, careful attention must be given to the design of the beam delivery system. In this scenario, beam properties such as wavelength, pulse energy, average power, energy density, beam profile and divergence are particularly relevant in order to achieve the best results.

Depending on the gas mixture in the laser head, excimers can be operated at several different wavelengths. As mentioned above, the most common being 193, 248, 308 and 351 nm. Commercial sources can produce pulse energies between approximately 10 mJ and 2 J. This is especially important because the wide range of energies allows the user to tailor the pulse energy to individual materials processing applications.

Energy densities in the beam vary between about 10 mJ/cm^2 and 500 mJ/cm^2 depending on the model. This value is quite significant for the beam delivery system, which in turn produces the proper energy density on the workpiece.

The beam profile is of rectangular symmetry. It is a flat top without peaks or holes and is well suited for processing material using imaging techniques. In the case of plane parallel resonator mirrors (often called stable resonator) the beam divergences are about 4 mrad horizontally and 1 mrad vertically. When focused, an elliptical rather than a circular area is filled with high beam intensity.

Curved resonator mirrors can be used to produce an unstable resonator that results in a beam divergence that is about ten times smaller. The negligible difference between horizontal and vertical divergence allows the design of focal point optical systems.

5.5.1 UV delivery systems – optics

Ultraviolet transmission

The first design challenge encountered when crossing into the ultraviolet region is the reduced number of optical materials that transmit efficiently. Most of the materials depicted in the upper half of the optical glass chart are lost at UV-A wavelengths (Fig. 5.7). These include the high index, high dispersion glasses typically used for the correction of chromatic aberration. At progressively shorter wavelengths, low index, low dispersion glasses become highly absorbent. For a xenon chloride laser operating at 308 nm, in the middle of the UV-B range, only fused silica, fused quartz and some crystalline halides remain as viable refractive materials.

Reflective lens forms

To avoid the material transmittance problem altogether, employ an all-reflective design. Kingslake (1978) has given a highly informative review of such mirror systems.

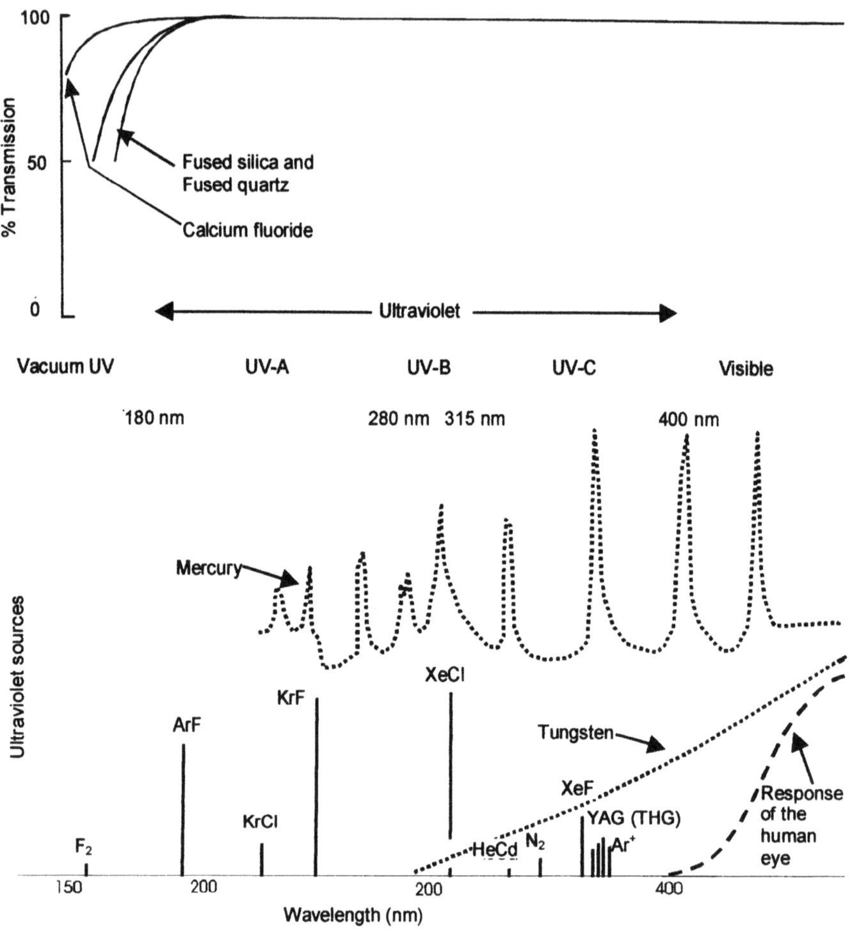

Fig. 5.7 Ultraviolet sources and material characteristics.

The Schwarzschild form is perhaps the most remarkable. By maintaining a specific ratio of the concentric mirror radii, primary spherical aberration, coma and astigmatism are all zero. Also, the absence of refractive elements frees the design from chromatic aberration. The Schwarzschild form is especially well-suited for microscope objectives as well as other small-field applications. General drawbacks of this form, and most other all mirror systems, are the small angular field, limited numerical aperture, curved image and central obscuration. The obscuration typically introduces a throughput loss of 10 to 50 % for a uniformly illuminated beam. In addition, the obstruction also changes the spatial frequency response. Compared to a system with equivalent f number and without obstruction, the obscured system mid frequency response is lower, while higher frequencies exhibit a slightly enhanced modulation response.

Refractive lens forms

Ultimately, however, some applications cannot be solved with an all-mirror system. For example, the economies of integrated-circuit production demand efficient throughput that is directly related to field size. Microlithographic lenses imaging the mercury I line at 365 nm provide submicron resolution over an image field approaching 30 millimeters in diameter. Likewise, an all-mirror system is generally not feasible at large numerical apertures; a practical limit seems to be an NA of 0.50. Yet refractive UV microobjectives with limited fields have been designed with NAs. as large as 0.90, yielding 0.4 μm resolution.

Refractive index

Pursuing a refractive lens form, the designer must next contend with index of refraction and dispersion characteristics. Measured indices for specific glass melts, supplied by glass manufacturers, are generally made at visible wavelengths, often far removed from the wavelength of interest. The use of formulae intended for interpolation, but actually used to extrapolate indices of refraction in the ultraviolet, can be unreliable. This is primarily due to the rate of change of index with respect to decreasing wavelength caused by nearby absorption bands.

To attain the optical performance required in systems with high numerical apertures, the designer must know the index of refraction to a high degree of accuracy (to a few parts in 10^6). For melt sensitive designs, it is common practice to review the "actual" index of the optical glass, as reported by the glass manufacturer, prior to the manufacture of the optical system. The optical designer then has the opportunity to alter nominal lens radii to re-optimise optical system performance for specific melt data. Of course, this approach fails to yield the intended performance level if the measured or calculated index is in error. The measurement of material index at UV wavelengths corrects this potential pitfall.

Dispersion characteristics

The dispersion of optical materials also increases in the ultraviolet. Initially, this appears to be a benefit toward gaining colour correction in the UV. However, the asymptotic behavior of the index curve near absorption bands is such that the relative dispersion of different materials becomes nearly equal. The impact on the refractive design is either the likely addition of more elements to compensate chromatic aberration or narrowing the bandwidth of radiation. Unfortunately, both options are undesirable in terms of system complexity and increased cost.

Fused silica versus fused quartz

When a refractive solution is required at wavelengths of 308 nm or lower, further scrutiny of material properties is required. The most common materials used, and often confused, are fused silica and fused quartz. Although both are composed of the same basic material, silicon dioxide (SiO_2), they are made by different processes and exhibit slightly different properties. Fused quartz is essentially that - natural crystalline quartz that has been ground and then fused at a high temperature. In contrast, fused silica is a synthetic material produced by the flame hydrolysis of silicon tetrachloride.

Fluorescence

Impurities in quartz will produce a blue fluorescence when illuminated with ultraviolet light, with the peak excitation occurring around 257 nm. The fluorescence persists up to 10 ns following excitation at the shorter UV wavelength. The fluorescence of quartz is particularly noticeable with KrF illumination at 248 nm. Fused silica exhibits significantly less fluorescence than fused quartz, because of its higher purity. However, individual samples from a single melt have been observed to fluoresce to varying degrees. Depending on the specific application, fluorescence can create serious performance degradation. This serves to reinforce the need for improved glass process control and special UV material inspection for critical applications.

In general, this effect degrades optical performance when the undesired light reaches the image plane. Specific examples include loss of contrast for optical inspection systems and decreased feature resolution in optical lithography. In the latter case, the fluorescing "output" of the optical component may contribute to uncontrolled exposure of the photoresist.

Fused silica is available in both UV and IR grades; the fundamental difference is the hydroxyl ion (OH⁻) concentration. The IR grade is "low OH silica" (<10 ppm) and has relatively low absorption in the 2.7 μm region. This grade of fused silica also exhibits a relatively strong absorption band similar to the 257 nm band in fused quartz; in this case, however, the fluorescence is red.

"High OH⁻ silica" (>800 ppm) has a less pronounced absorption band in the UV, at the expense of the IR region. At sufficiently high fluence levels, red fluorescence still appears in UV grade fused silica and must be taken into account, especially if a KrF laser will be used.

Fluorescence could be obtained by focusing the beam of a high energy KrF laser into the samples (Fig. 5.8). The fluorescence of the sample in the foreground results from an energy density approaching 800 mJ/cm^2.

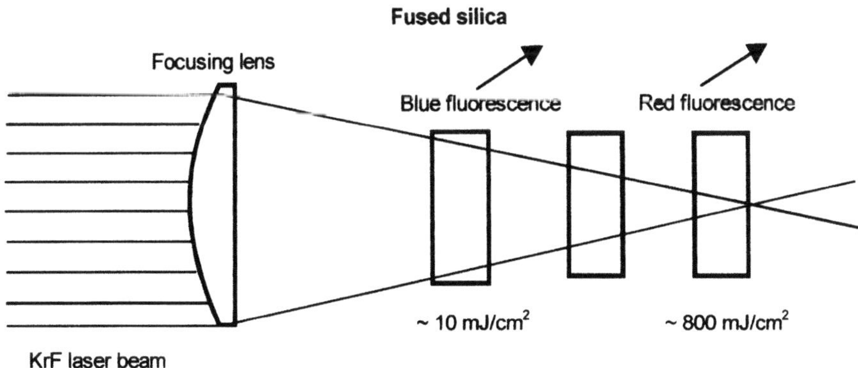

Fig. 5.8 Schematic of set up depicting red and blue fluorescence from fused silica.

Colour centres and solarisation

Escher (1988) has reported an overview of KrF induced colour centres in commercial fused silica. The conclusion supports the mindful selection of fused silica for use with KrF lasers. Further, changes in transmission of optical systems have been measured as a function of pulse energy density and number of pulses.

The change of transmission in optical materials is commonly referred to as solarisation. This condition was first observed in glass exposed to solar radiation, but the term is now extended to describe various optical glasses susceptible to this effect, irrespective of the light source. While the condition of solarisation has been noted (Setta et al., 1988). Continued effort on the part of glass manufacturers to arrive at suitably stable UV materials is necessary.

The impact of solarisation to an optical system can range from cosmetic discoloration to catastrophic loss of transmission. In the case of UV projection printing, the increased exposure time adversely affects throughput efficiency, or foreshortens the potentially useful lifetime of the light source.

In some instances, the solarisation of an optical system may be reversed. Elevated temperatures and illumination at different wavelengths have been employed to effect the reversal. Likewise, alteration of glass compositions has been suggested (Babcock, 1977) to reduce susceptibility to solarisation.

Crystal halides

At the shorter wavelength end of the UV-C region, fused silica and fused quartz are no longer useful. The majority of remaining materials are crystalline substances. Lithium fluoride (LiF), magnesium fluoride (MgF_2), calcium fluoride (CaF_2), barium fluoride (BaF_2), and sapphire (Al_2O_3) are the most common of these materials. The crystal halides exhibit several undesirable characteristics:- birefringence, colour centre susceptibility, scattering, sensitivity to thermal shock and fogging. The fogging of polished surfaces results from the hygroscopic nature of these materials and generally limits their useful lifetime in moisture-bearing environments.

Sapphire performs reasonably well in this spectral region. However, accurate material parameters (index of refraction, dispersion, homogeneity, and so on) for sapphire are not readily or routinely available. The same comment generally applies to crystal manufacturers, when the quality and quantity of material information is compared to that currently supplied by optical glass manufacturers. The rationale for this relative dearth of information may be the limited (to date) interest in these materials for high performance imaging applications.

Surface figuring

The challenges associated with the design effort are rivaled in the manufacturing processes. One direct consequence of decreasing wavelength is the diminishing allowable surface error. For example, a twentieth wave ($\lambda/20$) surface figure measured at 632.8 nm corresponds to an eighth wave error at the KrF wavelength. To maintain a $\lambda/20$ tolerance at KrF wavelengths, the measured error at the HeNe wavelength would have to be controlled to approximately $\lambda/50$. This degree of surface figure control surpasses the capability of many optical fabrication facilities.

One consequence of manufacturing an all mirror system is even more stringent surface figure control, because the optical path difference error doubles in reflection. In addition, mirror systems are highly susceptible to component distortion, which again induces a two-to-one contribution to wavefront error. These sources of wavefront error and other drawbacks, such as coating difficulties, make deep UV mirror systems particularly challenging.

Scatter

The manufacture of low scatter optical components requires the production of smooth surface finishes, generally free of smooth surface finishes, generally free of defects. In the UV, surface micro-irregularities predominate over other defects such as scratches, digs, contamination and bulk scatter (Bennett, 1978). At first glance, that assertion may seem surprising, but surface scattering increases exponentially with decreasing wavelength.

Surface finish depends not only on the manufacturing process but also on the optical material selected. For example, fused quartz has been measured with a surface finish as small as 2 to 13 Å (rms), while CaF_2 and sapphire range from 10 to 61 Å. In general, surface micro-irregularity for crystalline halides tends to be greater than for quartz and silica based optical glasses.

Relatively new "super polish" techniques may offer some improvement for UV systems. This capability is closely associated with laser gyro technology and is currently yielding optical surfaces with approximately 2 Å (rms) micro-irregularity. However, it remains to be seen whether these techniques can be successfully applied to the range of optical materials and component dimensions required in UV optical systems.

Related developments in optical metrology also merit attention with regard to surface finishing. Several firms currently offer products capable of high-resolution, non-contact surface profiling. While the relatively high price of these instruments may prevent widespread use beyond specific markets such as laser gyro optics and magnetic media, the technique offers a means of better quantified process development.

UV optical coatings

Many of the common dielectric and metal materials used for coating in the visible and near-infrared regions begin to absorb at ultraviolet wavelengths. However, several materials are still available for low index (SiO_2, CaF_2, MgF_2) and high index (Al_2O_3, HfO_2, Sc_2O_3) layers. Alas, multilayer coatings become less effective as the wavelength decreases, because the index ratio between high index and low index materials declines. Thus, UV dielectric coatings require more alternating high and low index layers to achieve performance comparable to similar structures at visible wavelengths. Unfortunately, these additional layers impart secondary disadvantages:- surface stress, absorption and scatter.

5.5.2 Workstation optics - beam manipulation

Imaging techniques

Nearly all industrial materials processing applications that employ excimer lasers require a uniform illumination of the workpiece. The first consideration is to ensure the uniformity of the laser beam. There are several alternatives. One is to employ better preionisation in the laser. Unfortunately, these types of lasers are not yet mature enough to be considered for most industrial applications.

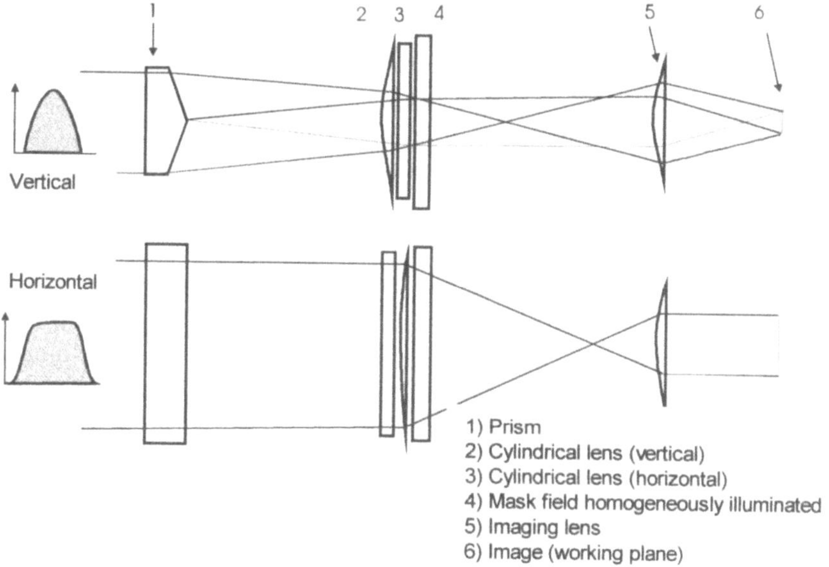

Fig. 5.9 Beam homogeniser.

Another way is to use a beam homogeniser. This technique produces a beam that is normally flat in the long dimension and approximately Gaussian in the other. A first step is to reshape the Gaussian dimension by using a set of prisms (Fig. 5.9). The simplest form of a beam delivery system is a mask imaged onto the workpiece. The target interaction area is also reduced at higher magnifications. A scan mask imaging technique is used in applications requiring larger areas to be processed, such as marking, or requiring higher fluences. The edge precision in marking, for example, is given by the resolution of the imaging lens, which for diffraction limited imaging is given by

$$R = 0.5 \frac{\lambda}{NA} \tag{5.7}$$

In this equation λ is the wavelength of the laser beam and NA the numerical aperture at the lens.

The short wavelength of an excimer laser can be used to produce structures of less than 1 μm. This cannot be achieved with standard commercial lenses. Because of their simple construction with spherical or planar surfaces only, image defects such as distortion and fuzzy edges become significant whenever the numerical aperture exceeds 0.1. The field lens is a significant component of the imaging process. Its purpose is to maintain a small value of the numerical aperture, reduce distortion and improve depth of focus to achieve higher resolution. A simple model calculation helps to quantify the use of a field lens. In this example we consider mask imaging with a demagnification factor of four. The imaging lens has a 50 mm diameter, a focal length of 200 mm, a central thickness of 5 mm and is constructed as a plano convex lens. The hypothetical object consists of n radiant points, emitting m rays into a certain angle. This angle is chosen to match the beam divergence of an excimer laser. With these parameters we can define quantitatively the imaging defect as:

$$B_i = \sum_{i=1}^{n} \sum_{j=1}^{m} \left[r_{ij}(z) - \bar{r}_1(z) \right]^2 \tag{5.8}$$

Here i runs over the n radiating points, j over the m rays originating from each point, and z is the coordinate along the optical axis. The imaging defect is the sum of the squares of all deviations from the centre of the image, and the distance from the optical axis is given by

$$\bar{r}_1(z) = \sum_{j=1}^{m} \frac{r_{ij}(z)}{m} \tag{5.9}$$

In this definition the image curvature is also considered an imaging defect. A field lens can reduce imaging defects by a factor of about 1000 and it also permits objects that are more than twice the size of the imaging lens to be produced with only small imaging defects. Another benefit is that a depth of field of about 1 mm with a resolution in the micron range is so large that the positioning of the workpiece does not present any technical problems.

UV optics testing

Modification of existing visible techniques accounts for much of UV optical system testing at present. In the case of reflective systems, wavelength independence permits direct use of visible techniques. However, when a system must be aligned and tested at its design UV wavelength, the method for viewing the image must be addressed.

For initial or coarse alignment of an ultraviolet system, something as simple and inexpensive as a fluorescent card will work. This technique can be extended for more detailed viewing with the use of a video camera. For more critical real time imaging, a CCD camera with a visible blocking filter should be used. Most CCD cameras have a layer of UV absorbing material over the detector elements and therefore are not recommend for direct UV use. However, CCD cameras modified by the addition of a scintillation or fluorescent layer are well suited for UV applications. These modifications extend the sensing range from a previous cut off of approximately 400 nm down to 185 nm or even shorter wavelengths.

Testing at different or longer wavelengths for some systems can be accomplished by merging computer design models with computer based metrology. For example, the transmitted wavefront error of an optical system can be measured off-wavelength and then expressed in terms of Zernike coefficients. The computer model can then subtract the error arising solely from the off-wavelength contribution, leaving a measurement of the system at the intended wavelength. However, a caveat to off-wavelength testing must be emphasised. After coating, surfaces tested off-wavelength will exhibit wavefront distortions stemming from phase changes upon reflection. This phase dispersion of dielectric multilayers must not be overlooked.

High precision

The short wavelengths of excimer lasers permit precision machining of materials. The normally nonlinear removal (within significant heat conductivity) above the high resolution already mentioned requires the use of corrected optics. With standard lenses, the surfaces of which are plane or spherical, the largest possible numerical aperture at which the lens defects are still smaller than the diffraction limit is well below 0.1. This means that an advance can be made into the 1 to 5 μm range with normal single lenses. The depth of field is then only about 50 μm, that is to say, the positioning of the workpiece requires great care. Precision in the sub μm range, however, can only be achieved with specially designed optics. Figure 5.10 shows an optics setup in a system used to make microperforations. The laser beam is first collimated and apertured to produce a small beam with uniform intensity. The beam is then focused by the $f = 150$ mm lens and the focus is imaged onto the surface of the workpiece.

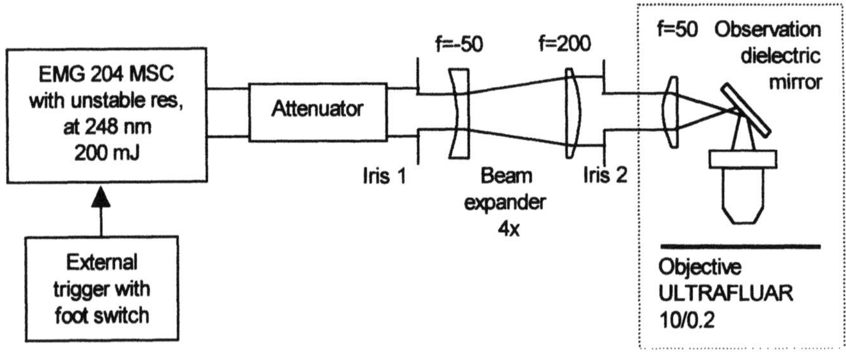

Fig. 5.10 Optics setup for microperforation system.

Mask selection

The important criteria in mask materials selection for imaging an excimer laser beam are low absorption of the UV, high thermal conductivity of the UV radiation and fabrication, producing the image or pattern in the material. Most metals strongly absorb UV radiation and this makes specification somewhat difficult. From a material property point of view, aluminium is the best choice. It has a low absorption and is a good heat conductor. However, the metal makes it difficult to pattern the image using standard mask-making techniques. The most common mask material employed in excimer laser systems is stainless steel, nickel silver and chromium on a quartz substrate.

The methods used to produce characters or features in the mask are punching, direct writing with an excimer laser and photolithography, which uses a chemical etching approach. In all three cases the thickness of the mask material is limited. For mask structures of 150 μm the thickness is limited to 0.1 to 0.2 mm.

These masks are limited to 1 J/cm^2 energy density, and they can withstand no more than 5 to 10 W average power before they become distorted due to thermal effects. Powers of 100 to 200 W can be sustained by these masks if they are water cooled. This, of course, adds to the complexity and cost.

Masks made with photolithography or chemical etch techniques are much more restricted concerning maximum energy and average powers. For example, a typical mask has a thickness of only a few microns and can withstand an energy density of no more that 50 mJ/cm^2. The metal coating expands as it heats up, and in time it separates from the quartz substrate.

Another method being developed is using a dielectric, high-reflectivity coating on a substrate. In other words, rather than any of the radiation being absorbed it is reflected. This allows energy densities of several hundred mJ/cm^2 and high average powers. The disadvantage to this approach is the high cost of dielectric coatings.

5.5.3 Summary

Many different techniques and optics are being employed in designing an excimer laser beam delivery system. The best, most practicable, approach is to start from the workpiece to be processed and work backwards, keeping in mind wavelength, energy density and average power in the selection of the proper optical elements, including the mask. Through careful material selection and more stringent optical processes, high-performance UV optical systems can be produced. As requirements for shorter wavelengths intensify, new optical materials and better characterisation and process control of existing materials will be required. Material stability under intense UV illumination will become increasingly important.

5.6 PULSE SHAPING

As was stated earlier, excimer systems are high gain systems. This is due to the nature of the participating energy levels, their energy separation and transition oscillator strengths. As such excimer lasers inherently operate in the pulsed mode, emitting ns duration pulses. Such pulse durations are typical for high gain systems, where a small number of passes (typically 1 or 2) are sufficient to extract most of the useful gain of the system, and for the ability of the power supplies to maintain cavity discharges in such systems. As a result, typical commercial excimer lasers provide pulse durations in the 10–60 ns regime. However, the demands of several applications require the employment of excimer laser pulses outside this time regime. There are no straightforward ways of experimentally controlling the pulse duration of such systems. The various techniques, applicable in most other types of lasers, are difficult or not applicable in the case of excimer lasers. To achieve longer pulse durations, one may simply superimpose and combine a number of suitably time delayed replicas of the original pulse. This method is acceptable and manageable for a relatively small number of such replicas. Commercially available units can be obtained, or constructed in house, for specific wavelengths and degree of pulse extension. A alternate solution is through the extensive modification of the power supply of such systems. In such cases, the added requirement on their design is to maintain the discharge for a longer time with the usual characteristics of even, uniform, non filamented discharge profiles in the cavities.

For the formation of short pulse durations, the picture is more complicated, as the known mechanisms of short pulse production face serious problems when applied in excimer lasers. Q-switching techniques suffer from the lack of suitable saturable absorber materials, for passive methods, and suitable UV transmitting crystals with reasonable damage thresholds for active methods. It should be noted however, that normal excimer laser formation is similar to Q-switching, as the giant pulse formed in a Q-switched system results from 1 or 2 passes through the gain medium and after the gain profile has been artificially increased (external means of lowering the Q of the cavity) to simulate a high gain medium. On the topic of the production of ps and sub ps pulse durations from excimer systems, the methods are again indirect. The active media have the necessary bandwidth and gain profile to produce short duration pulses, but the mechanisms and devices through which this bandwidth can be utilised (active and/or passive mode locking) lack the necessary materials. Additionally, even if such devices could be obtained, a major problem would remain from operational considerations. Normally mode locking techniques, active or passive, rely on maintaining gain in the cavity for a substantial number of round trips. This allows modes within the cavity to build up and adjust their phase relationships so that the output pulse is coherent and has the largest possible bandwidth. Clearly in high gain, few pass systems this build up cannot occur. Therefore a mode locked oscillator in the excimer wavelength regime is difficult to built. Nonetheless, short duration pulses can be readily amplified in excimer systems, with good gain amplification, short time profile and reasonable transverse mode characteristics.

5.6.1 Long pulse durations

The research efforts into the questions and problems surrounding the topic of long pulse duration excimer lasers has been substantial, as the goal is not only of theoretical interest but has many practical aspects that make such pulses desirable. A driving force behind the development of such long pulse excimer lasers, is the scaling of excimer lasers to very high average powers and increased pulse rate frequencies. Such scaling will benefit from any excitation scheme that increased overall system efficiency and reduces loads in the cavity. A longer pulse operation provides additional advantages like lower stresses on the electrical components, lower heat and acoustic loads on the cavity, increased efficiency, better and more controllable laser beam parameters such as divergence, polarisation and linewidth (McvKee, 1991). Such longer pulses will also be more efficiently transmitted in fiber delivery systems.

Power supply

The most widely used scheme of a discharge pumped excimer laser, based on the classical excitation schemes, like C-L-C circuits, will limit the laser emission to short pulses (10–70 ns) and relatively low efficiency (<2%). This is due to the limited ability of such power supplies to maintain a discharge in the cavity for longer time durations. Several methods to increase pulse width and efficiency have been studied, such as control of discharge parameters with classical C-L-C circuits (Long et al., 1989; Taylor and Leopold, 1989), double discharge (spiker-sustainer) (Hueber et al., 1992 and 1992) and stabilisation by segmented electrodes associated to resistors or inductors (Sze, 1983) or by e-beam pumping methods and other methods (Montagne et al., 1992; Fransceschini et al., 1992; Taylor, 1986). The double discharge power supply (spiker/sustainer configuration) is such a technique and an equivalent circuit block diagram, is shown in Fig. 5.11.

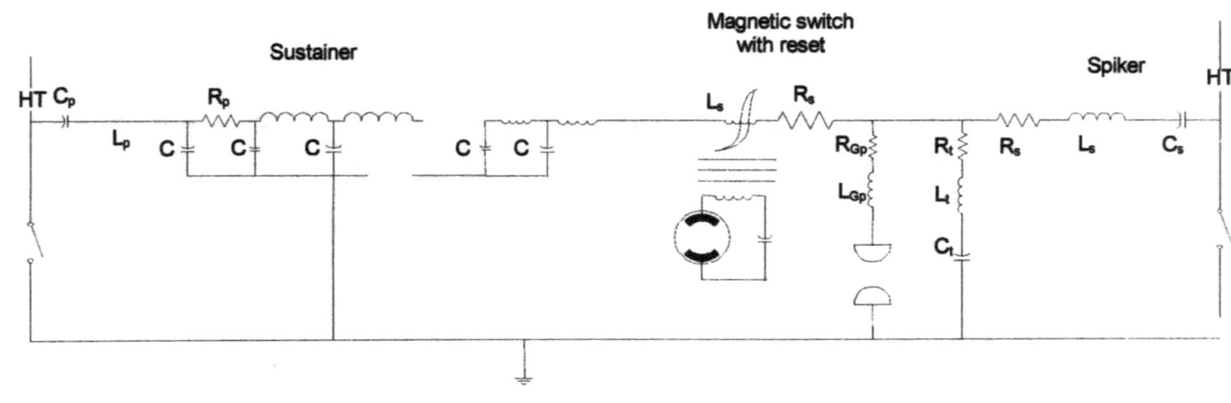

Fig. 5.11   Equivalent circuit schematic of a double discharge excimer power supply. (Hueber et al., 1991 and 1992.)

The excitation scheme provides a very fast high voltage spiker for avalanche discharge, a time delay between spiker and sustainer through the behaviour of the saturated magnetic switch and the discharge sustaining action of the sustainer section. The scheme has been successfully applied to the XeCl and KrCl laser with good results. Extraction efficiency has been notably increased (>3% and ~1% respectively) with associated pulsewidths of ~200 ns.

External methods

The same goal can be accomplished with a more simplistic approach, whereby a laser pulse is optically separated into a number of replicas, which are subsequently suitably delayed with respect to one another and finally combined in a single beam, of the required duration. The method just spreads the available energy of the original pulse in a longer time interval, and suffers from losses in the various optical elements that are used. It affords a semi continuous means of adjusting the pulse width, but the resulting profile is highly asymmetric, with a rather steep rise and slow falling edge. Commercial units exist that result in pulse durations up to 200 ns, with reasonable efficiencies and pulse profiles. The weakest component in such a system is the beam splitter and combiner, where most of the energy losses occur, and the limiting of the extending ability of the system occurs. Either a dielectric or a grating type beamsplitter combiner can be used.

Short pulse durations

Notwithstanding the lack of good experimental methods of mode locking the available bandwidth in an excimer laser, pulse durations in the ps and sub ps range can be easily obtained in the UV region by using the excimer cavity as a convenient high-gain amplifier. Such radiation is of great research interest in a number of areas, such as high intensity effects in laser-matter interactions, time resolved studies of surface dynamics and behaviour of atomic systems. Starting from laser pulse of the desired pulse duration, one needs to generate a seed pulse in the excimer wavelength of interest that can be amplified. The seed pulse can be generated in a number of mixing techniques, such as second harmonic generation, sum mixing, etc, and amplified in the excimer cavity, in a single or multipass arrangement. Significant output energies can be obtained from such systems (>100 mJ) in a sub ps pulse that can result in focused intensities of 10^{20} W/cm^2 (Nabekawa et al., 1992 and 1993; Taylor and Leopold, 1986).

The following figures 5.12 to 5.14, depict the schematic layouts of existing amplified systems of Xe and Kr, that have sufficient bandwidth so as to amplify short duration seed pulses, which in all cases are derived from dye laser systems. It should be noted that the amplifiers operate in the saturated regime, in most cases, so that the maximum energy will be extracted. Moreover in such a case, one can attempt to pulse compress the resulting pulses to shorter durations, due to the extra-bandwidth that the seed pulse may extract from the saturated amplification.

For the XeCl laser, the search is still ongoing to achieve mode locking operation, as early indications (Taylor and Leopold, 1986) observed microsecond duration optical pulses from a UV preionised laser. This long duration laser action, together with the available bandwidth in a XeCl cavity, could result in ps duration mode locked operation, provided a suitable "saturable absorber" can be demonstrated for such a system.

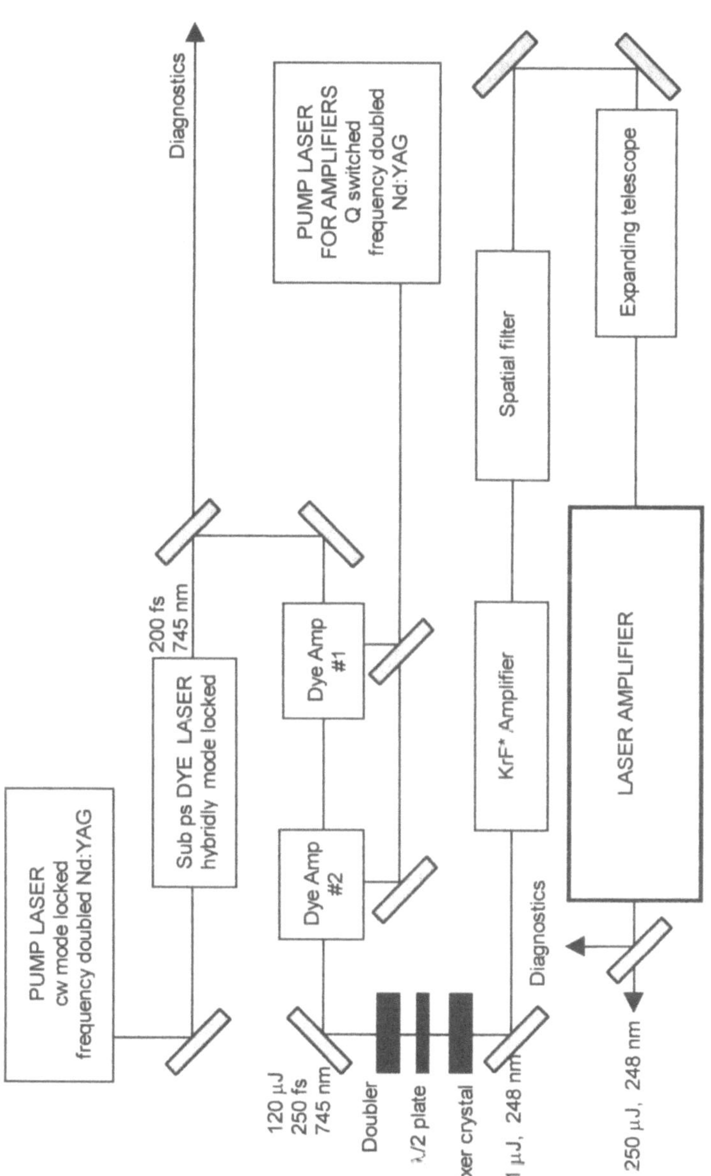

Fig. 5.12 Schematic diagram of the ultrahigh intensity **KrF** laser system. The laser amplifier produces subpicosecond (~600 fs) laser pulses.

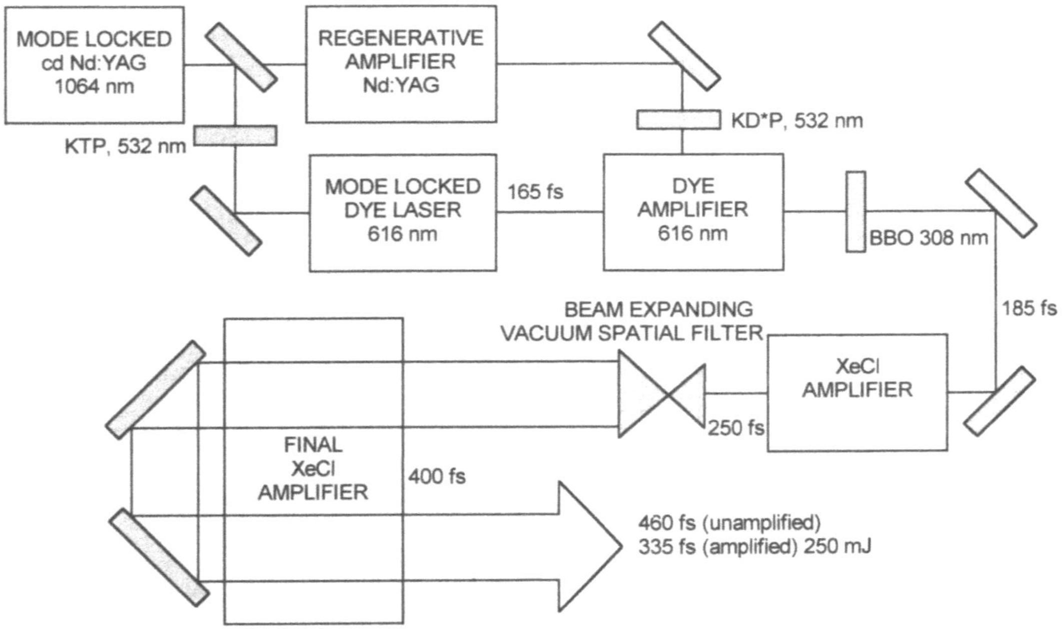

Fig. 5.13 Schematic layout of a XeCl short pulse laser amplifier system. Measured pulse durations are shown after each stage.

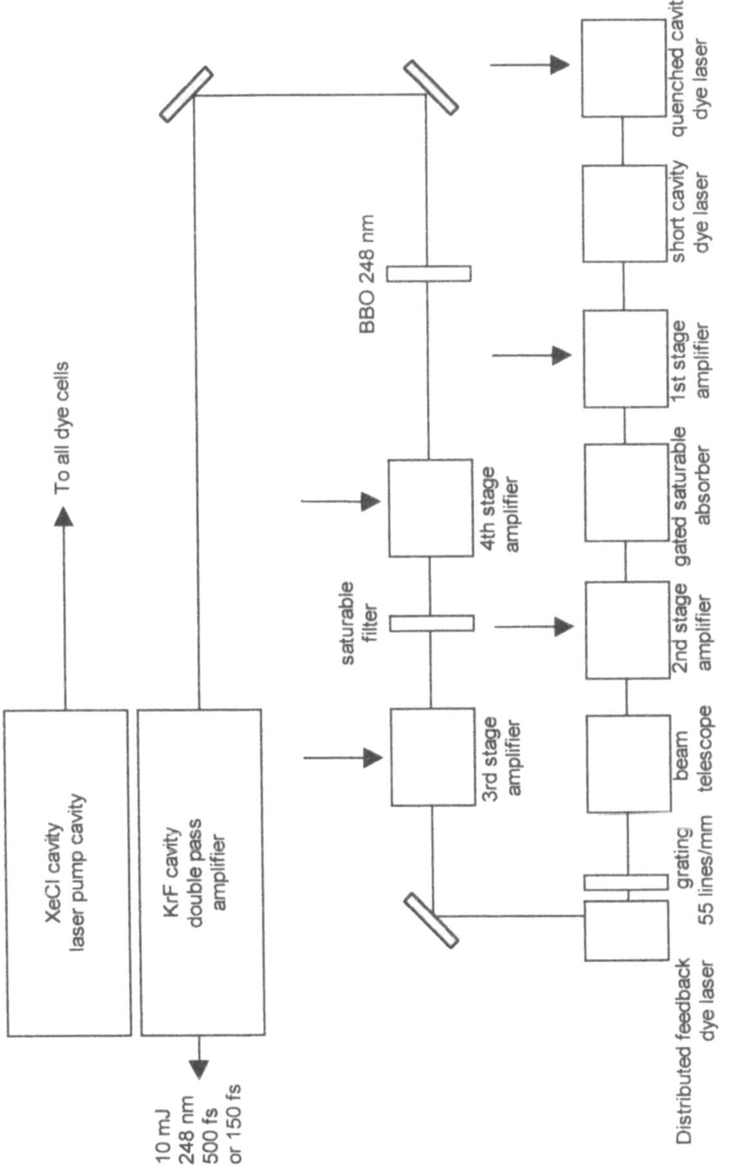

Fig. 5.14 Schematic layout of a DFDL based KrF amplifier system. Pulse durations of 150 fs can be obtained after pulse compression.

5.7 CURRENT STATUS AND FUTURE DEVELOPMENTS

There are five major issues which determine the performance and maintenance requirements of excimer lasers:

1. Gas mixture
2. Optics
3. High voltage components
4. Stability of Performance
5. Operational duty cycle

In all these issues there has been significant progress over the last few years. For example the gas lifetime, which is a major factor determining the running costs of the lasers has been dramatically increased by using external cryogenic gas processors, advanced materials for limiting the halogen degradation, automatic halogen injection and better gas handling systems. In a recent development a rare gas storage device, makes the laser more user friendly by eliminating the rare gas pressure tank handling (Rebhan, 1994). The high voltage components and the discharge efficiency have also been improved. The electrodes shape, the preionisation components and the electrical circuit adaptation to the discharge impedance are key issues in this respect. In general, the optimisation of efficiency depends on the use of the appropriate voltage for each gas composition for a given status of the electrical components, as for example the electrodes. Clearly, the stress on laser materials decreases with increasing laser efficiency.

Efforts for an extended gas lifetime and better high voltage performance have led to an improved excimer laser stability with a good beam profile. Currently, available commercial systems are capable of producing $\sim 3 \times 10^9$ laser pulses in stabilised outputs of 150 to 200 W. To give a better feeling of this lifetime, it should be considered that a 250 Hz XeCl laser yields 9×10^5 pulses in 1 hour of operation or 36×10^6 pulses in a 40 hour week. Considering that 1×10^9 pulses correspond to about 28 weeks, these excimer lasers can be good industrial tools in terms of maintenance intervals and two-shifts operation.

Finally, the ease of optics cleaning and replacement intervals should be considered. Optics cleanliness besides the laser output energy, also affects the beam quality. Careful selection of materials in the pressure vessel is important for minimising "dust" formation due to sputtering during the electric discharge and material deposition on the windows. Systems for minimising the exchange time of the laser windows are in operation (Rebhan, 1994).

The industrial lasers which are currently available include high average power system of either high peak energy and relatively lower repetition rate or systems of lower energy and high repetition rate. The best example of the former approach is a 1 kW XeCl laser, emitting pulses of 10 J energy at 800–1000 Hz repetition rate, which has been produced in France as a result of the EUREKA - EUROLASER EU 205 project. Within the same project a 200–500 W system of high repetition rate has been produced in France and Germany (Rebhan, 1994; Fontaine and Forestier, 1987; Lacour, 1992). In a similar industrial laser programme in Japan called AMMTRA (Advanced Material Processing and Machining System) several 200–500 W excimer laser systems have bee developed. It should be pointed out that most of these systems are XeCl lasers and the current trend is to improve the operation and meet industrial specifications for the KrF and ArF lasers. Improvement of the beam characteristics both in terms of homogeneity and plateau profile is another important goal in this respect.

5.8 APPLICATIONS

There is a great diversity of industrial, environmental and medical applications of excimer lasers. For example, ceramics, glass quartz and many plastics can be conveniently processed at the excimer laser wavelengths. Micromachining and thin film growth applications are particularly interesting in this respect. In lithographic applications, shorter wavelengths allow higher densities. In the medical field eye keratoplasty is at a mature level nowadays, while there are optical radars (LIDAR) used for monitoring the environmental pollution directly. Selected excimer laser applications, will be discussed, including a recent application in which excimer lasers are used in the conservation of painted works of art.

5.9 REFERENCES

Babcock, C.L. (1977) *Silicate Glass Technology Methods*, John Wiley & Sons, New York

Bennett, H.E. (1978) *Opt. Eng.*, **17**, p480-486

Escher, G.C. (1988) *Proc. of the SPIE*, **998**, p30-37

Fontaine, B.L. et. al., (1987) *CLEO Technical Digest* , p310

Franceschini, M.A. (1992) *SPIE*, **1810**, p435

Golde M.C. and Thrush, B.A. (1974) *Chem. Phys. Lett.* **29**, p

Golde, M.F. (1975) *J. Mol. Spectroscopy* **58**, p261

Hueber, J.M. et al., (1991) *SPIE* **1503**

Hueber, J.M. et al., (1992) *SPIE* **1810**

Kingslake, R. (1978) *Lens Design Fundamentals*, Academic Press, New York

Krauss M. and Mies F.H. (1979) in *Excimer Lasers* (ed. Ch. K. Rhodes), p5, Springer Verlag.

Lacour, B. et al., (1992) *SPIE* **1810**, (eds. C. Fotakis, K. Kalpouzos and Th. Papazoglou), p368

Long, W.H. Jr, Plummer M.J., and Stappaerts E., (1989) *Applied Physics Letters* **43**, p735

Luk, T.S., (1989) *Opt. Lett.* **14**, p1113,

McKee, T.J., (1991) *Applied Optics* **30**, p365

Montagne, J. E., Inglesakis, G. and Autric, M., (1992) *SPIE,* **1810**, p447

Nabekawa,Y. (1993) , *Opt. Lett.* **18**, p1922

Rebhan, U. et al., (1995) *SPIE* **2502**, p426

Rebhan, U., Nikolaus, B. and Basting, D. (1994) in *Excimer Lasers* (ed. L.D. Laude), p1, Kluwer Academic publishers

Ronki, M. et al., (1978) *IEEE, J. Quant. Electronics* **14**, p464

Searles S.K. and Hart, G.A. (1975) *Appl. Phys. Lett.* **27**, p243

Setta, J.J. et al., (1988) *Proc. of the SPIE*, **970**, p179-191

Stehle, M. (1993) in *Excimer Lasers*, (ed. L. D. Laude), NATO Series E-265,15, .

Sze, R.C. (1983) *Journal of Applied Physics,* **54**, p1227

Taylor, A.J. (1990) *Opt. Lett.* **15**, p39

Taylor, R.S. (1986) *Appl. Phys.* **B41**, p1

Taylor, R.S. and Leopold, K.E. (1986) *Applied Physics Letters* **47**, p81

Taylor, R.S. and Leopold, K.E. (1989) *Journal of Applied Physics*, **65**, p22

Velazco, J.E. and Setser, D.W. (1975) *IEEE J. Quant. Electron.* **11**, p708

6

Semiconductor lasers

E. Wintner

6.1 INTRODUCTION

At first glance, the semiconductor laser, or more precisely the semiconductor diode laser, is a special solid state laser, but for good reasons it is usually treated as a laser family of its own. Solid-state lasers in general are formed by a laser medium which is represented by a host material (of crystalline or glassy nature) doped by suitable ions (rare earths, transition elements). Generally, those dopants have clearly defined energy levels, which are more or less perturbed, i.e. broadened, by the binding fields of the host. However, in case of the semiconductor laser the lasing transitions take place between the valence and conduction bands as a consequence of the collective interaction processes of the electrons and the lattice as a whole. Therefore, the density of energy levels is extremely high in comparison to any other laser material. The excitation, represented by mobile electrons and holes, can be transported through the crystal to the recombination area which is called the active zone.

As a preliminary test set up, many laser pumped semiconductor lasers have already been demonstrated. In general, however, it is the forward biased diode which is referred to as a diode laser. This idea has led to the most efficient conversion of electrical into optical energy which nowadays can reach up to 50%. Thus, it is absolutely necessary to have available the relevant technology to form rectifying diodes of the desired materials. This has been relatively easily achieved with III-V or IV-VI compounds for many years. II-VI compounds, interesting for shorter wavelengths however, have caused a real hardship in this respect.

Historically, the development of the diode laser started soon after the first realisation of a laser (Ruby laser by Maiman, 1960) being part of the continous effort ever since in order to create new lasers of various materials and designs for any desired wavelength.

The first suggestion of a possible semiconductor diode laser was made by Basov et al (1961). Soon after that, the first laser of this kind was demonstrated in three laboratories at nearly the same time: by Hall et al, (1962) at General Electric, Nathan et al, (1962) at IBM and Quist et al, (1962) at MIT Lincoln Laboratory. For a long time the diode laser seemed to be far away from practical application because it could only be operated at cryogenic temperatures in a pulsed mode. The real breakthrough to overcome the enormous current densities was the idea of the heterostructure (Kroemer, 1963; Alferov and Kazarinov, 1963), i.e. that a layer of higher bandgap could form an effective potential barrier to carriers in order to keep them in a certain area of the device. In 1969 this idea was successfully realised for the first time. From then on an unprecedented series of improvements led to the present situation where the diode laser can be estimated to be the best version of a laser in many respects:

The diode laser combines the technologies of microelectronics and optics. Hence, such devices can be made extremely small and cheap. Very stable light emission in cw or pulsed mode can be achieved. The laser can be modulated internally by varying the drive current allowing efficient and simple modulation at very high frequencies (more than 10 GHz). Single frequency emission as well as tunable output can be realised in a single mode. Nowadays, many wavelengths are covered, starting in the blue-green spectral region, encompassing the 800 nm spectral area of the most common emissions, and reaching via the optical communication wavelengths (1300 and 1550 nm) far into the infrared regime. Diodes emitting only milliwatts are extremely cheap today and produced in tens of millions for various applications. High power devices can emit any absolute value of total noncoherent optical power. More informative parameters in order to characterise these devices are intensity (power per emitting area) and, even better, brilliance (intensity per solid angle). Those specifications are limited by factors like heat conduction and waveguide characteristics. Typically, up to 100 W cw power usually around 800 nm can be generated.

These properties and characteristics of diode lasers have led to manifold applications in various fields from physics and optics to engineering and numerous medical applications: Efficient collimation offers alignment and pointing operation via pilot beams, the ultrafast modulation capability and miniaturisation allows optical communication and optical data storage. High power diode lasers are effective pump sources for other lasers, especially solid state. Medical applications are numerous, stretching from low power biostimulation to diode laser pumped solid state laser treatments in dentistry, surgery etc. With the blue-green diodes now under development, a new generation of compact disc devices as well as new full-colour displays can be envisioned. Surface emitting lasers might allow effective beam steering and potentially replace the electronic picture tube. Quantum cascade lasers might fill the gap of devices in the infrared, nowadays between 4 and 11 μm, having reasonable performance characteristics even for commercial use (Faist et al., 1996).

6.2 MATERIALS AND EPITAXIAL GROWTH TECHNIQUES

6.2.1 Composite direct semiconductors

For 90% of all semiconductor applications silicon is the best candidate material: it can be refined to the highest purities, it has a suitable bandgap, can be doped relatively easily, offers good carrier mobility and forms a very highly insulating oxide - a very practical and important aspect. There is one drawback, Si is an indirect semiconductor, i.e. due to different locations of the relevant minima and maxima of the energy bands recombination of carriers must involve phonons (see Fig. 6.1). This leads to an increase of carrier lifetime into the microsecond range as compared to the nanosecond lifetimes of direct semiconductors making recombination less efficient and destroying phase relations due to the statistical nature of the phonons involved.

III-V compounds are the nearest relatives to group IV; they are isoelectronic and their lattices are similar (sphalerite or wurtzite versus diamond structure) but their bandgaps are direct. This is the reason why GaAs was the first successful semiconductor laser material.

In order to fabricate a functional diode laser the availability of a good substrate is essential like in all electronic device technology. Single element crystals can be realised most perfectly (Si: $10^{12}/cm^3$ pure and free of defects). Single crystals of binary compounds are more complicated to grow because stoichiometry is an additional problem to solve. As long as the constituents are not too different in chemical physical behavior (e.g. volatile) single crystals of reasonable qualities can be grown.

In practice, GaAs and InP are the III-V substrates to be used having purities of approximately $10^{15}/cm^3$. A binary compound has a certain band gap corresponding to a specific wavelength to be emitted.

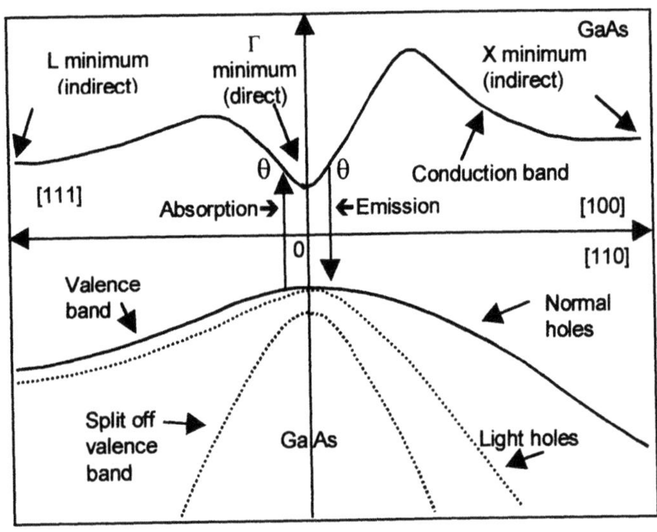

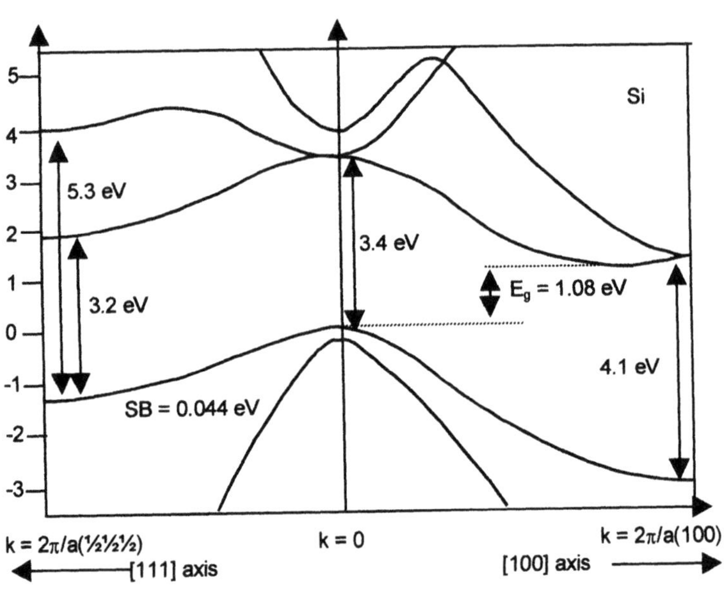

Fig. 6.1 Direct band structure of GaAs (top) and indirect band structure of Si (bottom).

For two reasons there is a necessity to vary the gap energy:

1. to generate wavelengths as needed.
2. to form heterostructures for carrier confinement as it will be discussed later.

It is useful to illustrate possible material combinations with the help of a diagram showing band gaps versus wavelengths (Fig. 6.2). The binaries are represented by dots in this diagram. It is, of course, possible to form more complicated compounds like (III_1/III_2)-V or (III_1/III_2)-(V_1/V_2). These ternaries are localised on the connecting lines, the quaternaries can be positioned within the four connecting lines of the respective binary combinations. Quaternary compounds allow full flexibility with respect to band gap and lattice constant, i.e. one can both match them to a substrate and choose a wavelength. The most useful and common III-V compound semiconductor laser materials are shown in Table 6.1.

Table 6.1 The most important semiconductor laser materials

Compound	Substrate	Wavelength region (nm)
$Al_x Ga_{1-x} As$	GaAs	780–880
$In_x Ga_{1-x} As$	GaAs	880–1100
$Al_x Ga_y In_{1-x-y} P$	GaAs	630–690
$In_{1-x} Ga_x As_y P_{1-y}$	InP	1100–1600
under development		
$In_x Ga_{1-x} N$	Al_2O_3	390–620
$In_x Al_y Ga_{1-x-y} As_z Sb_{1-z}$	GaSb	750–6200

AlGaAs is the most sucessful ternary group so far yielding highest effiency and allowing the highest variation of diode laser designs realised so far including the high power array lasers. It represents a special case of a series of ternary materials with nearly constant lattice parameter. Therefore, in fact all combinations are lattice matched to GaAs. Unfortunately, for Al contents > 43 % the material is indirect.

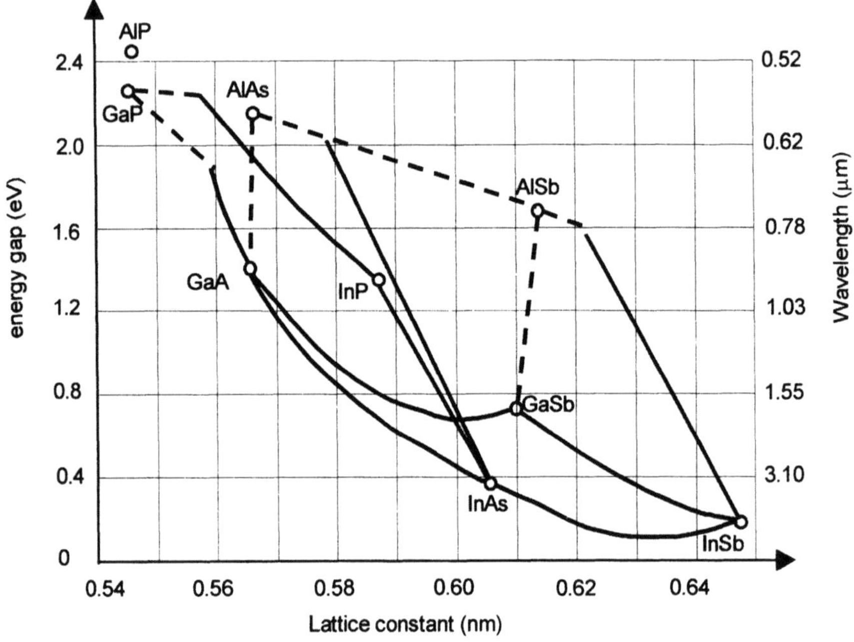

Fig. 6.2 Bandgaps versus lattice constants of III-V compounds.

InGaAsP has been developed with large industrial efforts to cover the needs of the optical communication business, i.e. the wavelengths of the lowest losses in optical silica fibers at 1.3 µm and 1.55 µm. After the solution of problems of contact deterioration and temperature behaviour in the 1980s, devices made out of this material have matured and been commercialised.

In recent times, the demand of covering the shorter wavelength half of the visible spectrum has increased. AlGaInP can at least cover the shorter red wavelengths. This allows to replace the HeNe laser in a number of applications. Furthermore, the first high power array diodes came onto the market in early 1995 (e.g. Applied Optronics, 700 mW cw) still showing difficulties of performance and lifetime. They are interesting for pumping of broad-band solid-state lasers.

For many years, attempts have been made to cover the green, blue and even the near UV region with semiconductor diode lasers. None of the efforts has been very convincing so far and has not yet allowed to bring such diodes on the market.

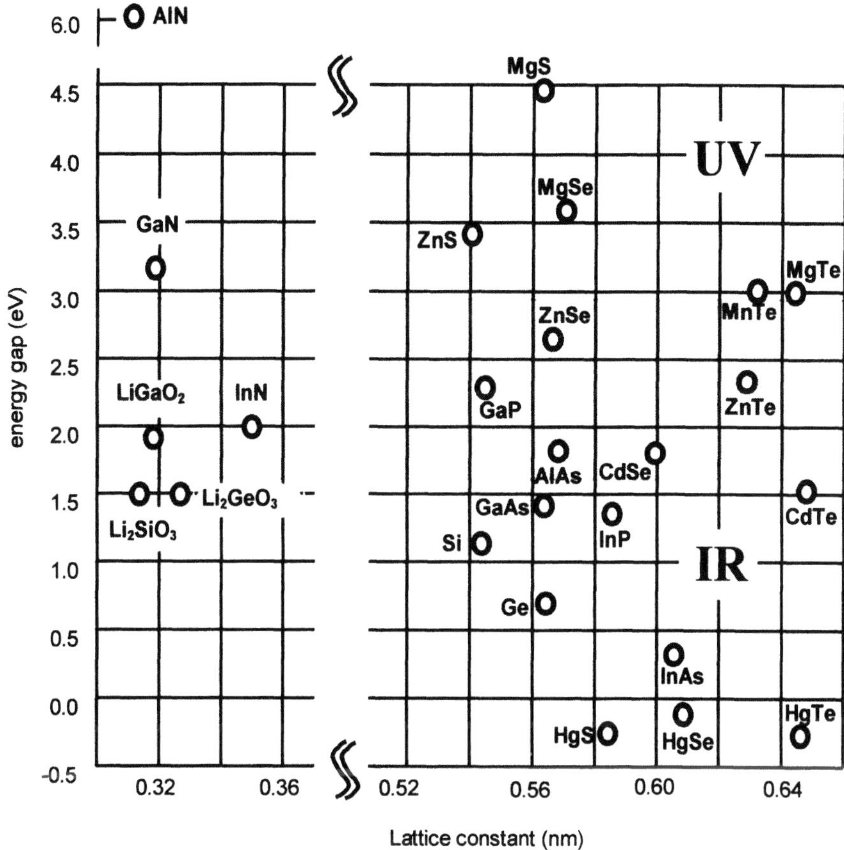

Fig. 6.3 Bandgaps versus lattice constants of II-VI compounds including a few other semiconductors for comparison.

There are two candidate systems:

1. For a number of years, people have worked on II-VI compounds which in general have larger and direct bandgaps (Fig. 6.3). They seemed to be the logical continuation of the idea of isoelectronic semiconductors. Unfortunately, problems of two different natures arise and could not be fully overcome: i) doping of donors and acceptors is very hard to be achieved simultaneously together with satisfying carrier mobility. This is, however, indispensable for the formation of p-n-junctions. ii) The more the constituent elements are apart from group IV the more ionic is the nature of the bindings. This also relates to the mechanical hardness and stability of the materials negatively influencing the durability of contacts etc. Such a property, however, is part of their

nature and cannot be changed. Therefore, it is still doubtful if II-VI compounds like the mostly favored ZnCdSe system (see e.g. Gunshor and Nurmikko, 1995) eventually can reach device specifications comparable to the afore mentioned III-V compounds. So far, green 508 nm emission at room temperature for 1 hour cw at the maximum has been achieved. 483 nm pulsed operation of mWs at 5% duty cycle has also been reported (Laser Focus World, 1995).

2. A very new option for blue diode lasers has recently been considered, namely GaN on $LiAlO_2$ or $LiGaO_2$ substrates. The following ideas led to this development: In 1994 the commercial production of GaN LEDs was announced by Nichia Chemicals of Japan with a light yield 100 x that of SiC LEDs. Laser action of GaN on sapphire (Al_2O_3) has failed so far due to the poor lattice matching.

GaN has a wurtzite structure, the two obvious related substrate compounds would be SiC and ZnO. Both of them are not easy to grow. But now research has revealed the above mentioned Li crystals as ideal candidates. Research work to realise the first devices is in progress (Chai, 1995).

Worth mentioning is the large group of IV-VI lead chalcogenides or lead-tin chalcogenides (PbS, PbTe, PbSe; $Pb_{1-x}Cd_xS$, $PbS_{1-x}Se_x$, $Pb_{1-x}Sn_xTe$, and $Pb_{1-x}Sn_xSe$) for the far infrared. They have been known for a long time and partially highly developed although they never reached substantial commercial importance. This is due to i) operation at cryogenic temperature and ii) few applications existing in the regime of 3 to 30 μm which can be totally covered. They usually can be tuned over a large spectral region via changes of temperature, pressure or magnetic field; typical parameters are (Herman and Melngailis, 1974):

$$\partial E_g / \partial T = 5.10^{-4} \; eV / K$$

$$\partial E_g / \partial H = 10^{-3} \quad eV / T$$

$$\partial E_g / \partial p = -5.10^{-11} \; eV / Pa$$

This allows emission of wavelengths with high accuracy (10000 x better than the resolution of a grating spectrograph) for calibration purposes etc. Typical are Microwatts to Milliwatts with lasing thereshold currents of 100 A/cm^2 at temperatures < 40 K.

As has been shown so far, semiconductor properties can be varied over wide ranges by forming compounds with varying stoichiometry. Another possibility has been discovered since modern layer growth techniques, as they will be described in section 6.2.2, have been developed allowing the precise realisation of even extremely thin (2–10 nm) layers, very often in a repetitive alternating sequence. If the thickness of a layer is made thinner and thinner eventually its dimension becomes comparable to the de Broglie wavelength (Planck's constant / momentum). Then the motion of injected carriers is restricted in the direction perpendicular to the thin layer, and the kinetic energy is quantised into discrete energy levels. Layers with such electronic properties are called quantum wells (QWs). If the active zone of a semiconductor laser contains only one QW it is called single quantum well laser. There may be also a larger number of quantum wells embedded in a diode laser comprising alternating layers with different bandgaps (the ones with higher bandgap are called quantum barriers). It is referred to as a multiple quantum well (MQW) laser.

In a QW structure, electrons and holes are not restricted in the parallel direction, but they are quantised perpendicularly. Thus, the density of states is modified in comparison to a normal semiconductor (Fig. 6.4). The QW shows a step like function for the density of states while the ordinary semiconductor is characterised by a parabolic function (dotted line). The step like density of states introduces some excellent new characteristics and possibilities to the laser material: the density distribution is narrower than normal, therefore gain in the diode can be higher and hence thresholds smaller. The emitted wavelength between the quantised states is shorter than the wavelength corresponding to the bandgap energy. In case of certain laser structures, other characteristics like spectral linewidth and modulation properties can be inproved. The temperature performance generally is better.

In many cases, structures similar to MQWs are used having larger spacings   (>10 nm). They are usually called superlattices. The quantum nature plays no role in such an arrangement. Otherwise, averaged over the dimensions of the usual wavelengths (like 800 nm) new semiconductor properties can be tailored.

Another degree of freedom is possible if the layers in superlattices or MQWs are lattice-mismatched. Generally, such a mismatch causes numerous crystal defects (e.g. dislocations) giving rise to enormously increased losses.

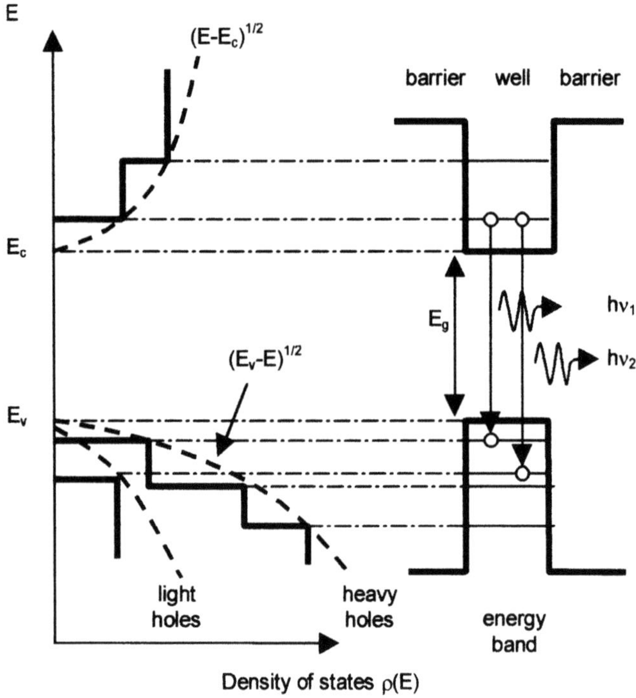

Fig. 6.4 Density of states and energy bands in a quantum well structure.

However, if the layers are sufficiently thin, the mismatch will only cause strain (usually uniaxially) and thereby the potential energy will be lowered (strain energy is proportional to volume, the energy stored in defects goes with the interface area). Usually, strained layers are stable if the thickness < 3 nm. The strain and the associated stress is enormous (e.g. 0.65% mismatch corresponds to 5 bar of stress). This effectively alters the bandstructure and allows the creation of semiconductors with new properties just as they may be desired. For diode lasers the most important feature are higher differential gain and smaller linewidth enhancement factors.

As a conclusion, laser material properties can be varied by:

1. formation of multiconstituent compounds
    (binary, ternary, quaternary, quinternary ......) with variable
    stoichiometry
2. formation of regular planar lattices of two alternating
    semiconductors (superlattices, multiple quantum wells) with
    lattice match or lattice mismatch

6.2.2   Epitaxial growth techniques

Electronic devices are usually built up on top of a monocrystalline substrate which should be as pure and free of defects as possible. The thin layers deposited also have to be monocrystalline. There must be a perfectly conducting contact which is achieved best if the lattices are matched. Such layers are called epitaxial. There are four well established methods for the formation of epitaxial layers:

1. **Liquid phase epitaxy LPE** This method was very successful right from its start. Melts of group III elements (Ga, In) are used as solvents for group V elements (at temperatures between $600°$ an $900°$ C). Such solutions are brought in contact with appropriate substrates. If the temperature is lowered slowly monocrystalline layers start to grow onto the substrate in an oriented fashion. The growth procedure takes place very close to thermal equilibrium. See Table 6.2 for a comparison of the advantages and disadvantages of this method compared to others.

2. **Vapour phase epitaxy VPE** The constituents are introduced into a reacting zone in a gaseous condition formed in two alternative ways: i) Trichloride method: $AsCl_3$ and /or PCl_3 flows over Ga, In or GaAs, InP thereby forming metal chlorides. ii) Hydride method: metal chlorides are formed via the reaction of HCl vapor with heated In or Ga. They are mixed with AsH_3 and/or PH_3. The mixture of the reacting gases is brought into a deposition zone of slightly lower temperature ($700°$ - $800°C$). The thermal equilibrium is changed thereby yielding the deposition of solid layers onto the substrate provided stoichiometry is defined by the partial flow rates.

3. **Metal-organic chemical vapor deposition MOCVD** This is one of the more modern techniques derived from VPE. The reacting constituents are group III-alkyles ($Ga(CH_3)_3$, $In(C_2H_5)_3$) and group V-hydrides (e.g. AsH_3, PH_3) introduced at certain exactly defined rates into the reaction chamber. There the substrate is located on a graphite holder heated via radio frequency. The gaseous reactants flow in a laminar fashion at rates of 10–20 l/min at a pressure of 0.1–0.5 bars and thermolyze in the hot vicinity of the substrate. The elements of groups III and V are deposited on the substrate surface (temperature 450–650 °C) as a result of this thermolysis. In contrast to LPE and VPE, MOCVD works well away from thermal equilibrium.

4. **Molecular beam epitaxy MBE** This is the newest of all epitaxial techniques with the highest technical complexity (e.g. ultrahigh vacuum!). Purest quantities of all elements involved are evaporated into the UHV and hit the substrate (temperature 450–600 °C) via molecular or atomic beams. The stoichiometry of the deposited layers is only defined by the beam flow rates modified by the adhesion coefficients.

MOCVD and MBE allow precisely defined layers to be produced which also may change repetitively by composition and/or thickness. This was the precondition to successfully grow quantum wells and superlattices. MBE is most flexible and therefore well suited for experimental conditions while MOCVD allows to grow larger layer areas at faster rates, a prerequisite for industrial production processes.

Table 6.2 Comparison of epitaxial growth techniques

	Advantages	*Disadvantages*
LPE	Relatively simple and cheap; excellent crystalline layer quality (high rates of radiative recombination); Al can also be treated; relatively high growth rate.	Layer thickness not very exact and constant; unstable stoichiometry; rather poor surface quality; small area samples; InP cannot grow on GaInAsP (melting points!).
VPE	Smooth and uniform layers of rather large size and constant thickness; suitable for industrial production; low background doping achievable; no problem with InP on GaInAsP.	Double hetero- and multiple layers are hard to grow (multichamber reactor); problems with Al and Sb because of their high reactivity; crystal quality not optimum; use of highly poisonous gases.
MOCVD	Low reaction temperature; Al and Sb may be used without problems; good material quality and purity yields highest mobility in e.g. GaAs; stoichiometry can be altered rapidly - well suited for hetero and quantum well structures; Sub-μm layers of constant thickness and constituency may be realised; good surface quality, large areas.	Growth rate relatively low (~2μm/h; layers may contain organic impurities, complex adsorbates may occur; highly poisonous gases employed.
MBE	Control of layer thickness excellent (monoatomic layers) constant stoichiometry; purest growth conditions; ideal for heterostructures, quantum wells and superlattices.	Extremley expensive and complex; very low growth rate (0.3–3 μm/h); small sample areas; problems of sublimation of layers; problems with P.

6.3 ELECTRONIC AND OPTICAL PROPERTIES

6.3.1 Diode laser fundamentals

In order to understand the basics of a semiconductor diode laser the fundamentals of a laser as well as of a semiconductor diode have to be briefly discussed:

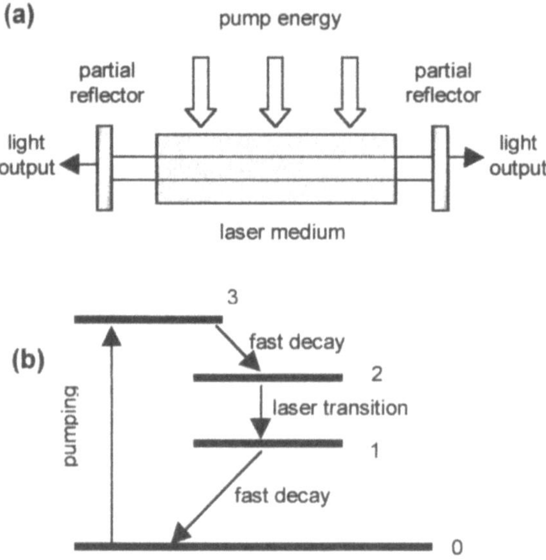

Fig. 6.5 a) Simple schematic of a laser   b) Four-level system of a laser.

A laser, in principle, is a very simple device (see Fig. 6.5a): It consists of an optical resonator, mostly of the Fabry Perot (FP) type, i.e. two mirrors of partial transparency adjusted in a way that light is captured within this arrangement for many bounces. Eventually, the light intensity will fade away because of two major loss mechanisms: i) light will leave the resonator, generally on both sides, a fact which is desirable and enables to make use of the beams. ii) Like any realistic resonator it is lossy (via absorption inside the FP, scattering and refraction off the axis) unless there is an amplifying medium inside the resonator to compensate for those losses - the laser medium. If this is the case the resonator behaves nearly like an undamped one. In case of light, amplification can be achieved by stimulated emission (Fig. 6.6). These physical phenomena are reflected by the acronym LASER meaning **L**ight **A**mplification by **S**timulated **E**mission of **R**adiation.

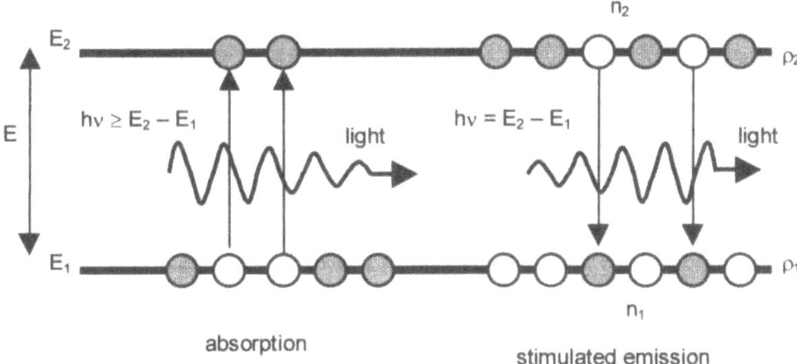

Fig. 6.6 Light absorption and stimulated emission.

In general, the laser emission characteristics like divergence, spectral distribution of intensity and polarisation correspond to the characteristics of the resonator while intensity, emission mode (continous wave cw or pulsed) and bandwidth are defined by the laser medium and its operating conditions. From practical optical experience usually there is no understanding for stimulated emission. The usual emissions to be seen in every day life are of a spontaneous nature (following similar rules of 'non-causality' like nuclear decay) having no classical analogue, e.g. all kinds of luminescence. Stimulated emission, however, is the real counterpart of absorption. In terms of a classical view of electrons elastically bound to the nucleus, levels of higher excitation of such a harmonic oscillator can be reached by the oscillating driving force of the fields of incident light (absorption). In the same way deexcitation can be yielded by external optical fields (of different phase) thereby causing the emission of optical energy (stimulated emission). Einstein (1917) had shown that the probability of absorption is equal to that of stimulated emission. Furthermore, the ratio of probabilities of spontaneous to stimulated emission is proportional to the cube of frequency making lasing at shorter wavelengths very hard to achieve.

An amplifying laser medium, of course, needs some energy supply - the pump. It is one of the most attractive features of a diode laser that the pump energy is introduced via direct conversion of electrical current through a diode into electronic excitation of electrons and holes which eventually recombine and thereby emit light with high probability.

In laser theory, different energy level systems were analyzed. The so-called four level system turned out to be the most favorable to get laser action at low pump powers (Fig. 6.5b). The basic idea is the separation of pump (0–3) and emission levels (2–1). If the pumping action can populate level 2 to higher density than level 1 the condition of inversion is achieved (a thermal non-equilibrium situation sometimes formally described by negative temperatures) and stimulated emission will dominate over absorption of the incident resonator radiation. In case of a semiconductor, there is a huge number of levels within the conduction and valence bands, i.e. many four-level systems are interleaved (Fig. 6.7).

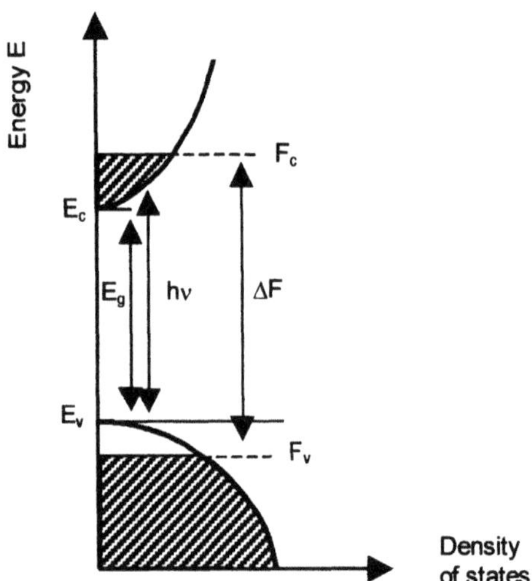

Fig. 6.7. Population inversion between the conduction and valence (heavy hole) bands of a direct semiconductor at 0 K. Energy versus density of states is depicted.

In order to get net amplification via stimulated emission in a semiconductor, necessarily the photon energies of the emitted light hf must remain smaller than the difference ΔF of the quasi Fermi levels F_c and F_v because the latter are defined by the corresponding population densities of electrons and holes, respectively. A sharp line of maximum energies being populated exists at 0 K, whereas at higher temperatures the levels being occupied are not abruptly terminated, but show a transition regime, whose energy widths lie in the order of kT.

The threshold condition for the onset of net stimulated emission simply reads $F_c - F_v > E_g$, whereby the inequality amounts to the order of magnitude of kT. Bearing in mind the quantitative relationship between the Fermi energies and the number densities of electrons or holes one gets the temperature dependent threshold concentrations of those carriers reflecting the necessary concentration of donors and/or acceptors resulting in values of the order of 10^{18} cm^{-3}. In the following some of the most important features of a semiconductor diode will be reviewed: such a device is based on the idea of a homojunction, as depicted in some detail in Fig. 6.8.

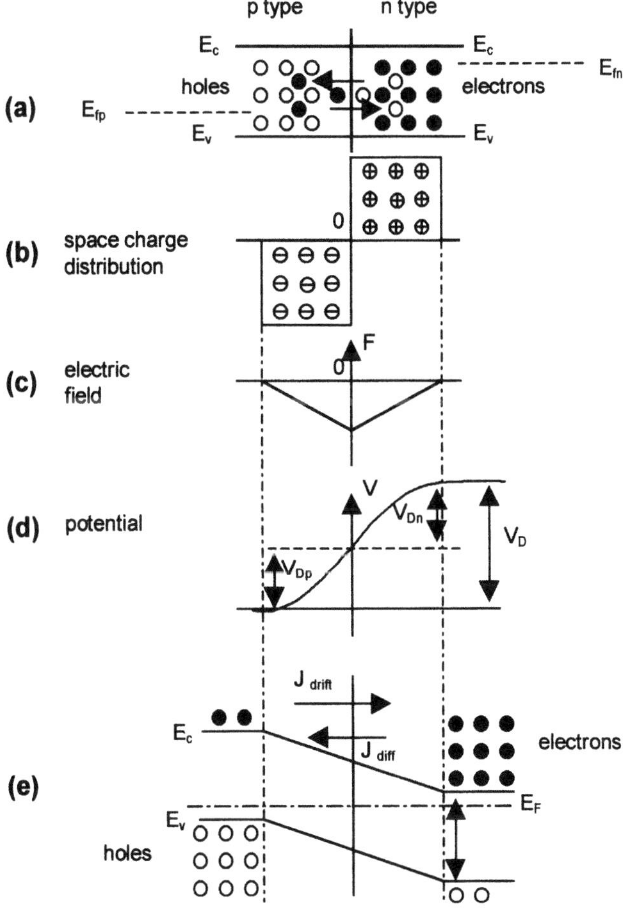

Fig. 6.8 Schematic homojunction in the thermal equilibrium state.

The density of mobile carriers in p and n doped semiconductors is different. If they are brought into electrical conducting contact (crystalline contact, epitaxial layers) diffusion will start in order to reduce the concentration difference. Thereby the immobile ionised acceptors and donors are left behind forming a space charge. This gives rise to a potential difference, called the diffusion voltage having a polarity corresponding to a backward bias, i.e. it prevents additional carriers from flowing across the junction. This process takes place at equilibrium and, consequently, the energy bands in the p and n regions are adjusted in a way that the Fermi levels are located at the same energy.

The situation under forward bias is shown in Fig. 6.9. The applied voltage effectively reduces the diffusion potential. As a consequence, diffusion still goes on and transports carriers to opposite sides of the junction thereby increasing the density of minority carriers by several orders of magnitude. Within a characteristic distance of the depletion zone called the diffusion length, carrier densities reach equilibrium again. Majority carriers from the contacts encounter the minority carriers and recombine. Therefore, those areas are called diffusion zones and eventually correspond to the active zone of a diode laser. Usually, p-n junctions are doped in an asymmetric way because of different mobilities of electrons and holes (e.g. for GaAs: μ_n = 8500 $cm^2V^{-1}s^{-1}$, μ_p = 400 $cm^2V^{-1}s^{-1}$ at 300 K). Hence, only one diffusion zone at the p doped side exists where laser action may take place.

Based on such considerations the first diode lasers were built (Fig. 6.10). The design was very simple: a resonator was realised by using the cleaved front and back sides of a monolithic piece of GaAs containing a p-n junction and providing ~40 % reflection via the natural step of the index of refraction of a semiconductor against air ($\Delta n \approx 3.5$). The current flow took place through the whole cross-section of the device giving rise to light emission over the whole facet width. Such diode lasers faced a lot of troubles when operated at conditions appropriate to generate the necessary carrier densities ~ $10^{18}cm^{-3}$ needed for inversion as mentioned above. Because the area of recombination is approximately 5 μm thick due to the value of the diffusion length very high current densities of around 50000 A/cm^2 are to be used causing enormous heat problems. At cryogenic temperatures the carrier density required is lower ($\propto T^{3/2}$) and the diffusion length smaller ($\propto T^{1/2}$). Hence laser operation at 77 K with current densities of 1000 A/cm^2 could be realised in a pulsed mode.

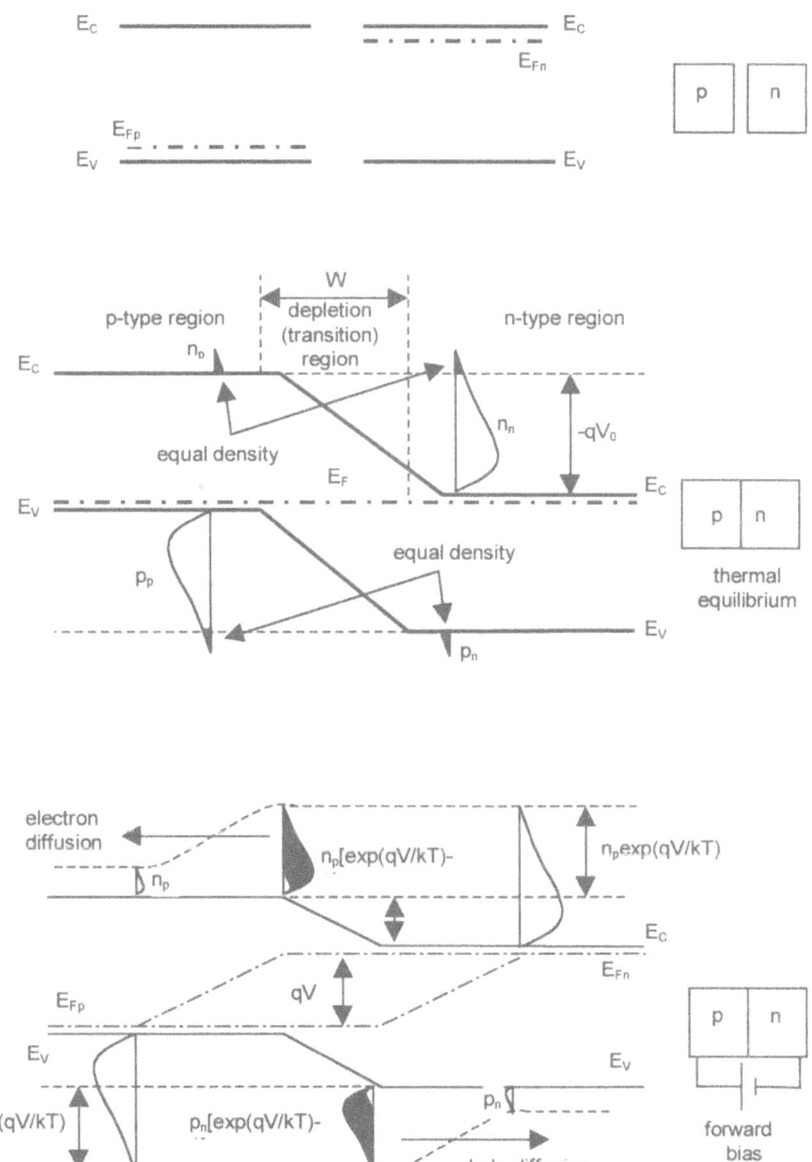

Fig. 6.9 Energy band diagram of a p-n homojunction and carrier density distribution in the thermal equilibrium state and in forward bias.

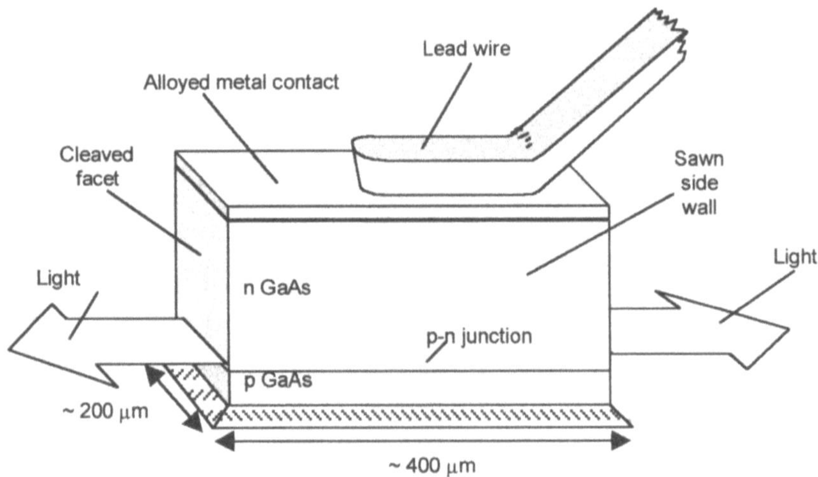

Fig. 6.10 Simple broad-contact semiconductor laser mounted on a heat sink.

At this state, such a device seemed not to be very practical. It took several years until substantial improvements were made: the hetero- and double-heterostructure.

A diode laser, of course, shows all features of a real laser: it faces a threshold as schematically depicted in Fig. 6.11a. If the current density is increased, starting from zero, there is no laser action at first. Nevertheless, incoherent and undirected light is emitted corresponding to the operation mode of a LED (light emitting diode). At a certain current density J_{th}, the threshold current density, the necessary carrier density n_{th} is reached like it was mentioned before allowing the start of the laser action. At any further increase of J no change of n takes place any more, i.e. any increase of pumping action increases the light output power as described in Fig. 6.11b. The slope of the current-light output power relation is a very important diode laser parameter called the slope efficiency η_S. It characterises the efficiency of the conversion of electrical power into light.

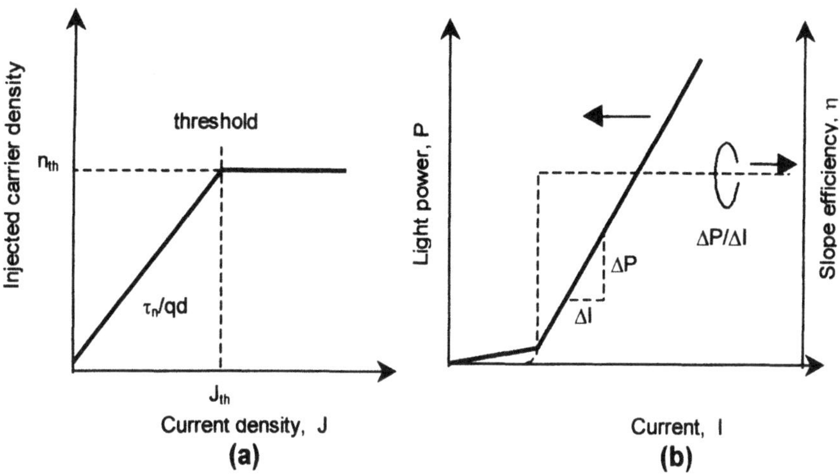

Fig. 6.11 a) Injected carrier density as a function of current
b) Current-light output power relation and slope efficiency η_s.

In theory, laser emission as a function of time at different pump levels can be calculated via the rate equations, i.e. two coupled nonlinear diffential equations for injected carrier density and photon number (equivalent to output power):

$$\frac{dn}{dt} = \frac{J}{qd} - \frac{n}{\tau_s} - \frac{gs}{V} \tag{6.1}$$

$$\frac{ds}{dt} = gs - \frac{s}{\tau_p} - \frac{\beta_{sp}Vn}{\tau_s} \tag{6.2}$$

where n means the carrier density, J the current density, q the electronic charge, d the active layer thickness, τ_s lifetime of the injected carriers, g stimulated emission gain coefficient, s photon number and V volume of the lasing mode, τ_p photon lifetime, β_{sp} spontaneous emission factor.

Such equations usually can only be solved numerically. Fig. 6.12 shows a typical general solution in the case where a step-current is applied to a diode laser. The damped oscillatory behavior is called relaxation. Interesting information about threshold, however, can be easily acquired under the assumption of stationary inversion (very small photon numbers) or steady state operation (constant photon number and carrier density) yielding information like depicted in Fig. 6.11.

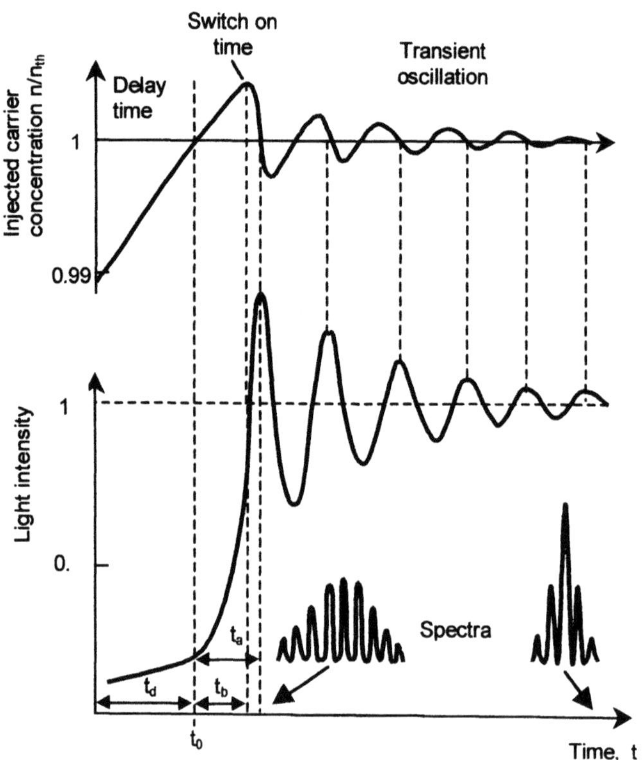

Fig. 6.12 Injected carrier concentration and output power versus time as a step-current response solved by rate equations.

6.3.2 Electronic confinement

It was already mentioned in the introduction that the realisation of the idea of a heterostructure based on the effective confinement of carriers by the potential walls of a larger bandgap represented the breakthrough for the usability of the diode laser. A homostructure contains a p-n junction based on one identical laser material with differently doped regions. In case of a heterostructure regions of two related semiconductors with different or even identical doping levels are in epitaxial contact. The steps in the conduction and valence bands are arranged in equilibrium condition in such a way that the Fermi levels are balanced. This usually leads to a larger potential barrier in the conduction band. To be effective, such a barrier should be ~ 100 meV, i.e. ~ 4 times the thermal energy at 300 K.

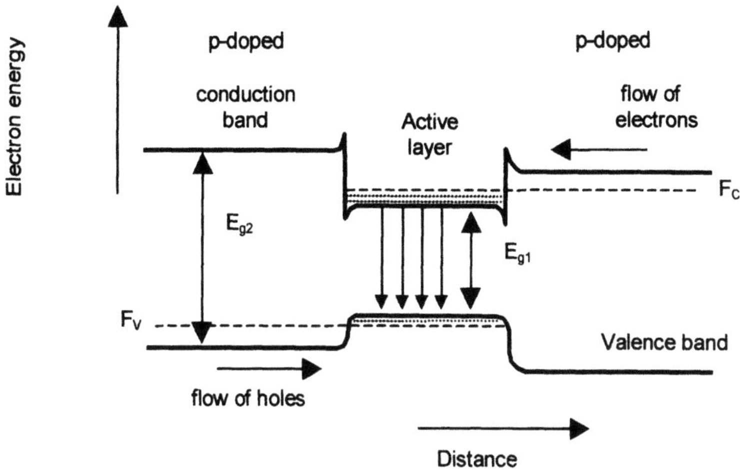

Fig. 6.13 The energy band diagram of a double heterostructure laser under forward bias.

Historically seen, the heterostructure was first introduced as a twofold p layer thereby erecting an effective potential step at the p side frontier of the active zone. In this way the thickness of the lasing area could be taylored as desired in order to reach threshold at much lower current densities. Such a single heterostructure allowed to reduce threshold currents at room temperature to 6000–8000 A/cm^2 but nevertheless faced severe drawbacks via the ineffective potential wall at the p-n junction and via the assymetry of the lasing mode. The light intensity stretches on one side into the n layer where no amplification but absorption takes place.

So, the next and most successful step was the building of a double-heterostructure diode laser which eventually allowed cw room temperature operation (Hayashi et al, 1970). Fig. 6.13 shows the energy band diagram of such a double heterostructure laser under forward bias. Within the active layer the electrons as well as the holes are confined effectively by potential walls and are forced thereby to recombine within this area. This leads to lasing under much lower current densities as compared with homojunction lasers. Typical values are J_{th} = 4300 A/cm^2 to J_{th} = 470 A/cm^2 for diode lasers at this stage of development.

Lateral confinement of injected carriers is also very important. For this purpose in the first place the stripe geometry of the upper metal contact of the laser was developed, thereby channelling the carriers to flow within a laterally loosely restricted area providing gain to the oscillating laser mode. Hence this structure is called gain-guided (Fig. 6.14a). Much better carrier confinement is achieved when introducing appropiate blocking layers on both sides of the active zone where no current flow takes place (e.g. proton bombarded, oxygen implanted, or sequences of layers similar to a blocked transistor) (Fig. 6.14b). This whole group of lasers is called stripe-contact lasers.

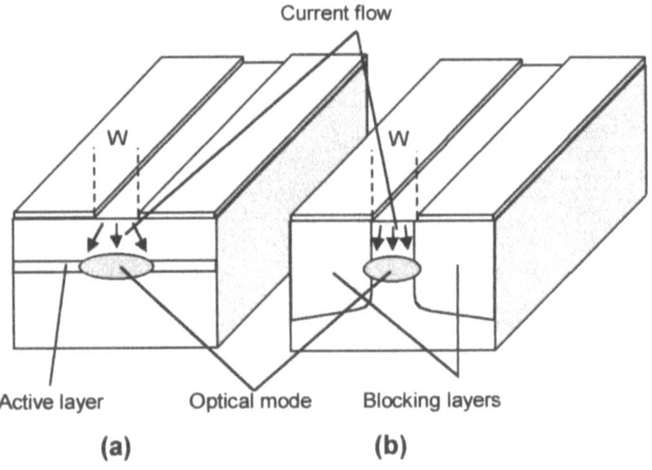

Fig. 6.14 Stripe contact lasers
  a) gain-guided laser with poor lateral confinement
  b) gain-guided laser for strong lateral confinement.

Technically more complicated, but a straightforward extension of the heterostructure concept is the type of laser named buried heterostructure laser. In this case the active area is just a stripe-like area of ~μm thickness totally buried within the hetero-material. Hence, the potential barriers are effective in all four directions perpendicular to light emission (Fig. 6.15). There are many types and modifications realised on the basis of this concept carrying names like V-groove, mesa... referring to the expitaxial designs steps involved.

Most diode lasers face a limitation of their output power via the degradation of the mirror surfaces. The very high intensities (~100 MWcm^{-2}) associated with heat can easily destroy the facets. The surface generally is the largest defect of a crystal. Hence, there the operating conditions are different to the bulk area - surface recombination currents turn the potential gain area into an absorption area. The ultimate realisation of the heterostructure concept leads to the fabrication of thin hetero-material windows in place of the original active material in order to avoid absorption. In such case, the active area is surrounded by the larger bandgap material from all sides. Such lasers are called window lasers.

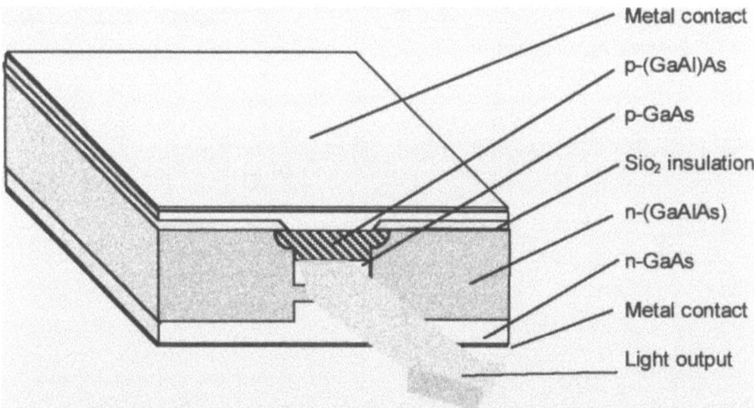

Fig. 6.15 Buried heterostructure laser with confinement of light and injected carriers in both transverse directions.

6.3.3 Optical confinement

Microwave oscillators need cavities with reflecting surfaces in all directions because the oscillating wavelength has the same dimension as the resonator. For lasers, in general, reflecting surfaces are just necessary at both ends of the beam propagation axis having such a curvature that the radiation is kept within the cavity even after several bounces to and fro. It is most easy to think of a Gaussian beam having a certain curvature of the wavefront at any position along the progagation axis. This beam is sustained if the mirrors have surface curvatures identical to that of the wavefront at their specific position (Fig. 6.16).

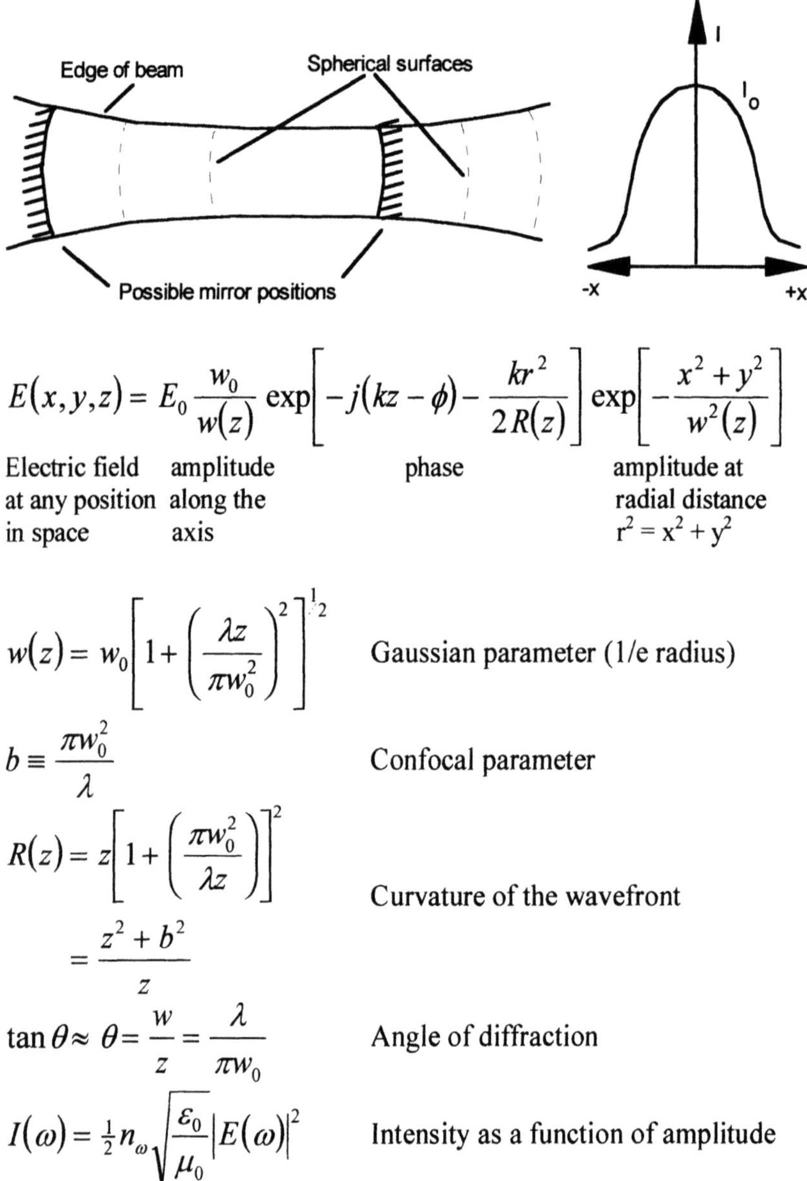

$$E(x,y,z) = E_0 \frac{w_0}{w(z)} \exp\left[-j(kz - \phi) - \frac{kr^2}{2R(z)}\right] \exp\left[-\frac{x^2 + y^2}{w^2(z)}\right]$$

Electric field amplitude   phase   amplitude at
at any position along the   radial distance
in space   axis $r^2 = x^2 + y^2$

$$w(z) = w_0 \left[1 + \left(\frac{\lambda z}{\pi w_0^2}\right)^2\right]^{\frac{1}{2}} \qquad \text{Gaussian parameter (1/e radius)}$$

$$b \equiv \frac{\pi w_0^2}{\lambda} \qquad \text{Confocal parameter}$$

$$R(z) = z\left[1 + \left(\frac{\pi w_0^2}{\lambda z}\right)\right]^2 \qquad \text{Curvature of the wavefront}$$

$$= \frac{z^2 + b^2}{z}$$

$$\tan\theta \approx \theta = \frac{w}{z} = \frac{\lambda}{\pi w_0} \qquad \text{Angle of diffraction}$$

$$I(\omega) = \tfrac{1}{2} n_\omega \sqrt{\frac{\varepsilon_0}{\mu_0}} |E(\omega)|^2 \qquad \text{Intensity as a function of amplitude}$$

Fig. 6. 16 Gaussian beam within a general laser resonator and its mathematical description.

Diode lasers, however, are of microscopic dimensions and usually have flat mirrors achieved by cleaving. Therefore, the radiation within the resonator has to be guided to keep the beam concentrated along the active zone and enforce a stable and reasonable beam diameter. Unfortunately, the beam of a diode laser is elliptical corresponding to its formation within an approximately rectangular volume (Fig. 6.17). This fact causes some problems when focusing such a beam, e.g. for pumping of solid-state lasers. In this context, the detailed knowledge of the behavior of Gaussian beams of the most general elliptical case is absolutely necessary.

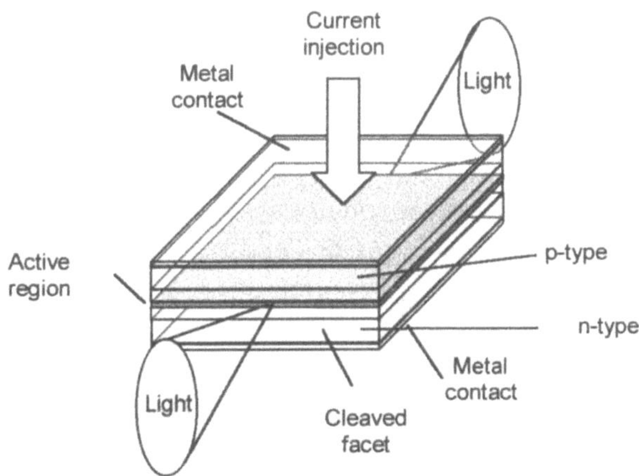

Fig. 6.17. Schematic illustration of a Fabry-Perot diode laser. The light is emitted from two partially reflecting cleaved facets of the diode.

For the diode laser, the waveguide of interest is the dielectric slab waveguide, i.e. a structure containing 3 optical layers with the central one having the largest index of refraction μ_I (Fig. 6.18). This structure usually is realised simultaneously together with the formation of double heterostructures (mostly the material with larger bandgap has the lower index of refraction). The crucial guiding mechanism is total reflection. If the angle θ of the beam with the reflecting interface is smaller than the critical angle θ_C guiding via such reflections takes place (e.g. 25% difference of Al concentration in AlGaAs heterostructures corresponds to $\theta_C \approx 15°$).

$$\theta = \cos^{-1}\left(\frac{\mu_2}{\mu_1}\right) \; if \; \mu_2 < \mu_1 \tag{6.3}$$

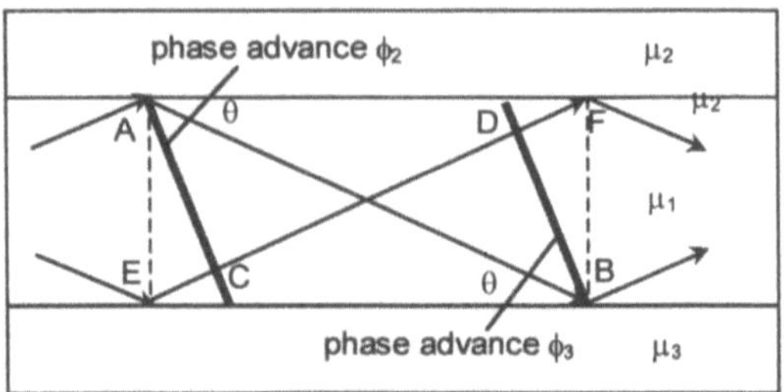

Fig. 6.18 Ray directions in an optical dielectric slab waveguide

In the case of 3 layers there are two critical angles. If $\mu_3 < \mu_2$ the limiting critical angle is θ_{c12} because it is smaller than θ_{c13}. Fig. 6.18 also illustrates the superposition of the propagating reflected waves which in all stable cases must remain in phase (phase difference $n.2\pi$). According to the geometrical drawing a wavefront AC may propagate towards BD on two different pathways: i) directly via CD or reflected twice along AB. Phase shifts ϕ_2 and ϕ_3 have to be considered in case of total reflection. It can be seen from the geometry of the wave propagation that

$$AB - CD = 2DF = \left[n + \left(\frac{\phi_2 + \phi_2}{2\pi} \right) \right] \frac{\lambda_0}{\mu_1} \tag{6.4}$$

where n is an integer and λ_0 is the vacuum wavelength. The angle of propagation θ of the mode corresponding to a specific n can be calculated

$$\sin\theta = \frac{DF}{d} = \left[n + \left(\frac{\phi_2 + \phi_2}{2\pi} \right) \right] \frac{\lambda_0}{2\mu_1 d} \tag{6.5}$$

where d is the thickness of the central layer. The wave is guided if $\theta < \theta_{c12}$.
At θ_{c12} the phase shift $\phi_2 = 0$. Hence, the criterion for guidance is

$$n < \frac{2d\delta\varepsilon^{1/2}}{\lambda_0} - \frac{\phi_3}{2\pi} \tag{6.6}$$

where $\delta\varepsilon^2 = \mu_1^2\mu_2^2$ corresponding to the smaller step of the index of refraction. If d is reduced consistently, fewer and fewer modes are guided. Eventually, even the zero order mode is forbidden because of ϕ_3 having a finite value. Only in case of a symmetric waveguide ($\mu_2 = \mu_3$) $\phi_2 = \phi_3 = 0$

at the critical angle. Hence, the zero order mode is guided even at smallest values of n.

Figure 6.19 shows some examples of the distribution of optical intensity across the thickness of a symmetric slab waveguide. For relatively thick slabs several modes are guided (top figure), for very thin waveguides (typical for quantum wells) only one mode is maintained. In this case, guiding is very weak according to the long exponential tails of the so-called evanescent field on both sides. In the centre region the distribution is cosine-like.

In some cases, like a single QW, guiding is enforced by adding two additional heterolayers, just for the sake of optical confinement thereby yielding a much better transverse mode structure (Fig. 6.20). The additional layers are inserted between the active layer and the normally doped cladding regions. Such a structure is called a separate confinement heterostructure (SCH). A further expansion of this idea is realised in case of high power laser diodes (array lasers, see section 6.4). In order to increase the area of reflection at the surfaces, separate confinement is realised via two graded index layers with a parabolic index profile (e.g. by systematically varying the Al content of AlGaAs layers). This structure is called a GRINSCH (graded index...) and offers the following advantages:

1. a small fraction of the cross-section of the wave propagates in an area where absorption may take place;
2. low J_{th} and high η_{diff} because of the small active QW volume;
3. weaker temperature dependance;
4. λ may be determined via the potential well and the shape of the GRINSCH layers. The guiding of modes in such a structure is related to the properties of a gradient index optical fiber.

In the case of a stripe laser, the confinement on both sides of the active layer would also act as a slab waveguide. As opposed to the previous considerations the dimensions are much wider and therefore the optical confinement effect with respect to a stable mode structure is not so much pronounced. Figure 6.17 gives a realistic impression of the emitted beams. The narrow width in direction of current flow gives rise to much more diffraction. Therefore, the ellipse of the beam cross section is much wider in this direction.

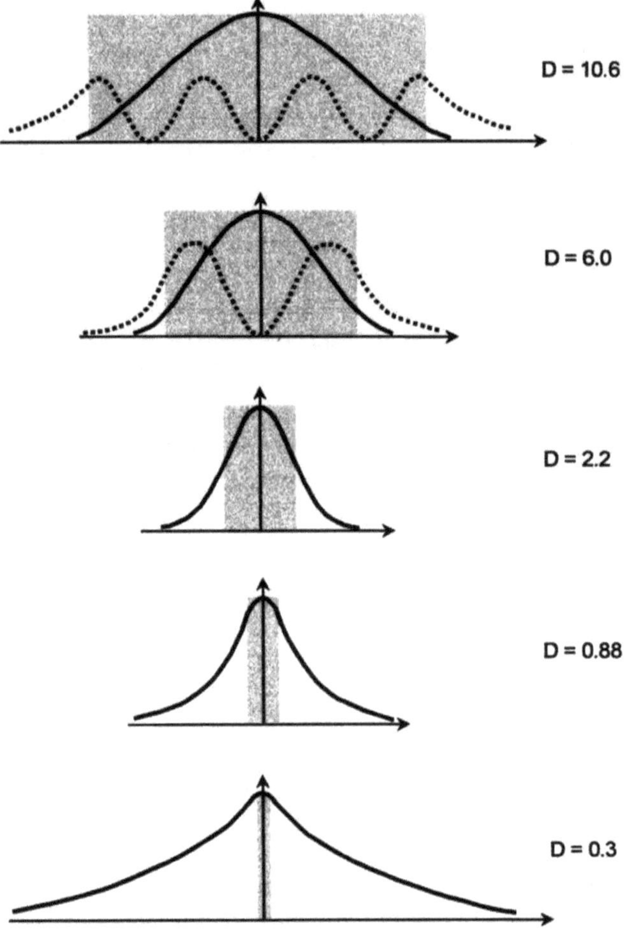

Fig. 6.19 Distribution of optical intensity across the width of a symmetrical slab waveguide for values of the normalised width $D = (2\pi/\lambda)\delta\varepsilon^{1/2}d$ of the centre layer between 0.3 and 10.6, shown shaded in grey. Zero order transverse modes shown by continuous curves and highest order transverse modes shown by dashed curves where applicable. (after Thompson, 1980).

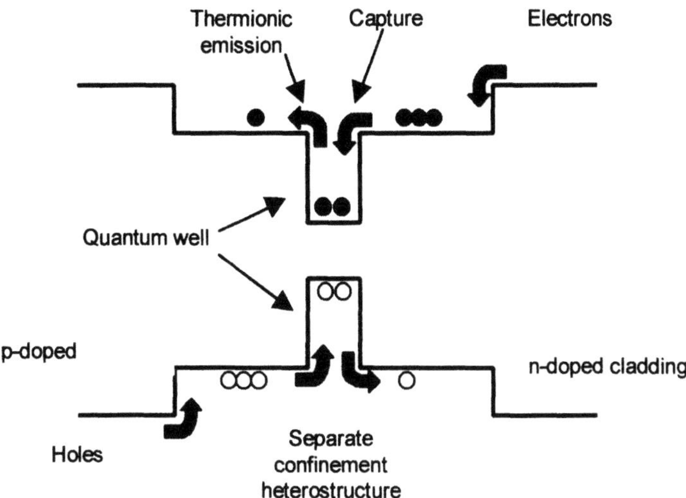

Fig. 6.20 Schematic of carrier transport in a single separate confinement heterostructure.

Of course, gain guided lasers also take advantage of optical confinement. Via the Kramers-Kronig relation, the absorption (i.e. also gain) properties of an optical material are closely related to its dielectric properties (i.e. index of refraction as a function of wavelength). Figure 6.21 illustrates this mutual dependence.

The modes discussed above are called transverse modes because they describe the time independent distribution of the optical fields in a transverse direction of a laser resonator.

In the general case of a full size Fabry-Perot resonator the transverse mode structure is a two dimensional group of characteristic field functions (Hermite-Gaussian functions). In most cases one is interested in only one single, fundamental mode corresponding to a Gaussian beam as described above. Such a laser is referred to as a single mode laser.

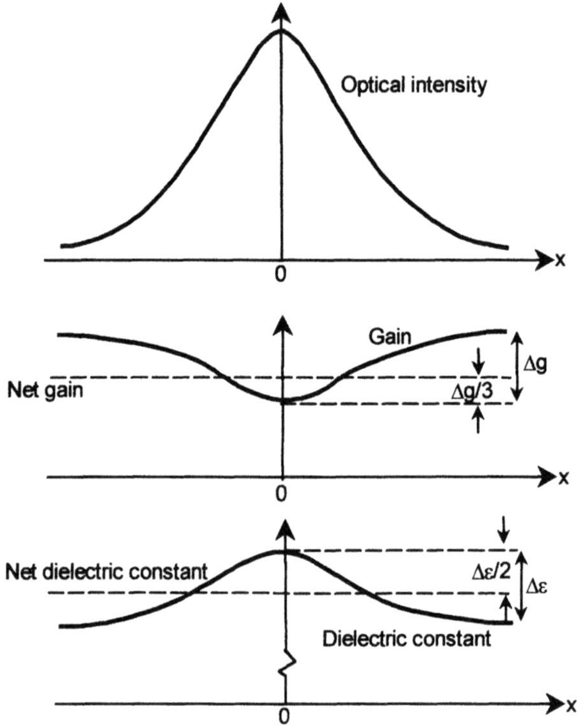

Fig. 6.21 Distribution of optical intensity, gain, and dielectric constant for idealised self-focused filament.

6.3.4 Spectral and temporal properties

There are also, as is commonly known, longitudinal modes, i.e. standing waves resulting from the superposition of travelling waves in both directions within the resonator. Their wavelength λ_m depends on the number of half waves contained within the length of the resonator

$$\lambda_m = \frac{2\mu L}{m} \tag{6.7}$$

where L is the length of the resonator, μ the index and m an integer.
In terms of angular frequency, equation (6.7) becomes

$$\omega_m = \frac{m\pi c}{\mu L} \tag{6.8}$$

The difference in frequency between adjacent modes is

$$\delta\omega = \frac{\pi c}{\mu L} \tag{6.9}$$

It is important to remember that in a diode laser the refractive index μ varies with the frequency ω (material dispersion) and therefore, strictly speaking, also the intermodal spacing, $\delta\omega$. m is of the order of 10^6 for most full scale lasers. For typical diode lasers, however, with $L \approx 200\text{–}500$ μm, $m \approx 400\text{–}1000$. Only a fraction of them will experience sufficient gain to lase. Figure 6.22 shows a typical emission spectrum of a cw AlGaAs diode laser containing approximately 25 longitudinal modes with a spacing of ~0.35 nm corresponding to ~100 GHz.

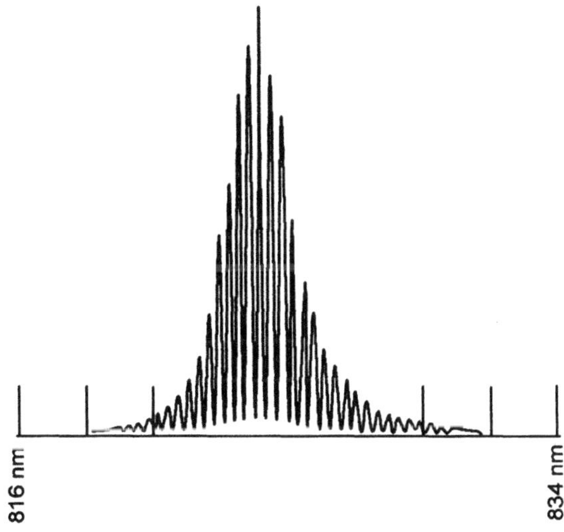

Fig. 6.22. Emission spectrum of a cw-GaAlAs diode laser at room temperature.

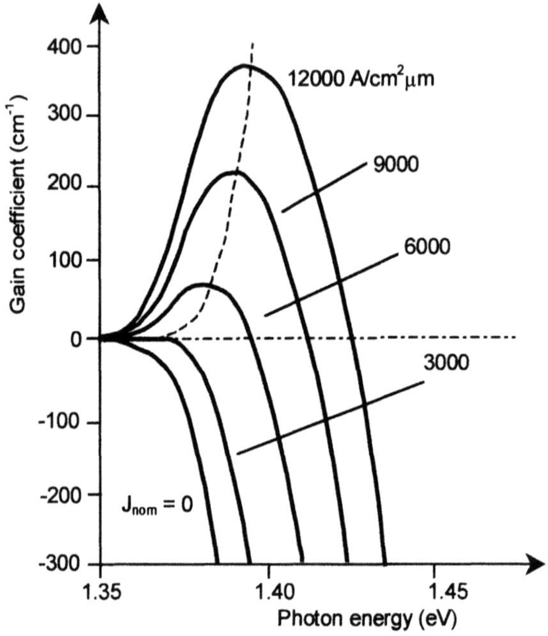

Fig. 6.23 Calculated dependence of gain coefficient on photon energy for GaAs at 297 K for several values of the injected current density.

The gain spectrum of a diode laser normally is very wide (Fig. 6.23). It depends, of course, via the density of injected carriers and the corresponding quasi Fermi levels (compare Fig. 6.7) on the current density. Hence a considerable number of modes can be amplified above threshold at a sufficiently high current. On the other hand, it may be intended to allow only one mode to oscillate in order to realise a single frequency laser. In such a case, very short (50–100μm) diodes are used so that the mode spacings become very large. Then, at moderate currents only one mode may lase. However, shortening of the laser cavity reduces the available optical gain. Therefore, to achieve lasing, the active layer must have very high gain and the optical feedback must be maximised. Short cavity lasers with enhanced facet reflectivities (R_1, $R_2 > 0.85$) can operate under cw conditions in a single mode.

It is also interesting to analyze what determines the spectral width of a single longitudinal mode of a diode laser: Generally, in a laser technical noise arising from vibrations of the mirrors and various fluctations of optical parameters within the resonator is the dominant origin of the linewidth to be measured. Much more stable, down to kiloHertz, are monolithic solid state laser arrangements (in fact, the diode laser is also monolithic). The source of the prevailing finite linewidths is, roughly speaking, quantum fluctuations associated with spontaneous emission. This is described by the Schawlow-Townes formula (already derived in 1958!)

$$\delta f = \frac{\pi h f \Gamma^2}{P} \tag{6.10}$$

where Γ is the linewidth of the 'cold' resonator (arising from damping) and P is the power of the mode. This formula gives the right numbers for most lasers (e.g. $\delta f \approx 0.01$ Hz for a HeNe laser) but is very wrong in case of the diode laser. The calculation yields 1 MHz, in reality $\delta f \approx 50$ MHz. This fact has led to specific investigations of the specific properties of the diode laser. Qualitatively, the following facts have to be considered:

1. Spontaneous emission does not only affect the phase of the laser radiation (considered by the Schawlow-Townes formula) but also the amplitude. Hence continuous relaxation oscillations take place increasing the bandwidth.
2. Diode lasers have extremely small dimensions (the smallest are $0.1 \times 1 \times 10 \ \mu m^3$) involving only $\sim 10^5 - 10^6$ electrons into the lasing process. Hence, statistical fluctuations of this number are non-negligible.
3. Together with the density of electrons, the refractive index changes via two interdependences:
   (a) The interaction with free carriers (plasma effect) reduces ε.
      $\delta \varepsilon \approx 0.02 \ (\sqrt{\varepsilon} = \mu)$.
   (b) Amplification means change of absorption coefficient, i.e. via the Kramers-Kronig relation μ is also changed. This effect also causes a reduction of μ which is approximately 2 x the value of (a).
4. Modulation of intensity originating from spontaneous emission noise yields side-bands which in case of diode lasers are accessible for measurement (spectral distance typically 2 GHz).

The temporal behavior of light emitted from diode lasers is a very important aspect, especially in optical data transmission. It is one of the major advantages that direct modulation is possible via the change of the drive current. As explained already in Fig.6.12 there is a damped resonant relation between optical intensity and carrier density which presents an upper limit for pulse repetition frequency and, associated with it, pulse duration. This resonance frequency is given by

$$f_r = \frac{1}{2\pi}\left(\frac{I/I_{th}-1}{\tau_n\tau_p}\right)^{1/2} \tag{6.11}$$

where I and I_{th} are the actual current and the threshold current, respectively, τ_n is the carrier lifetime at threshold and τ_p is the photonic lifetime. τ_m is typically ~ 1ns, $\tau_p \approx$ ps for a usual diode (having a length of ~ 300 µm). As one can see, f_r is proportional to the square root of power. It can reach values up to 10–40 GHz.

In optical communications applications of modern type, both single frequency and high repetition modulation are desired. Such a requirement can only be fulfilled in specially equipped diode lasers containing internal filtering, i.e. distributed Bragg reflector and distributed feedback lasers (see section 6.4).

Another very important technique for the generation of short pulses out of lasers is mode locking. In this context the idea can only be discussed in brief: according to the bandwidth of a laser several (hundreds) of longitudinal modes can operate simultaneously having a constant frequency spacing, as pointed out, but no phase relation. However, if the phase difference between all the modes can be kept constant the temporal interference between all of them can yield very short pulses with intervals of the cavity round trip time. Pulse duration depends on the number of modes and hence on the bandwidth. For diode lasers pulses of picoseconds can be realised. The method to enforce a constant phase relation is called the mode locking technique. There are various possibilities which cannot be described within the scope of this chapter.

6.3.5 Loss mechanisms and temperature behavior

Optimum operation of a diode laser would be the case if every electron - hole pair would recombine via stimulated emission emitting a coherent photon which finally could leave the oscillator through the facet of choice. This is obviously never achievable. There are various sources of carrier and of optical losses:

1. Spontaneous emission causes the reduction of carrier density without increasing the oscillation intensity of the laser resonator. However, it is needed to allow the start of the lasing emission because stimulated emission only allows amplification.
2. Carriers may leave the active zone jumping over the potential wall by thermal activation and recombine elsewhere.
3. Carriers may recombine nonradiatively via
    recombination centres or
    Auger recombination or
    surface recombination

    (The Auger process is a three particle interaction. Two of them, an electron and a hole, recombine and transfer their energy to another carrier which thereby is excited to a higher energetic position within the respective band. See Fig. 6.24).
4. Optical losses usually face three contributions, namely free carrier absorption, band-edge absorption and scattering losses.

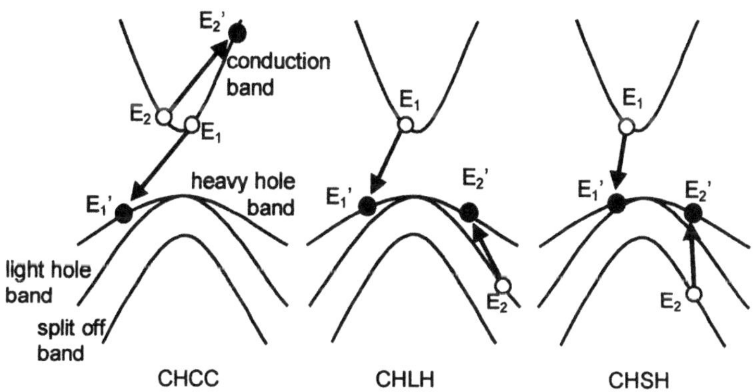

Fig. 6.24 Band-to-band Auger processes.

Free carrier absorption arises from a direct interaction between the optical waves and the electrons and holes. The absorption is directly proportional to the carrier concentration and approximately proportional to the square of the wavelength. For instance, in GaAs with an electron concentration of 10^{18} cm^{-3} the absorption coefficient at the emission wavelength is ~3 cm^{-1}. It seems that the reduction of carrier concentration could reduce this loss, but paradoxically, gain increases at a higher rate with n. In general, free carrier absorption is not important in GaAs but can become significant with longer wavelengths, e.g. GaInAsP.

Band-edge absorption takes place most strongly in the central part of the active layer where it is of no significance as it just regenerates the injected carriers. However, if an optical wave propagates with a high proportion of its cross-section within the unpumped medium where absorption cannot generate carriers to be used again this effect becomes non-negligeable. In any kind of heterostructure laser, however, the adjacent regions to the active zone are of sufficiently greater band-gap so that no band-edge absorption should take place.

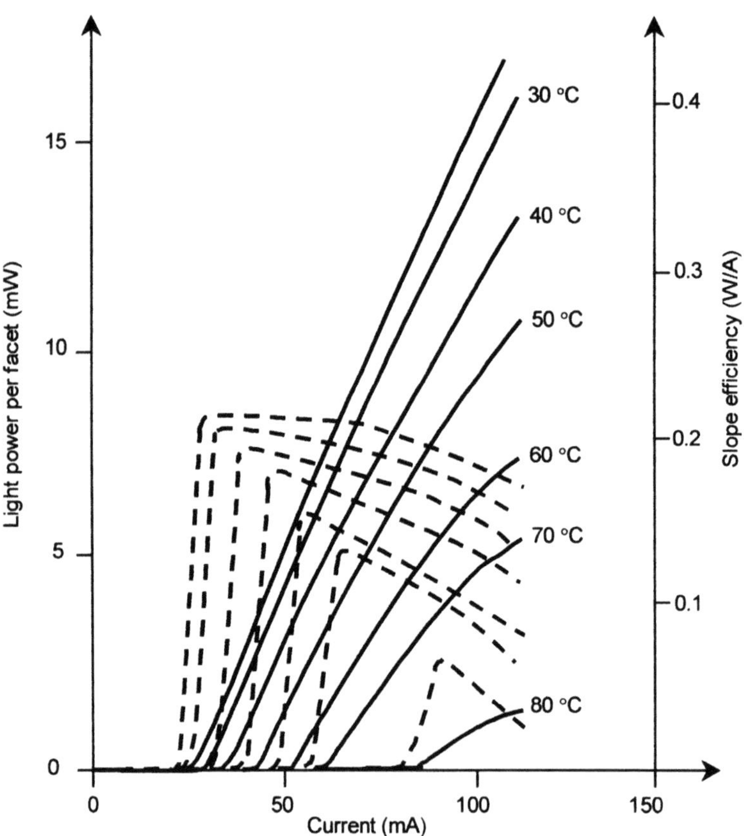

Fig. 6.25 Temperature dependence of current-light output power characteristics in a 1.3 μm, InGaAsP/InP BH laser, $T_0 \sim 60$ K. Light output solid lines, slope efficiencies dotted lines (after Fukuda, 1994).

The threshold current and the efficiency of a diode laser depend on temperature as it can be seen when measuring the output power versus current (see Fig. 6.25). Empirically, an exponential law is observed:

$$I_{th} = I_{th(0)} \exp\left(\frac{T}{T_0}\right) \tag{6.12}$$

where T is the temperature and T_0 a characteristic coefficient. T_0 is different for the various laser materials: e.g. $T_0 \approx 250$ K for GaAs, but $T_0 \approx 90$K for GaInAsP. At closer inspection, even T_0 can be temperature dependant (in GaInAsP it varies from 90 K ($T < 260$ K) to 30 K ($T > 340$ K)). This can be explained by the different loss mechanisms activated at different temperatures.

6.4 DIODE LASER DESIGN

There is a huge variety of different designs of diode lasers which to describe or depict in detail is far beyond the scope of these lecture notes. In the following the most important types of lasers should be covered.

6.4.1 Stripe lasers

This type of laser is most commonly used at rather low power levels for optical data storage and for optical communication. For such applications, beam and spectral quality are rather important. Hence, index guiding is the best solution to the problem. Figure 6.26 gives a overview of a number of laser designs which with two exceptions (a and b) all are index guided.

When compared to gain guided lasers, such as planar-types, BH lasers show some excellent characteristics. In addition to a stable transverse mode, BH lasers have a low threshold current, high device efficiency, the capability of high output power operation, and so forth. These characteristics are favourable for use in equipment and systems, especially optical fiber transmission systems. Those excellent characteristics are, of course, introduced by the BH structure (compare Figs. 6.26 e,g,h). The current path and optical confinement region are strictly determined by the structure. The InGaAsP active layer is surrounded by InP layers and the refractive index distribution is similar to an optical fiber. The propagating light is confined to the active layer. Consequently, the transverse mode is stable and fundamental, even at high output power if the dimension of the active layer is suitably set. Usually, the stripe width of the active layer is around 2 µm or less.

The current flow is limited to the active layer by current blocking layers (burying layers). The p-n junction in the blocking layer is reverse biased when a forward bias is applied to the active region, and thus the current is confined to the active layer by the reverse-biased p-n junction. Recently, a highly resistive layer, such a Fe-doped InP, has been used as a current blocking layer instead of the p- and n-type layers. As a result of the strict current confinement, the threshold current can be decreased (although the current density increases).

However, the structure and fabrication processes of a BH laser are complicated. During fabrication new causes of degradation could be introduced. The rate of eventual degradation is very dependent on the fabrication process and on the structure and thus, various degradation modes are known.

Distributed Bragg reflector and distributed feedback lasers:

DBR lasers employ in-plane periodic structures to provide distributed frequency-selective feedback. The built-in grating leads to a periodic perturbation in the refractive index, and feedback occurs by Bragg diffraction. In contrast to DFB lasers, where a grating is incorporated in the region of the active layer, DBR lasers use gratings etched outside the active region, near the ends of the cavity as shown in Fig. 6.27. These unpumped corrugated end-regions act as frequency-selective mirrors. Here optical gain and wavelength tuning are provided by the active section and the Bragg section, respectively. The passive phase-control section can be used to ensure single-mode operation. A problem inherent in DBR lasers is that when the unpumped active material is used to etch the gratings, optical losses inside the DBR region are high and the resulting DBR reflectivity is poor. This problem can be overcome by using a material for DBR region that is relatively transparent at the lasing wavelength.

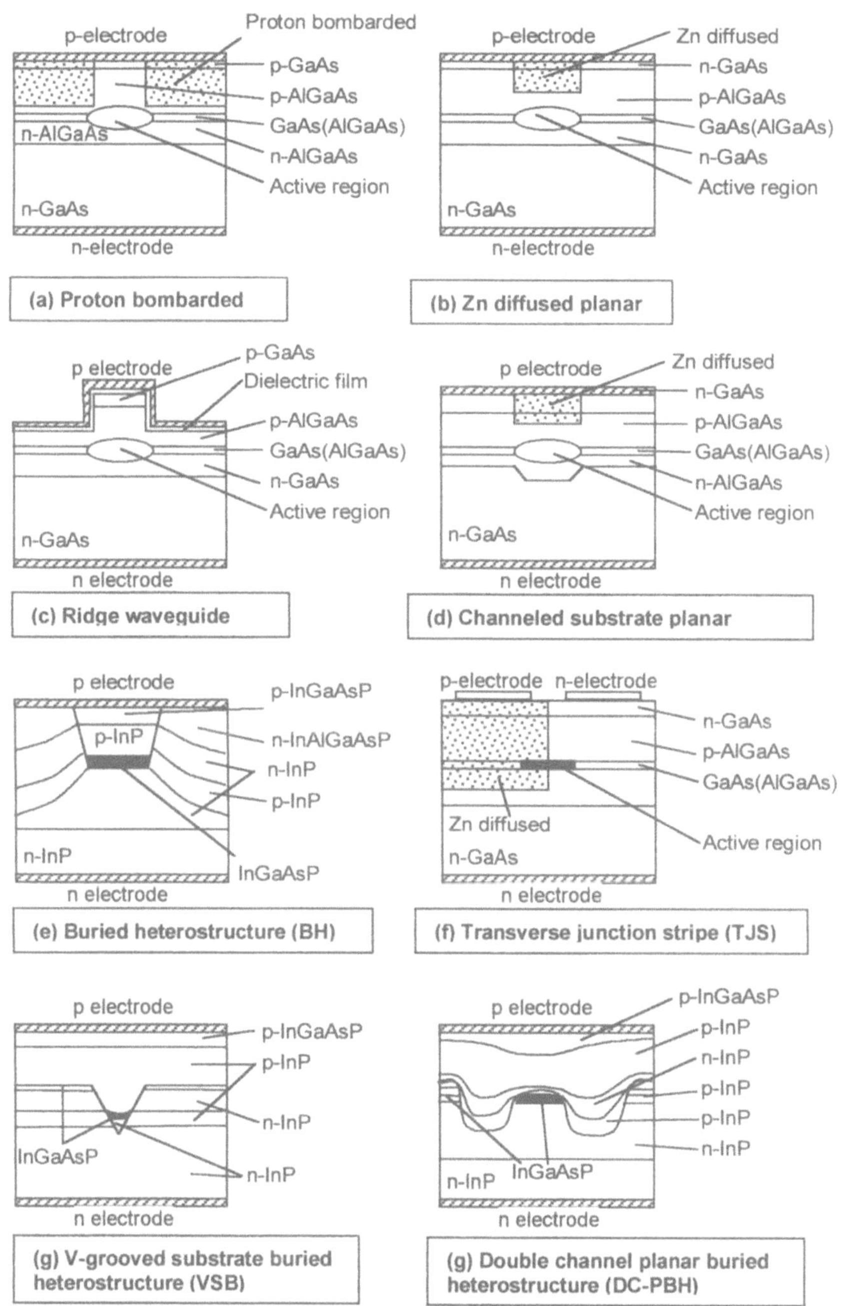

Fig. 6.26 Popular structures for various lasers (after Fukuda, 1980).

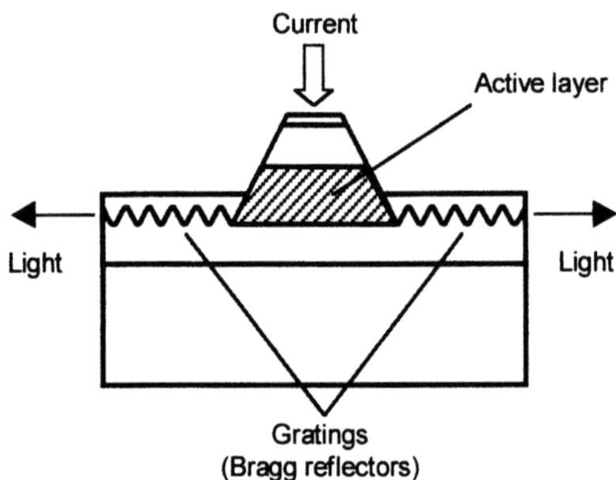

Fig. 6.27 Schematic view of a distributed Bragg reflector laser.

The first-order gratings are generally formed in DBR regions with a coupling coefficient of approximately 100 cm^{-1}. The longitudinal mode closest to the Bragg wavelength has the lowest threshold gain. The Bragg wavelength λ is related to the pitch of the grating Λ by

$$\Lambda = \frac{m\lambda_B}{2\mu_m} \tag{6.13}$$

where μ_m is the mode refractive index and the integer m represents the order of the Bragg diffraction. For example, a first order grating has a pitch of 0.23 μm at a wavelength of 1.5 μm and a typical value $\mu_m \cong 3.4$.

A DBR laser is usually designed in such a way that the dominant mode has a threshold gain 8-10 cm^{-1} lower than the neighbouring modes. These modes are typically suppressed by ~ 30 dB. In contrast to FP lasers, the longitudinal modes of a DBR laser are not equispaced. Because of its spectral purity, DBR diode lasers are promising for applications in optical fiber communications.

The feedback in a DFB laser is provided by a grating that runs along the active region. Two DFB laser structures are shown in Fig. 6.28. Periodic perturbations in the refractive index along the laser cavity provide frequency selective feedback. In DFB devices, an optical wave traveling in one direction is reflected by the grating into a wave traveling in the opposite direction and vice versa. Coherent coupling between the counter-propagating waves occurs only for wavelengths that satisfy the Bragg condition given by equation (6.13). However, DFB lasers with uniform gratings and cavities do not lase at the Bragg wavelength. Instead, two-mode lasing can occur on either side of the stop-band. Asymmetries in the cavity or at the facets and variations of the carrier distribution along the cavity will determine the dominant mode. But the phase and residual facet reflectivity vary considerably even among lasers cleaved from the same wafer, and the carrier distribution along the cavity varies under different drive conditions. Therefore, even if the gain margin between the modes is sufficient to ensure a high yield of lasers with single-mode operation under CW conditions, stronger mode selectivity is required for lasers that are subject to direct modulation.

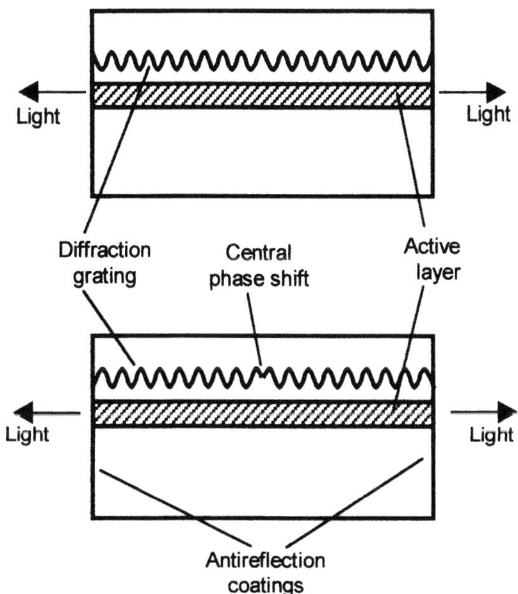

Fig. 6.28 Longitudinal cross-section of two DFB diode lasers (a) without and (b) with a phase-shift (after Vasilev, 1995).

In order to avoid two-mode emission, an additional phase shift of $\pm\pi$ for the roundrip gain must be introduced. This condition implies that both counter-propagating waves should suffer a $\pi/2$ or $\lambda/4$ phase shift. Phase shifting the grating by $\lambda/4$ in the middle of the laser moves the lasing mode to the Bragg wavelength. If antireflection facet coatings are used in addition to the phase shifiting, much higher gain margins can be achieved, ensuring dynamic single mode operation. Alternatively, an uncorrugated section of a typical length of 10 µm can be introduced into the centre of the laser cavity such that the optical phase shifts occur during wave propagation along this region. However, $\lambda/4$ phase-shifted DFB lasers suffer from longitudinal mode spatial hole burning. Here, lasers with strongly coupled gratings experience increased optical intensity in the region of the phase shift at the centre of the cavity. This leads to a reduced carrier concentration at the centre of the laser, compared with that at the facets (the spatial hole).

Other DFB structures have been proposed to give a higher side mode suppression ratio than that achieved using uniform gratings, and without the longitudinal mode spatial hole burning associated with $\lambda/4$ phase-shifted lasers. The operation of lasers with two $\lambda/8$ phase shifts has been successfully demonstrated. Gain coupled DFB lasers, which are still in the early stages of development, are expected to have improved mode selection and a flattened longitudinal-mode intensity.

6.4.2 Array lasers

Phase locked arrays of diode lasers have been studied extensively over the last 15 years. Such devices have been pursued in the quest to achieve high coherent powers (> 100 mW diffraction limited) for applications such as space communications, blue-light generation via frequency doubling, optical interconnects, parallel optical-signal processing, high-speed, high-resolution laser printing and end-pumping solid-state lasers. Conventional, narrow-stripe (3-4 µm wide), single-mode lasers provide, at most, 100 mW reliably, as limited by the optical power density at the laser facet. For reliable operation at watt-range power levels, large-aperture (≥ 100 µm) sources are necessary. Thus, the challenge has been to obtain single spatial mode operation from large aperture devices, and maintain stable, diffraction limited beam behavior to high power levels (0.5–1.0 W).

By comparison with other types of high -power coherent sources (master oscillator power amplifier (MOPA), unstable resonators), phase locked arrays have some unique advantages: graceful degradation; no need for internal or external isolators; no need for external optics to compensate for phasefront aberrations due to thermal and/or carrier induced variations in the dielectric constant; and, foremost, beam stability with drive level due to a strong, built in, real index profile. The consequence is that, in the long run, phase-locked arrays are bound to be more reliable than either MOPAs or unstable resonators.

The four basic types of phase-locked arrays are shown schematically in Fig. 6.29: leaky wave coupled, evanescent wave coupled, Y junction coupled and diffraction coupled. Leaky wave coupled devices are arrays for which the lasing modes have the major field intensity peaks in the low index array regions, so called leaky array modes. The first phase locked array to be published was a gain guided array, which by its nature is a leaky wave coupled device. Evanescent wave coupled devices use array modes whose field intensity peaks reside in the high index array regions. They have been the subject of intensive research over the 1983–88 time period. Since operation of evanescent type arrays was difficult to achieve in the in-phase mode (i.e., single central lobe), Y junction coupled devices and diffraction coupled devices were proposed to select in-phase operation via wave interference and diffraction respectively.

Up to 1988 the results were not at all encouraging: maximum diffraction limited single lobe powers of ~ 50 mW or coherent powers (i.e. fraction of the emitted power contained within the theoretically defined diffraction limited beam pattern) never exceeding 100 mW. Thus, the very purpose of fabricating arrays, to surpass the reliable power level of single element devices, was not achieved. The real problem was that researchers had taken for granted the fact that strong nearest neighbour coupling implies strong overall coupling. In reality, as shown in Fig. 6.30, nearest neighbour coupling is "series coupling", a scheme plagued by weak overall coherence and poor intermodal discrimination. Strong overall interelement coupling occurs only when each element couples equally to all others, so-called "parallel coupling". In turn, intermodal discrimination is maximised and full coherence becomes a system characteristic.

Furthermore, parallel coupled systems have uniform near field intensity profiles, and are thus immune to the onset of high order mode oscillation at high drive levels above threshold.

Parallel coupling can be obtained in evanescent wave coupled devices, but only by weakening the optical mode confinement, and thus making the devices vulnerable to thermal and/or injected carrier induced variations in the dielectric constant. For both, full coherence and stability, it is necessary to achieve parallel coupling in structures of strong optical-mode confinement (e.g., built in index steps ≥ 0.01). Only strongly guided, leaky-wave-coupled devices meet both conditions. A stable, parallel-coupled source is highly desirable for systems applications since it has graceful degradation; that is, the failure or obscuration of some of its components does not affect the emitted beam pattern.

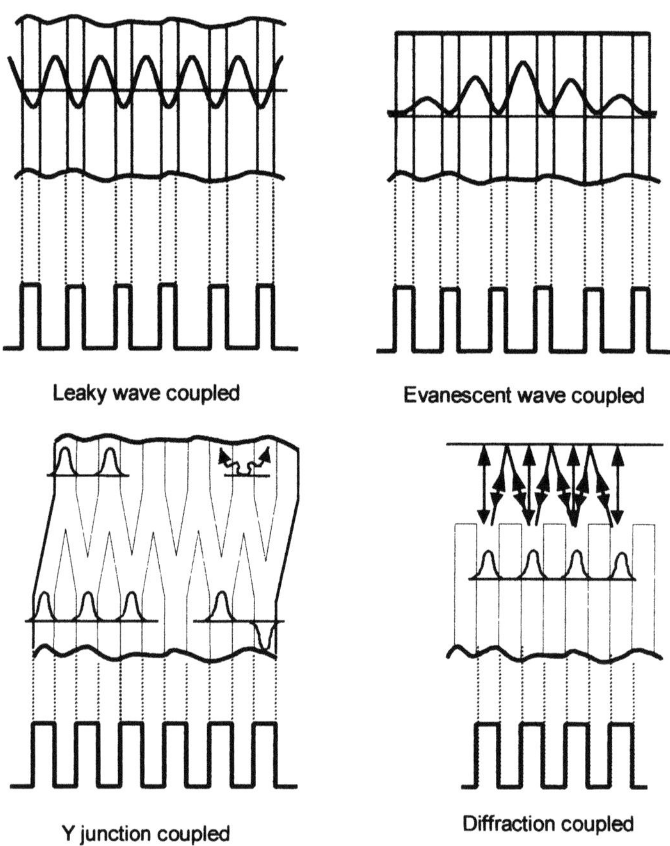

Fig. 6.29 Schematic representation of basic types of phase-locked linear arrays of diode lasers. The bottom traces correspond to refractive-index profiles. The top traces indicate the overall amplitude.

Only strongly guided, leaky wave coupled devices meet both conditions. A stable, parallel coupled source is highly desirable for systems applications since it has graceful degradation; that is, the failure or obscuration of some of its components does not affect the emitted beam pattern.

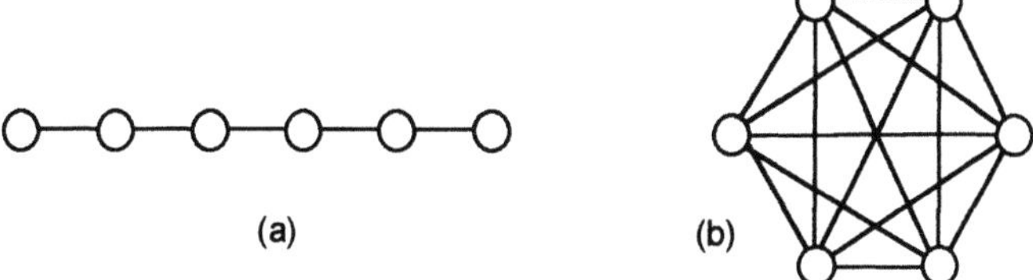

Fig. 6.30 Types of overall interelement coupling in phase-locked arrays:
(a) series coupling (nearest-neighbor coupling, coupled-mode theory);
(b) parallel coupling.

Table 6.3 gives a brief history of phase-locked array research and development*. In 1978 Scifres *et al.* reported the first phase locked array: a five element gain guided device. It was a further eight years before Hadley showed that the modes of gain guided devices are of the leaky type. Gain guided arrays have generally operated in leaky, out of phase (i.e. two lobed) patterns with beam widths many times the diffraction limit due to poor intermodal discrimination. Furthermore, gain guided devices being generated simply by the injected carrier profile, are very weakly guided and thus vulnerable to thermal gradients and gain spatial hole burning. The first real index guided, leaky wave coupled array (i.e. so called antiguided array) was realised in 1981 by Akley and Engelmann. While the beam patterns were stable, the lobes were several times the diffraction limit, with in phase and out of phase modes operating simultaneously due to the lack of a mode selection mechanism. Positive index guided arrays came next in array research (1983-88). In phase, diffraction limited beam operation could never be obtained beyond 50 mW output power. Some degree of stability could be achieved in the out of phase operating condition such that, by 1988, two groups reported diffraction limited powers as high as 200 mW.

---

* All references referred to below are given in Botez et al., (1994).

Table 6.3 Brief history of phase locked diode laser array research and development

Time period	Array type	Preferred array mode	Overall inter-element coupling	Single mode selectivity	Max. diffraction limited power
1978-	Gain-guided	Leaky in-or out-of-phase	Series	Poor	-
1981	Antiguided	Leaky in-or out-of-phase	Series	Poor	-
1983-88	Positive-index-guided	Evanescent out-of-phase	Series	Moderate	0.2 W
1988	Antiguided	Leaky in-or out-of-phase	Series	Moderate	0.2 W
1989-	In-phase resonant antiguided	Leaky in phase	Parallel	Excellent	2.0 W

Note: The dashes in the final column imply insignificant value.

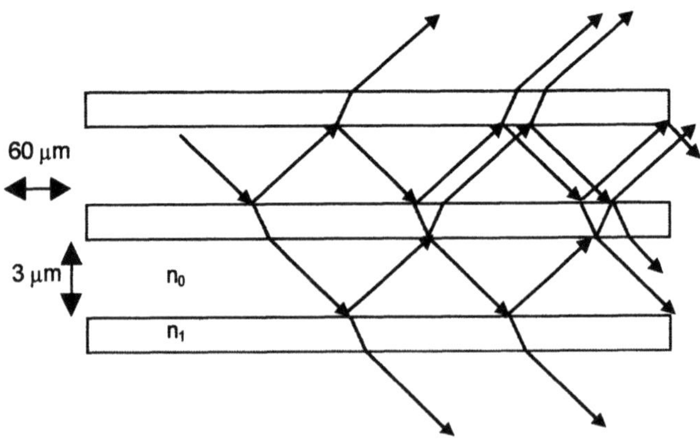

Fig. 6.31 Ray optical representation of leaky wave fan-out in a resonant array of antiguides. Only one original ray from top left is shown for clarity.

In 1988 antiguided arrays were resurrected and, from the first attempt, researchers obtained 200 mW diffraction limited in phase operation. Hope for achieving high coherent powers from phase locked arrays was rekindled, although there was no clear notion as to how to obtain single lobe operation. The breakthrough occured in 1989 with the discovery of resonant leaky wave coupling, which, as shown in Table 6.3 allowed parallel coupling among array elements for the first time, and thus the means of achieving high-power, single lobe, diffraction limited operation. The experimental coherent powers quickly escalated such that, to date, up to 2 W has been achieved in a diffraction limited beam.

Looking at Table 6.3 it is apparent that positive index guided devices were just on stage in array development, and that what has finally made phase-locked arrays a success has been the discovery of a mechanism for selecting in phase leaky modes of antiguided structures.

6.4.3 Surface-emitting lasers

In the mid 1980s, the demonstrated performance of diode lasers suggested that they could be used for extensive applications beyond fiber optics, compact optical discs, and optical recording. They could replace flashlamps as solid state laser pumps, provide optical interconnects between integrated circuits or within computers and possibly even replace large gas and solid-state lasers in high power, high coherence applications such as satellite communication and laser machining and welding. These new possibilities caused an increased interest in surface emitting geometries for semiconductor lasers in an effort to find the best device configuration for a given application or performance level. One expectation was that surface-emitting approaches would allow combining the power of hundreds or thousands of low-power, grain of salt sised devices into a monolithic, coherent high power array of semiconductor lasers while maintaining the efficiency and spectral properties of the individual cleaved-facet semiconductor lasers. Results obtained in the last few years are validating the hoped for performance of surface-emitting semiconductor lasers.

In addition to the useful features that make them attractive as replacements for conventional, cleaved facet semiconductor lasers in some applications, surface-emitting lasers can provide a basis for the use of optics in a number of technologies which cannot easily use cleaved-facet lasers.

A salient feature of the surface emitters is that they can be grown, fabricated, tested, and used in a monolithic planar geometry which is similar to the geometry used for electronic integrated circuits. For example, conventional cleaved facet lasers cannot be easily integrated into an optoelectronic-integrated circuit (OEIC) since the act of cleaving separates the laser from the rest of the chip. This problem can be partically reduced by using etching or micro-cleaving techniques to define the laser facets on the circuit. However, in many integration applications, it is desirable to have the laser connected to an on-chip waveguide, or to have the laser communicate to another chip. Unfortunately, an etched or microcleaved facet is not readily coupled to a monolithic waveguide. In addition, unless the laser is located on the edge of the circuit, the facet emission is difficult to access. Surface emitting lasers, however, can be located anywhere on an OEIC chip with any orientation, and the light can be easily accessed for external use. Another important feature of all the surface-emitters is that the light is emitted normal (or near-normal) to the surface of the wafer for use in a variety of applications. Of further note, light produced by some types of surface emitters can, in addition, be simply guided around the wafer and used to optically interconnect coplanar lasers and other optical devices such as switches, modulators and detectors.

Surface emitting technology also makes the fabrication of monolithic, two-dimensional arrays of semiconductor lasers possible. Phase-locked arrays are obtained by optically interconnecting the lasers on the chip using either on chip or external means of optical coupling. Such two-dimensional arrays offer the promise of very high powers with narrow beam divergences.

Historically*, surface emission dates back almost to the beginning of semiconductor lasers when Melngailis (1965) of MIT's Lincoln Laboratory reported on what has since become known as a vertical cavity structure. The concept (Kogelnik and Shank, 1972) and demonstration (Kogelnik and Shank, 1971) of distributed feedback lasers led to grating surface-emission in semiconductor lasers largely because of the difficulty of fabricating first-order feedback gratings which required periods of about 0.1 μm at wavelengths around 0.85 μm. Second, third, and fourth order distributed feedback or distributed Bragg reflection gratings were much easier to fabricate but also coupled light out of the surface in lower orders.

---

* The following references are contained in Evans and Hammer (1993) *Surface Emitting Semiconductor Lasers and Arrays*.

As a result, initial demonstrations of grating-surface-emission were reported simultaneously by groups from Xerox, (Burnham *et al.*, 1975), the A. F. Ioffe Physico-Technical Institute, (Alferov *et al.*, 1975), and IBM (Zory and Comerford, 1975). A few years later, results were published from Bell Northern, (Springthorpe, 1977) on surface-emitting lasers using 45° corner turning mirrors which were etched into the structure. These early demonstrations of surface emission remained relatively dormant until the 1980s, primarily because of fabrication difficulties and more pressing issues related to the performance of conventional semiconductor lasers. Three basic types of surface emitting lasers are briefly described in the following:

Vertical cavity surface-emitting laser

At present there are three basic configurations of surface-emitting lasers. One is the vertical cavity structure, in which the feedback mirrors are parallel to the top and botton surfaces of the semiconductor wafer as shown in Fig. 6.32(b). The active region can be as thick as several microns, or can be as small as a few tens of Angstrom units.

The vertical cavity surface emitting laser (VCSEL) has been extensively and continually developed since 1977 at the Tokyo Institute of Technology. In the last four or five years, many additional researchers from around the world have contributed to the development of vertical cavity lasers. Threshold currents of <1 mA have been reported along with packing densities of about one million lasers/cm^2. Because of their vertical emission small area, and low threshold, these devices are ideal for optical interconnects with low power consumption. VCSELs have also been considered for generating high power, and two dimensional arrays have been demonstrated (Orenstein *et al.*, 1991).

Grating outcoupled surface-emitting laser

An illustration of a grating-outcoupled surface-emitting (GSE) laser is shown in Fig. 6.32 (c). In these devices, the grating provides in-plane reflection for feedback for laser oscillation and also allows enough transmission to additional elements to provide optical coupling. The analysis of such gratings is quite complicated, and is discussed usually by coupled mode theory.

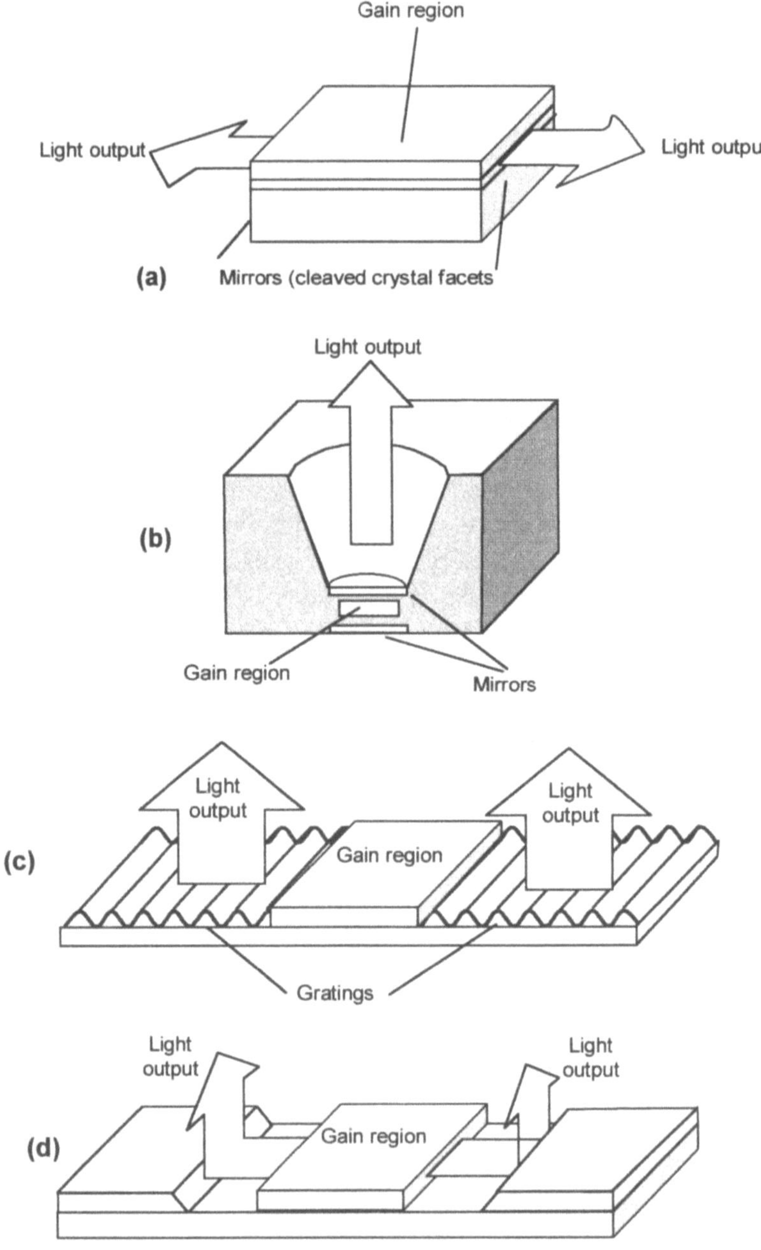

Fig. 6.32 (a) Conventional laser with edge emission; (b) vertical cavity surface emitting laser; (c) grating outcoupled surface emitting laser; (d) integrated deflector surface emitting laser (after Evans and Hammer).

In recent years, many research groups have pursued the development of several versions of GSE lasers. Because of a common, uninterrupted waveguide in all sections of a GSE laser wafer, monolithic integration into a coherent 2D array or with other planar optoelectronic devices is straightforward. In addition, a large fraction (>50%) of the two-dimensional surface can be optically emitting. Continous-wave powers of more than 3 W and peak powers of over 30 W have been reported for GSE arrays. Steering of the surface emitted beam has been demonstrated by electronic phase adjustment and by wavelength tuning.

Integrated beam deflector surface-emitting lasers

The final type of surface-emitting laser is known as an *integrated beam deflector laser* or *folded cavity laser*, and one version of this device is shown in Fig. 6.32 (d). In a common version of this device one or both perpendicular cleaved facets are replaced by an etched perpendicular facet and an etched integrated beam-deflecting mirror. This technology for surface emitters has been developed extensively during the last five years. The use of the mass-transport process unique to the InGaAsP system, resulted in the first high quality beam-deflecting mirrors and provided device performance equivalent to cleaved edge-emitting lasers. (Liau and Walpole, 1985). Although the major thrust in this area has been to fabricate incoherent arrays, a coherent two-dimensional array of etched facet lasers was demonstrated using an external dye laser for the master oscillator (Jansen *et al.*, 1989).

6.5 CONCLUSIONS AND OUTLOOK

The development of the diode laser started very slow soon after the realisation of the very first laser at all. After the invention of the heterostructure a highly successful period of improvement started. Eventually, very small devices with lowest thresholds, high efficiencies and long lifetimes were developed ideally suited for communications applications. Following the needs of lowest possible losses when propagating through silica fibers new versions of diode lasers were developed derived from the first successful GaAlAs/GaAs system, employing new material combinations even of quaternary or quintenary nature. The advent of molecular beam epitaxy equipment allowed highest precision and flexibility when realising superlattices or quantum wells.

Incorporating them into diode laser designs allowed to improve the specifications in several respects. Single mode operation could be demonstrated with substantial powers. Diode arrays allowed to make such lasers the most effective converters of electrical energy directly into optical power being at least very narry band although mainly incoherent.

Table 6.4 gives a summary of the state of the art of laser performance in different wavelength regimes based on different design approaches:

Table 6.4 Cw output power of matured semiconductor lasers (state of the art 1996)

Material and Substrate	Emission Wavelength	Transverse Single-Mode	100 μm Aperture	1 cm Bar
InAlGaP/GaAs	630 nm	60 mW	1.4 W	6 W
InAlGaP/GaAs	680 nm	295 mW	1.7 W	90 W
AlGaAs/GaAs	850 nm	300 mW	5.2 W	120 W
InGaAs/GaAs	980 nm	500 mW	5.0 W	
InGaAsP/InP	1.3–1.55 μm	380 mW		
InGaAsP/InP	1.8–2.0 μm	30 mW	0.5 W	8 W

High power laser radiation associated with good coherence properties can be yielded from resonant antiguiding arrays. Alternatively MOPA arrangements gained stability and performance when fabricated in a monolithic way containing DBR, active laser and power amplifier sections yielding up to 2 W cw (D.Welch, Spectra Diode Labs). Recent highlights have been i) the successful realisation of a GaN laser emitting $\lambda = 410$ nm, with room temperature pulsed operation at 50% duty cycle. ii) The demonstration of quantum cascade lasers between 5-8 μm, at room temperature under pulsed operation. iii) Extremely efficient VCSELs have been published at $\lambda = 800$-1100 nm, having wallplug efficiencies up to 57 % and maximun power of 40 mW (Ebeling, 1996).

The future of diode laser will be very bright manifested by a vast multitude of different designs offering steadily improved spectral or temperal, spatial or coherence properties covering a wavelength regime eventually stretching from the near UV to the middle IR regions.

Derived from such superb performance more and more applications will be possible. In a number of cases diode lasers will replace conventional lasers, diode pumped solid state lasers will be indispensable in material processing. Medical applications will stretch from soft biostimulation to dentistry or surgery. Endoscopic treatments with lasers will cause an increased need for well suited fibers in a wider wavelength regime than the silica fibre can offer today.

Communication lines, fibre bound or via free space propagation will be unthinkable without the dominating role of diode lasers as the optical sources. Data storage will yield even higher densities to be realised in an even further developed generation beyond the 1996 introduced DVD standard based on new blue diode lasers derived from the recently demonstrated GaN laser.

In summary, in a foreseeable future diode lasers will be introduced successfully in a very wide range of technological, metrological, medical and biological applications.

6.6 REFERENCES

Alferov, Zh. I. and Kazarinov, R. F. (1963) Author's certificate 1032155/26-25, USSR

Basov, N.G., Kroklin, O.N. and Popov, Y. M. (1961) Production of negative temperature states in p-n junctions of degenerate semiconductors. *Pis'ma Zh. Eksp. Theor. Fiz.*, **40**, p1879. Also, *Sov. Phys. JETP*, **13**, p1320 (1961)

Chai, B. H. T. (1995) private communication, June 1995

Einstein, A. (1917) Zur Quantentheorie der Strahlung, *Physikalische Zeitschrift*,**18**, p121

Faist, J. *et al.* (1996) *Appl. Phys. Letts.* **68,** p3680

Gunshor, R. L. and Nurmikko, A. V. (1995) Blue-green laser-diode technology moves ahead, *Laser Focus World*, March 1995, p97.

Hall, R. N. *et al.* (1962) Coherent light emission from GaAs p-n junctions. *Phys. Rev. Letts.* **9,** p366

Hayashi, I. (1970) Junction lasers which operate continously at room temperature, *Appl. Phys. Letts*. **17,** p109

Herman, T. C. Melngailis, I. (1974) Narrow Gap Semiconductors in *Applied Solid State Science Advances in Materials and Device Research,* **4,** (ed.Wolfe, R.) Academic Press, pp1-94, New York, London.

Kroemer, H. (1963) A proposed class of heterojunction injection lasers, *Proc. IEEE,* **51,** p1782

Maiman, T. (1960) *Nature,* 187, p493

Nathan, M. I. et al (1962) Stimulated emission of radiation from GaAs p-n junctions, *Appl. Phys. Letts.,* **1,** p62

Quist, T. M. *et al.* (1962) Semiconductor maser of GaAs, *Appl. Phys.Letts.,* **1,** p91.

Books and reviews recommended for further information:

Agrawal, G. P. and Dutta, N. K. (1986) *Long Wavelength Semiconductor Lasers,* Van Nostrand Reinhold, New York 1986

Botez, D. and Scifres, D.R. (1994) Diode Laser Arrays in *Cambridge Studies in Modern Optics,* **14,** Cambridge University Press (ISBN 0-521-41975-1)

Buus, J. (1991) Single Frequency Semiconductor Lasers in *Tutorial texts in optical engineering,* **TT5,** SPIE Optical Engineering Press, Bellingham (ISBN 0-8194-0535-3)

Eberling, K. J. (1989) *Integrierte Optoelektronik,* Springer, Berlin

Eberling, K.J. (1996) *Semiconductor Lasers: Recent Advances and Prospects,* Tutorial talk at ECLEO, Hamburg

Evans, G.A. and Hammer, J.M.(1993) Surface Emitting Semiconductor Lasers and Arrays in *Quantum Electronics - Principles and Applications,* Academic Press, Boston (ISBN 0-12-244070-6)

Fukuda, M. (1991) *Reliability and Degradation of Semiconductor Lasers and LEDS,* Artech House, Boston, (ISBN 0-89006-465-2)

IEEE Journal of Quantum Electronics, **29,** June 1993 (Special Issue on Semiconductor Lasers)

IEEE Selected Topics in Quantum Electronics, **1**, June 1995 (Special Issue on Semiconductor Lasers)

Mroziewics, B. Bugajski M. and Nakwaski, W. (1991) *Physics of Semiconductor Lasers*, North-Holland

Pearsall, T.P. (1982) *GaInAsP Alloy Semiconductors*, John Wiley & Sons, Chichester (ISBN 0-471-10119-2).

Petermann, K. (1988) *Laser Diode Modulation and Noise*, Kluwer Academic Publishers, Dordrecht

Proceedings of 14th IEEE International Semiconductor Laser Conference (Maui, Hawaii, USA), 1994, Institute of Electrical and Electronics Engineers 1994

Thompson, G.H.B. (1980) *Physics of Semiconductor Laser Devices*, John Wiley & Sons, Chichester, (ISBN 0-471-27685-5)

Vasil'ev, P. *Ultrafast diode lasers*, Artec House, Boston, 1995 (ISBN 0-89006-736-8).

Welch, D., Spectra Diode Labs was cited by Eberling (1996)

Zory, P.S. Jr. (editor), (1993) *Quantum Well Lasers*, Academic Press, New York

7

Safety

J. M. Green

7.1 INTRODUCTION

Lasers find a wide usage in general industry and engineering, from measurement applications to materials processing. When discussing safety in this context it is useful to refer to the combination of a laser, beam delivery optics, workpiece manipulation equipment and the guarding, fume extraction etc. associated with this integrated system as making up a **Laser Processing Machine**.

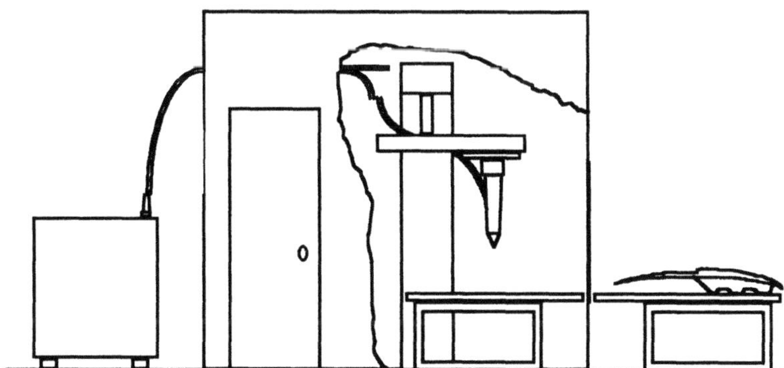

Fig. 7.1 Laser Processing Machine.

The essential safety philosophy for all machines is embodied in the essential health and safety requirements of the EC Machinery Directive. Standards have been written to satisfy these requirements for certain classes of machine, including one for Laser Processing Machines. A basic requirement of the Laser Processing Machine standard EN12266 is that ***the possibility of exposure of people to levels of radiation exceeding the AEL of Class 1 shall be eliminated during production (normal or otherwise)***, thus ensuring that when the machine is in normal use the user need take no further measures to address the laser radiation hazard. The Class 1 requirement is normally taken to mean that the laser radiation is totally enclosed.

Laser Processing Machines also need to be designed to address the other hazards associated with their use, which may include toxic fume generated during processing, electrical hazards associated with the laser source, mechanical hazards associated with the workpiece manipulation and high pressure gas hazards associated with the laser processing head. All of these hazards are commonly encountered in the industrial environment and there are standards and techniques/ equipment to help deal with them.

As with all machinery, routine maintenance and occasional servicing is required to keep the equipment safe. Safety management is usually dealt with by the user appointing a Laser Safety Officer to deal with the approval of servicing procedures and ensure that regular safety inspections and testing of equipment takes place. Occasional servicing of the laser and beam delivery components, which can involve operation of the laser with guarding removed, is often sub-contracted to the laser supplier.

Most of the larger integrated laser machines are one-off devices and this fact more than any other accounts for why most laser manufactureres do not get involved in such systems integration. In any event, the purchaser may choose to commission a company specialising in workpiece handling and automation to be the systems integrator, and where this happens, there is a danger of lack of sufficient attention to laser safety requirements, particularly the enclosure of the laser radiation which is scattered from the workpiece. This can lead to costly modifications later.

In this chapter, firstly the hazards associated with the use of high laser powers for materials processing are reviewed. Secondly, the requirements of current laser standards and legislation are outlined, and finally, some general guidance for the industrial laser user is provided.

7.2 BIOLOGICAL HAZARDS

7.2.1   Introduction

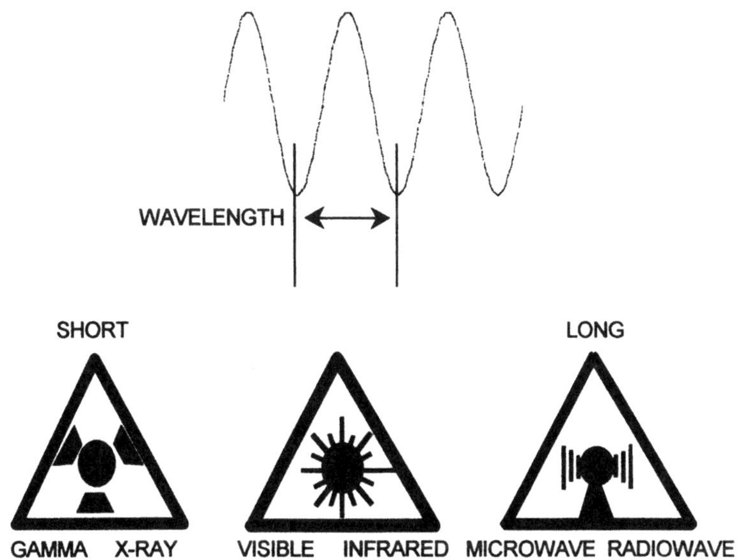

Fig. 7.2 Laser radiation hazard.

The laser is just one of many types of man-made sources of electromagnetic radiation. Electromagnetic radiation includes Gamma and X-Rays, ultravolet radiation, visible radiation (which we call light), infrared radiation, micro- and radio- waves. The feature which distinguishes these different types of radiation is wavelength.

There are many different kinds of laser. The full range of wavelengths that these different types of lasers emit is limited to ultraviolet, visible and infra-red. This is the so called non- ionising part of the Electromagnetic spectrum. Starting at the shortest laser wavelengths encountered in industrial devices and progressing to increasingly longer wavelengths: Excimer lasers emit UV radiation, Copper Vapour Lasers and doubled Nd:YAG emit visible radiation, otherwise known as *light*, the Nd:YAG laser emits near-infrared radiation while the CO_2 laser is a source of far-infrared radiation. All these wavelengths are strongly absorbed by the skin, which protects all the bodily organs except for the eyes. There are three principal distinguishing features of laser radiation:-

1. Laser radiation is said to be monochromatic. The so-called **conventional** sources emit radiation with a much greater spread of wavelengths than lasers do.
2. Laser radiation is coherent. Lasers emit as if the source of the radiation is an intense point source. Because of this
    (a) laser radiation can travel as a pencil-like beam over great distances
    (b) laser radiation is capable of being focused to produce very high intensities.
3. Some lasers produce very intense pulses of laser radiation of very short duration.

7.2.2 Radiation hazards in general

Lasers which emit ultraviolet radiation can cause photochemical damage to the skin and eyes. Lasers emitting visible or infrared radiation can cause only thermal (burn) damage. For damage to occur the radiation must be absorbed in the tissue. he case of the eye, the cornea, aqueous humour, lens and vitreous humour protect the retina except in the visible and near infra red regions. The retina is at risk only to wavelengths from 0.4 - 1.4 µm.

Ocular exposure to ultraviolet radiation

UV radiation causes photochemical damage. Sunburn is an example of such damage. When it occurs to the cornea of the eye it is known as arc-eye by welders and as snow blindness to skiers! The injury can be very painful, but damaged cells on the cornea are replaced and for mild over-exposure the healing process is complete in about two days. However, UV radiation at some wavelengths can penetrate to the jelly-like lens of the eye where photochemical damage can produce a cataract. The Xenon Chloride Excimer laser at 308 nm is particularly hazardous in this respect.

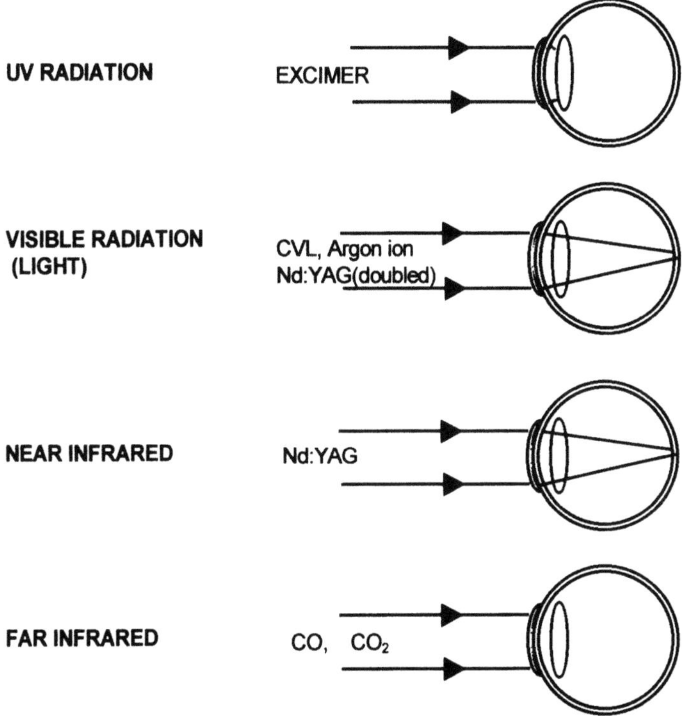

Fig. 7.3 Ocular exposure.

Ocular exposure to visible and near infrared radiation

Radiation in the 0.4–1.4 μm wavelength range is focused by the eye onto the retina. The sun and some powerful arc lamps can produce retinal burns, even for short exposures, but even low power laser radiation in this wavelength region is a hazard

The eye has a natural aversion response to protect it from bright light. The sun produces a retinal image of brightness about 100,000 times that of outdoor daylight. If one looks into it accidentally a protective involuntary response to bright light causes you to turn your head away. This blink reflex takes less than 1/4 second.

Another protective involuntary but relatively slow response to bright light is the reduction in pupil size. It can decrease to about 2 mm diameter, but no further.

Ocular exposure to medium and far infrared radiation

Radiation of wavelength longer than 1.4 µm is fully absorbed in the cornea. Thermal burns to the cornea can therefore result from over-exposure. By contrast to the photochemical damage, thermal damage to the cornea does not accumulate between periods of exposure but is immediate. Although corneal grafts can be performed, such ocular damage must be considered to be serious.

Skin damage

Skin damage from radiation is something that everyone has experienced, if only as sun burn from photochemical damage by UV and thermal burns to the legs from infrared radiation by sitting too close to a fire.

The skin is highly absorbent: the very thin (10-20 µm thick) outer layer of dead cells (the epidermis) strongly absorb long wavelength infrared, such as is emitted from a carbon dioxide laser, shorter wavelengths down to about 0.7µm (red) penetrate increasingly further, but less than 1 mm, and for yet shorter wavelengths the penetration depth again decreases until in the ultraviolet the outer layer of dead cells absorb virtually all the radiation

7.2.3 Laser radiation hazards

Everything said so far applies equally to the radiation from lasers as from conventional sources. Just as the use of UV and infrared lamps needs to be carefully controlled to prevent skin burns, the radiation from Excimer and CO_2 lasers poses a similar hazard to the skin and the cornea of the eye, the distinction being that the radiation from the laser is more intense and localised and poses a hazard over a much greater distance because it travels as a beam. Unlike UV and far infrared radiation, visible and near infrared radiation reaching the eye passes through the cornea and is focused by the lens onto the retina. Because of the unique focusing properties of laser radiation, the eye hazard presented by visible and near-infrared lasers is orders of magnitude worse than it is for conventional bright light sources.

The effects described above are referred to as threshold effects. It need hardly be pointed out that the skin and cornea of the eye provide insufficient protection against high power laser radiation! The particular characteristics of laser radiation in this context, some of which were mentioned above, are discussed below.

Hazard range

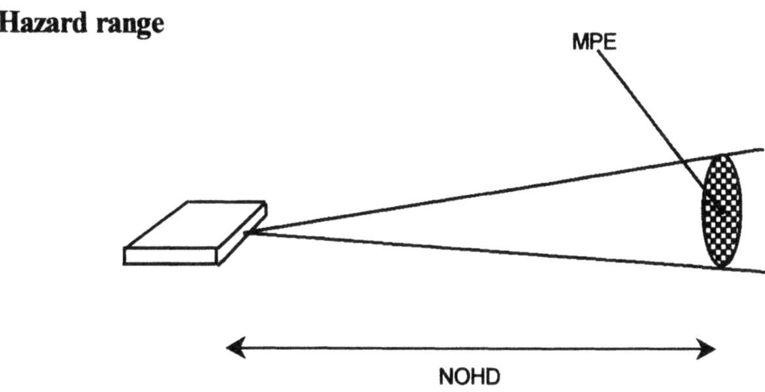

Fig. 7.4 Nominal ocular hazard distance (NOHD).

Laser beams normally have a low divergence. Consequently the hazard to eyes and skin is approximately independent of distance, although of course as the beam will slowly expand in cross-section the hazard will eventually decrease. This contrasts sharply with the rapid decrease with distance of **skin and corneal hazards** associated with conventional sources, caused by the radiation spreading out in a cone of wide angle.

Intensity

Lasers used for materials processing can emit radiation powers of several kilowatts confined within a beam diameter of, say, 20 mm. Focusing the radiation produces a tremendous increase in intensity, but even unfocused the intensity of the radiation within the beam of even a quite modest laser is much greater than that produced by all but the most intense conventional source. An example of this is the laser pointer which puts out less than 1 mW of red laser light, yet the red spot on a screen can be clearly seen even in the presence of bright room lighting. This consideration is particularly important in relation to skin and corneal hazards, while brightness discussed below is strictly the more appropriate parameter in relation to the retinal hazard.

Brightness

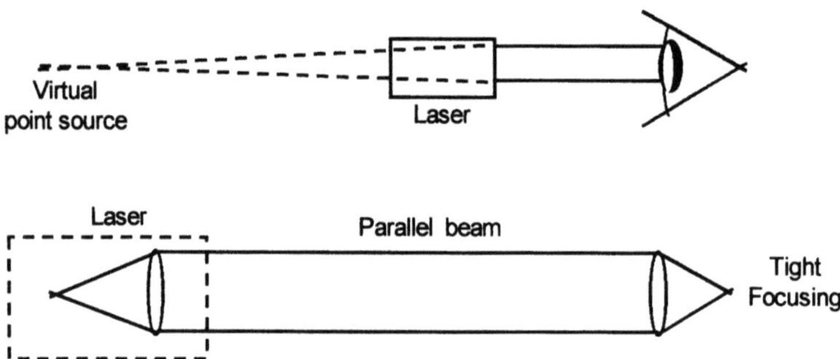

Fig. 7.5 Lasers are point sources.

Lasers can cause localised skin and corneal burns, but for lasers emitting radiation in the wavelength range 0.4–1.4 μm the main threat is retinal damage. In this range the eye acts as a focusing system. It has already been noted that the property of coherence means that laser beams appear to originate from point sources, as long as the eye can accommodate to the radiation a retinal image about 20 μm diameter will be formed - the smallest image the lens of the eye is capable of producing. The combination of intensity and focusability make up the parameter known as brightness: lasers are the brightest known sources of radiation.

Light is radiation in the wavelength range 0.4–0.7 μm. Radiation from 0.7–1.4 μm reaches the retina but is not directly sensed by the eye, not even as a pain sensation. Nd:YAG laser radiation lies in this most hazardous wavelength range in which damage can be done without the victim being aware.

The brain automatically orientates to image light from the object of principal interest in the field of view onto the fovea and so it is this region of the retina which is most at risk. Only the fovea produces a clear retinal image. Damage to the fovea cannot be compensated for by other parts of the retina, and a person with a laser damaged fovea could therefore be registered blind in one eye. On the other hand, damage to other parts of the retina will only affect peripheral vision and the brain can quickly compensate for such damage. (Diabetics receiving laser lesions to the retina to help combat glaucoma do not suffer noticeable visual impairment as a result). Damage to the blind spot (optic nerve) could cause total blindness.

The blink reflex of the eye does not provide adequate defence against:

- purposeful staring (as happens during the watching of solar eclipses)
- very bright sources.
- sources of radiation in the 700–1400 nm region: the retina is at risk but does not sense the radiation directly, nor does it sense pain from injury.
- pulsed laser radiation.

If radiation is coherent it can be imaged as a point, but if radiation is diffusely scattered (e.g. off a rough surface) its coherence is destroyed. Examples are visible laser beams producing spots or patterns on screens or passing through smoke, as used in entertainment and display applications. Specular reflections off a shiny surface do not destroy the laser beam's coherence. Mirrors and lenses can cause the beam to expand rapidly, thus decreasing its hazard range, but not its coherence.

Pulsed operation

Laser pulses are typically so short in duration, 1 ms or less, that whatever damage they inflict singly is done before the victim can respond. If sufficient energy is contained within the laser pulse the rapid heating of the tissue which absorbs it will produce strong mechanical forces and thermomechanical damage can occur. This phenomenon is particularly important in the context of the focused beam on the retina of the eye where a micro explosion which causes damage over a much greater area than is irradiated by the laser.

Principal Nd:YAG laser radiation Hazards

- Risk of loss of sight
- >1,000,000 x Safe Threshold are possible
- The Radiation is invisible
- The Radiation is not felt as it damages the retina

Radiation in the 0.7 - 1.4 μm wavelength range is particularly dangerous*

7.2.4 Associated hazards

Fume hazard

Over the last few years much progress has been made in understanding the nature of the fume created during laser materials processing. A characteristic of the particle content of such fume is a relatively high concentration of small particles, of order 0.1 - 1µm diameter, which can be inhaled deep into the lungs and retained. Chemical composition and particle size distribution and morphology have been quantified for some of the wide range of laser/assist gas/material/process combinations of industrial interest but the assessment of toxicity is another matter. The current evidence points to the need for effective fume extraction, especially during the processing of organic materials.

Metals Apart from the shift to small particle size, the fume produced during laser cutting and welding of metals appears very similar to that produced during other forms of thermal processing, such as flame cutting and MMA welding. Metal fume fever has for many years been a very common disease among welders, who conventionally have their heads close to the process, whereas at least in laser processing the operator is at a greater distance. TLVs exist for almost all metals but among these Chromium and Nickel, as common elements and putative carcinogens, deserve particular mention.

Inorganics In some applications, such as the cutting of glass, small inert fibrous dust can be produced which, by analogy with Asbestos, must be regarded as a potential carcinogen. Dust generally is a problem when cutting such materials as alumina, quartz, marble and limestone.

Organics This heading includes, in particular, wood and plastics. Despite the extensive work undertaken to date, it is still not possible to make accurate predictions of fume composition and concentration in the majority of cases. The fume often contains benzene and other aromatics which are known carcinogens, together with a large number of other bi-products. As well as carcinogenic concerns, the laser cutting of halogenated polymers, especially PVC, produces copious quantities of hydrogen chloride. This toxic gas is also a severe eye and throat irritant and will rapidly rust surrounding laser equipment. The general conclusion is that some materials, in particular PVC, should not be processed by laser.

Fume extraction and filtration Efficient local extraction of fume is not always a viable option. In the case of multi-axis robotic control of the laser beam delivery, for example, peripheral extraction, involving much larger volume flow rates of air is then needed. Special vacuum cleaners are available to remove potentially toxic heavy particulate (dust) that may collect on the laser equipment and on the floor.

It is arguable whether or not filtration of the laser fume is always necessary on environmental grounds where the extracted air is vented to atmosphere. However, the issue must be addressed in cases where the user requires the air to be recycled in order to reduce drafts and heating bills, or as in the food industry, where factory areas are sealed.

A wide range of filter types, including HEPA and electrostatic filters in wet and dry forms are available to process fume. Some filters are self cleaning and designed for maintenance with toxic products in mind, but proper disposal of the collected contaminated material can still be a problem. While solutions can be found for most individual cases, and occasionally certain classes of materials, there is no one form of filter that will deal effectively with all types of fume.

The filtering of plastics fume is a particular problem. Prefilters are required to achieve a practical lifetime for HEPA filters, while electrostatic filters quickly build up a coating which is then difficult to remove. In addition, water baths or cooling to remove any condensable vapours and selective chemical reactions to deal with non-condensable components of the plastics fume may be needed.

Electrical

Lasers are inefficient devices, the most efficient of the industrial lasers being the CO_2 laser at about 5% efficiency. 100 kW power supplies are therefore not uncommon in the industrial laser context. Depending on the laser type, the power supply may hold large amounts of stored energy at high voltage and proper earth connections for the supply and an earth stick for local earthing of components during servicing may be essential.

Other

There are other hazards which may need to be considered in the industrial laser context. These include:

Mechanical hazards, ranging from the sudden movement of robot arms to traps associated with XY tables.

High pressure gases to assist cutting and welding processes or supplied to gas lasers (including halogens in Excimer lasers, carbon monoxide and hydrogen in CO_2 lasers).

Fire from any flammable material exposed to high power laser radiation

Secondary radiation including the intense UV and visible radiation produced during processing.

All of these hazards are well known and little more needs to be said in the current context.

7.3 HAZARD CLASSIFICATION AND MPES

Two important terms

MAXIMUM PERMISSIBLE EXPOSURE (MPE)
That level of radiation to which, under normal circumstances,
persons may be exposed without suffering adverse effects.

ACCESSIBLE EMISSION LIMIT (AEL)
The maximum accessible emission level permitted within
a particular class

What constitutes a Safe level of exposure?

1. Base the assessment on the total output of the laser.

2. Choose a time basis of 30000 s generally, but 100 s for unintentional
   viewing of radiation provided it is not UV (i.e. < 400 nm)

Example: cw CO_2 laser radiation
$\quad\quad\quad$ (MPE(30000 s) = 1000 W/m^2; limiting aperture = 11 mm)
$\quad\quad\quad$ safe level of emission = 1000 x π $(0.011)^2/4$ = 0.1 W

$\quad\quad\quad$ This is a Class 1 acessible emission limit (AEL)
$\quad\quad\quad$ for a CW CO_2 laser

In general:
$\quad\quad\quad$ Class 1 AEL = MPE x Area of limiting aperture

7.3.1 Laser Hazard Classes

Class 1

Class 1 lasers are safe under reasonably foreseeable circumstances, either
because the output of the laser is so low or because of engineering
safeguards. For example, the accessible power of a CW Class 1 laser in the
visible region must not exceed 0.4 µW.

CLASS 1   NO RISK

Safe under reasonably foreseeable conditions

inherently safe (very low power radiation)
or
safe by engineering design (total enclosure)

Can apply to all wavelengths, pulsed and CW. The AEL is such that the
30000 s or100 s. MPE is reached at the focus of a lens
having a 50mm Ø clear aperture

Most commercial industrial laser systems are sold as Class 1 products: safe by engineering design. They contain Class 4 lasers, but all the laser radiation is sealed in. Such a system is referred to as an embedded laser, Class 1 by engineering design.

Class 2

CLASS 2   LOW RISK

Eye protection is afforded by the blink reflex

Can apply only to visible wavelengths and repetitively pulsed or CW lasers
The AEL is 1mW. The 0.25 s MPE is reached at the focus of a lens
with a 50 mm Ø clear aperture

Class 2 lasers are low power devices emitting visible (400-700 nm) radiation. They are not intrinsically safe but protection is afforded by the blink reflex. For CW lasers the power must not exceed 1 mW.

Class 2 is the only Class which applies only to CW and repetitively pulsed lasers and only to visible lasers. The Nd:YAG and carbon dioxide lasers cannot be Class 2 because they produce invisible beams.

Class 3A

CLASS 3A   LOW RISK

Safe provided that external optics are not used to concentrate the laser radiation. For visible lasers, eye protection is afforded by the blink reflex.

Can apply to any wavelength, pulsed and CW,
though principally to visible CW lasers.
The AEL is made up of two parts (1) The appropriate MPE and (2) 5 times the Class 1 or Class 2 AEL, e.g. 5mW and 25.4 Wm^{-2} for visible CW laser radiation.

Class 3A lasers may emit visible and/or invisible radiation. The requirement for Class 3A is that hazardous levels of radiation are not present in the unfocused beam. But unlike the situation for Class 1 and Class 2 lasers, there may be a radiation hazard **if the radiation is focused**. The laser is therefore safe, either by virtue of the blink reflex if the radiation is visible or intrinsically safe if not, provided the radiation is not concentrated into the eye, as would be the case of a person were using a telescope or binoculars at the time of exposure.

Class 3B

CLASS 3B   MEDIUM RISK

Direct viewing is hazardous
Only diffuse reflections of the unfocussed beam is safe to view
(for distances > 13cm, time < 10s)

Applies to all wavelengths, pulsed and CW.
The AEL is 0.5W for CW laser radiation,
except in UV B and C.

Class 3B lasers mark the border between medium power and high power. In qualitative terms the beam from a visible Class 3B laser is generally safe to view off a diffuse surface for times less than 10 seconds and distances greater than 130 mm, but the direct beam is hazardous even allowing for the blink reflex. The power of a Class 3B CW laser emitting visible laser radiation must not exceed 0.5 W. The heating effect of 0.5 W CW laser beam on the hand can be painful, but such a laser beam should not set fire to a piece of paper, even if focused.

Class 4

Class 4 lasers are devices exceeding the limits for Class 3B. There is no upper power limit for Class 4. Such lasers can be a fire hazard. Even diffuse reflected radiation from a Class 4 visible laser can be hazardous to the eyes.

CLASS 4   HIGH RISK

Even diffuse reflections may be hazardous.

Applies to all wavelengths, pulsed and CW. There is no upper limit to this Class and hence no AEL.

All lasers powerful enough to be used for materials processing are Class 4, although the laser may be embedded and the product made Class 1, as discussed above.

Limitations to the classification scheme

The laser hazard classification scheme can at best be a crude guide to safe laser use. The classification scheme relates only to the safety of the product in regard to laser radiation emitted during normal operation. Embedded laser products can operate in a higher classification during maintenance and servicing operations, and of course to say a laser is **Class 1** says nothing about the non-laser hazards of the product.

The main benefit of making a product Class 1 is that such a product can be used without implementing laser radiation safety precautions. The product could, for example, be used on a factory floor without those around the laser having to use eye protectors.

7.3.2 Maximum permissible exposure

Maximum Permissible Exposure (MPE) is a term used to describe the maximum safe exposure to laser radiation. It ***is that level of laser radiation to which, under normal circumstances, persons may be exposed without suffering adverse effects***. Laser Safety Standards tabulate MPEs, and the same values are broadly accepted world-wide. The values cover exposure of the skin and eyes to laser wavelengths from the UV to the far Infra red, from 180 nm to 1 mm wavelength.

MPEs relate only to acute effects, such as photochemical and thermal burns which become apparent during or shortly after exposure. The effects of exposure are limited to 30,000 seconds, an eight hour working day. Chronic (i.e. long term, cumulative, multiple exposure) effects such as might give rise to cataracts and the blue light hazard are ignored in setting MPE values.

Because the interaction of radiation with tissue in the wavelength range applicable to lasers is so complex, MPE values vary with both wavelength and exposure duration. For both eye and skin exposure, within the context of what may be regarded as **normal conditions** the worst case is assumed in the MPE tables. Raw damage threshold data is plotted on a probate curve. The 50% probability point is taken and then a safety factor introduced (for example, the MPE for 0.25 s exposure at visible wavelengths is 25 Wm^{-2}; this corresponds to about a probability of injury of about 10^{-4}, rising to 50% at 125 Wm^{-2}). It is also worth noting that the severity of injury upon exceeding the MPE is not the same at all wavelengths. For example, in the UV and far infrared regions the damage to the eye is to the cornea and is reversible whereas in the visible and near infrared the damage is to the retina and can result in permanent visual impairment.

The MPE at a particular laser wavelength depends on the duration of exposure, and in the case of a pulsed laser to the duration and pulse repetition frequency of the laser. Table 7.1 overleaf gives some representative values.

Table 7.1 Examples of MPE values for industrial lasers

Laser output	Exposure conditions + laser pulse duration and PRF where applicable	MPE eye exposure	MPE skin exposure
Excimer 248 nm (Krypton Fluoride)	Single pulse 10 ns pulse duration	30 Jm^{-2} per pulse	30 Jm^{-2} per pulse
Excimer 248 nm (Krypton Fluoride) repetitively pulsed	3×10^4 sec exposure 100 Hz, 10 ns duration	10 μJm^{-2} per pulse	10 μJm^{-2} per pulse
Helium-Cadmium CW (325 nm)	$3\ 10^4$ sec exposure	10 Wm^{-2}	10 Wm^{-2}
Argon Ion CW (514 nm)	0.25 sec exposure	25 Wm^{-2}	7800 Wm^{-2}
Nd:YAG CW	10 sec continuous exposure	51 Wm^{-2}	2000 Wm^{-2}
Nd:YAG (normal operation)	single pulse, 1 ms duration	0.5 Jm^{-2} per pulse	2000 Jm^{-2} per pulse
Nd:YAG (Q switched)	single pulse 0.1 μs duration	0.05 Jm^{-2} per pulse	200 Jm^{-2} per pulse
Nd:YAG Repetitively pulsed (mode locked)	$3\ 10^4$ sec exposure 100 Hz, 0.1μs duration	0.0012 Jm^{-2} per pulse	4.8 Jm^{-2} per pulse
Nd:YAG Repetitively pulsed (mode locked)	$3\ 10^4$ sec exposure 10 kHz, 0.1μs duration	0.004 Jm^{-2} per pulse	0.2 Jm^{-2} per pulse
CO_2 CW	10 sec continuous exposure	1000 Wm^{-2}	1000 Wm^{-2}
CO_2 TEA repetitively pulsed	10 sec exposure 1000 Hz, 10 μs duration	1 Jm^{-2} per pulse	1 Jm^{-2} per pulse

7.4 STANDARDS AND LEGISLATIVE REQUIREMENTS

7.4.1 Summary of the current situation

There is no legislation specific to the general operation of lasers. There is, however, legislation which has a direct bearing and there are laser product safety standards, one of which includes user guidance. The USA has a specific user standard, ANSI Z136.1 **American national standard for the safe use of lasers**, but this standard does not have an international or European standard counterpart.

European Directives and the regulations which implement them are beginning to have a major impact on the approach to industrial safety, including laser safety. Commercial laser equipment, like most electrical equipment, now has to conform with the Low-Voltage Directive and the EMC Directive. Laser Processing Machines, like all machines, have also to conform to the Machinery Directive. Compliance with standards is the common way of satisfying the requirements of Directives, but in the case of Laser Machines the standard, EN 12266 **Safety of machines using laser radiation to process materials**, is not yet in force.

The principal laser safety standard is EN 60825-1:1994 **Safety of laser products - Part 1: Equipment classification, requirements and user's guide** which introduces a number of important changes to its predecessor (EN 60825: 1992 **Radiation safety of laser products, equipment classification, requirements and user guide**), not least of which is additional user guidance related to high power lasers and an informative annex entitled "High power laser considerations particularly appropriate to high power lasers " which sets out the common causes of errant laser beams and appropriate preventative measures as well as some basic issues relating to the design of protective housings for high power lasers.

7.4.2 Legislation summary

Employers

European Health & Safety Directives have to be enacted in the form of regulations in each member state by an agreed date. The exact naming of regulations may vary from state to state, but in the UK the main regulations that affect laser use are:

1. **Management of Health and Safety at Work Regulations: 1992**. These regulations set out a formal procedure of risk assessment.
2. **Provision and Use of Work Equipment Regulations: 1992**. These regulations include requirements for guarding and control systems for machinery.
3. **Electricity at Work Regulations: 1989**. These regulations deal with the electrical supply to the laser.
4. **Control of Substances Hazardous to Health (COSHH) Regulations: 1988**. These regulations apply to materials to be processed, solvents for cleaning and chemicals needed by the laser.
5. **Personal Protective Equipment at Work Regulations: 1992**. These regulations cover the supply and care of laser protective eyewear.

For manufacturers

The main regulations which influence the design of laser equipment are:

1. **Supply of Machinery (Safety) Regulations: 1992**. These regulations set out general safety requirements for machines and includes dealing with radiation hazards.
2. **Low Voltage Electrical Equipment (Safety) Regulations: 1989**. These regulations set out general safety requirements for equipment operating with > 50 V electrical supply.
3. **Electromagnetic Compatibility Regulations: 1992**. These regulations set out limits for the emission and response to electromagnetic radiation.
4. **Personal Protective Equipment Regulations: 1992**. These regulations set out general requirements for the testing and marking of laser protective eyewear.

The Machinery Directive came into force on 1 January 1993 and the transitional arrangements, which exempt manufacturers from the requirements of the Directive, ended on 31 December 1994.

The principal requirements of the Machinery Directive are:

1. The machinery must satisfy the Essential Health and Safety Requirements specified in the Directive.
2. A Technical File must exist or be capable of being put together quickly.
3. The Responsible Person must issue an EC Declaration of Conformity.
4. The CE mark must be fixed to the machine (which implies that the machinery must satisfy all other relevant directives).
5. The machine must be safe.

7.4.3 Summary of standards

Below are listed the three standards dealing directly with laser radiation. There are many other potentially relevant standards dealing with general aspects of electrical and mechanical safety and electromagnetic compatibility:

1. **EN60825-1: 1994 Safety in laser products - part I: Equipment classification, requirements and user's guide**. This is the current European safety standard setting requirements for laser products and offering guidance for users.
2. **EN207 Eye Protection Against Lasers (laser safety eye protectors)**
3. **EN208 Eye Protection Against Lasers (laser adjustment protectors).** These standards apply only to laser safety eyewear and will probably be called up by the PPE directive. They do not apply to viewing windows in laser enclosures.
4. **EN 12266 Safety of Machines using Lasers to Process Materials.** This draft standard is the implementation of the Machinery Safety Directive as it applies to laser processing machines.

In addition, a range of relevant **harmonised** standards have been produced by CEN and CENELEC to assist in the interpretation of the Essential Health and Safety Requirements of the Machinery (Safety) Directive. The basic standard in this context is EN 292 -Safety of Machinery - Basic concepts, general principles for design+. In contrast to EN 60825-1, EN 292 deals openly with the concept of risk assessment and risk reduction.

7.4.4 EN 60825-1 Safety of laser products - Part 1: Equipment classification, requirements and user's guide

Manufacturers requirements

EN60825-1: 1994 should be seen as a **highway code** for the laser user and manufacturer. It replaces EN60825: 1991 and is the current Euronorm on laser safety. Major changes introduced by this new standard are:

1. Inclusion of LEDs
2. Measurement conditions for Classification: 100s period, minimum distance for measurement, extended sources, relaxed definition of accessible radiation.
3. Emission indicators at remote apertures
4. Requirements for walk-in workstations
5. Revised MPE tables: relaxation for mid-infrared, change to limiting apertures, new treatment of extended sources
6. Guidelines for protective housing for high power lasers.

The standard contains three main sections plus annexes:
Section one: General.
Section two: Manufacturing requirements.
Section three: User's guide.

The three major Annexes give examples of MPE and hazard range calculations, describe the biological effects of laser radiation and summarise considerations for protective housings for high power lasers.

Manufacturer requirements of EN 60825-1

Manufacturer

The individual or organisation that assembles the product at the highest level of integration.
*A **user** significantly modifying a previously classified laser product becomes a manufacturer*

To comply with the standard, a manufacturer or supplier must provide the following:

1. A **protective housing** to enclose the laser.
2. **Interlocking** of any removable panels on the protective housing. The design of the interlock depends on the hazard and the situation. Positive break switches and captive key operated overrides are common.
3. **Remote interlock connector**
4. **Key control**
5. **Laser radiation emission warning**, generally a light of fail safe design on the laser which comes on when the laser is in a state ready to operate. Emission indicators at remote locations are required for Class 3B and 4 lasers.
6. **Beam stop or attenuator**
7. **Labelling**. There are a total of four forms of labels and signs that are used in general:
    The universal laser radiation warning sign.
    A general classification and explanatory rectangle
    Warning labels at apertures and panels.
    Radiation output information.

To illustrate the application of the standard the manufacturers requirements for a Class 1 (totally enclosed) **walk-in** laser workstation incorporating a Class 4 laser are considered.

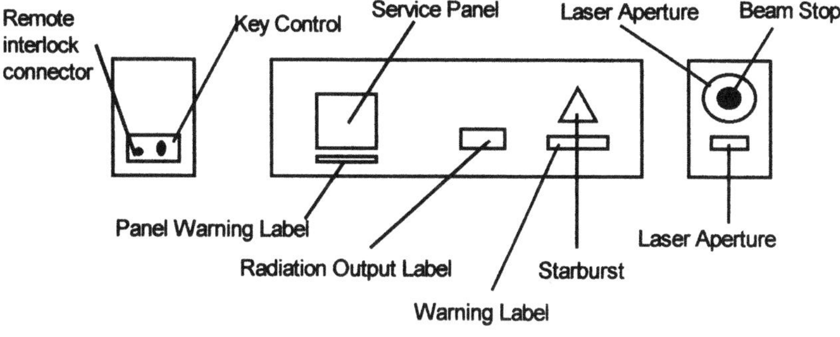

Fig. 7.6 Class 4 laser product.

Although more details of the workstation would be required for an accurate interpretation of the standard, typical requirements are:

1. **Protective housing** (in this case, including the workstation walls and ceiling) to enclose the laser product and prevent the escape of laser radiation in excess of Class 1.
2. **Safety interlock** on the door(s) and any access panels on the workstation which give access to laser radiation levels in excess of Class 3A.
3. **Remote interlock connector** on the Class 4 laser.
4. **Key operated master control** on the Class 4 laser.
5. **Beam stop or attenuator** - generally a safety shutter in the beam line.
6. **Laser radiation emission warning** - generally a light of fail safe design which comes on when the laser is in a state ready to operate.
    This or some other such device must provide adequate warning to a person entering the housing.
7. **Safety Shutter key** or some other means for someone entering the housing to prevent unintentional activation of the laser.
8. **Alignment aid** - a safe means of beamline alignment maintenance, generally a collinear low power visible laser beam.
9. **Labelling**. The following signs and labels would be expected:
    A **CLASS 1 LASER PRODUCT** label on the outside of the workstation.
    A radiation output and standards information explanatory label on the Class 4 laser.
    A **INVISIBLE LASER RADIATION/AVOID EYE OR SKIN EXPOSURE TO DIRECT OR SCATTERED RADIATION/CLASS 4 LASER PRODUCT** label on the Class 4 laser.
    A **LASER APERTURE** label, most likely at the end of the beamline where the laser beam leaves to meet the workpiece.
10. **User information** including instructions for maintenance.
11. **Servicing information** including instructions for adjustments and procedures.

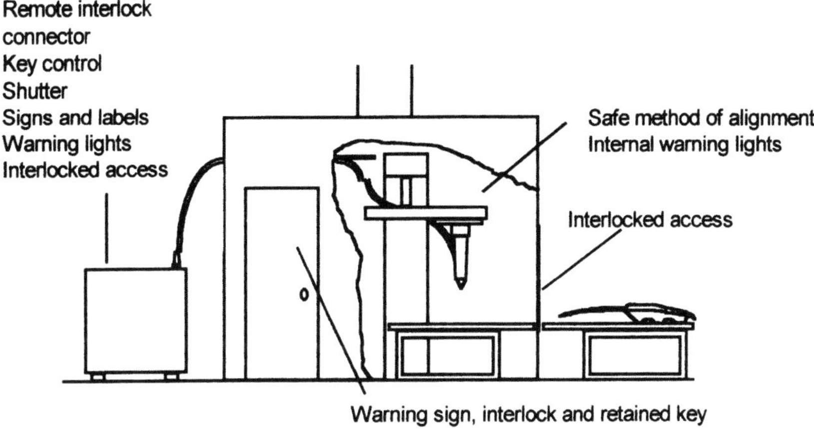

Fig. 7.7 Key features of EN60825-1

User guidance

The User Section is divided into four clauses:

1. Safety precautions
2. Hazards incidental to laser operation
3. Procedures for hazard control
4. Maximum permissible exposure

The first user clause

This deals with the general aspects of user precautions for laser radiation protection.

1. Appointment of a laser safety officer (LSO).
2. List of precautions/ control measures by class of laser.

The second user clause

This deals superficially with associated hazards.

The third user clause

This is a combination of hazard evaluation and implementation of control measures.

Controlling the hazard zone of open beams

1. Beam not at eye height (N.B... chairs, steps)
2. Beams not directed at personnel or doorways
3. Secure laser and beam steering components
4. Terminate all primary & secondary beams
5. Enclose where possible
6. Use Interlocks and barriers wherever possible
7. Beam path clear of reflecting surfaces

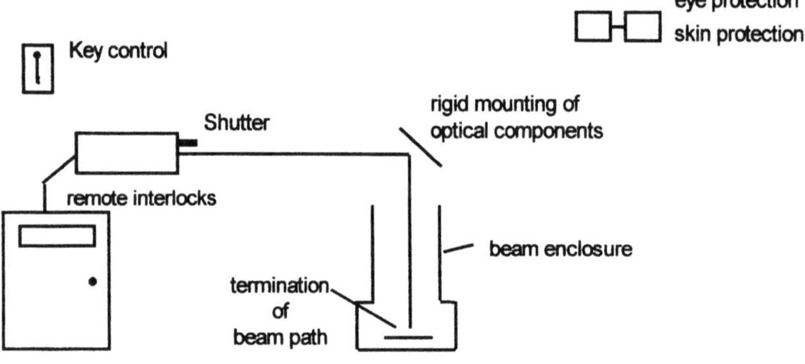

Fig. 7.8 Users' guide safety precautions – a combination of some of the above.

The fourth user clause

This provides tables and graphs for calculation of MPEs.

In normal operation a Class 1 laser product poses no laser radiation hazard and so, it could be argued, EN60825-1 has little if anything to offer by way of user guidance. Class 1 laser processing machines do, however, contain Class 4 lasers and the first part of the User Guidance section calls for the appointment of a Laser Safety Officer **for installations where lasers of a class greater than Class 3A are operated**. In addition, the User Guidance does address what in the present context falls under the heading of **servicing** operations and provides an approach to dealing with safety in general R&D where high power lasers are used. The specific guidance for these **open beam** situations is described below.

Well designed laser machines allow most alignment and replacement operations to be accomplished without access to laser radiation. Where this cannot reasonably be achieved an administratively controlled area must be established. This will probably include the erection of opaque screens with a labyrinth entrance, a warning light and laser hazard warning sign. It would be advisable if not essential to have an additional person inside the control area holding some form of hold-to-open emergency stop or safety shutter de-energiser, especially if a stray or unanticipated reflection of the laser beam could be strong enough to burn through the screen.

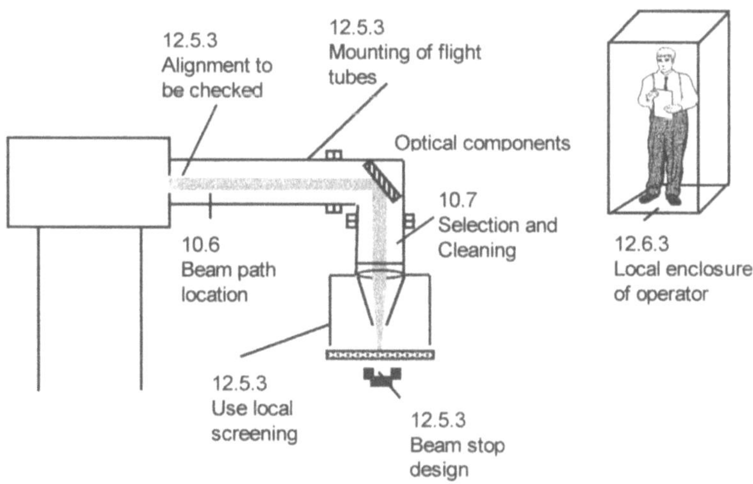

Fig. 7.9 New user guidance in EN60825-1 for high power laser use.

Protective laser eyewear must be worn by all those inside the controlled area but the use of protective skinwear is not so clear. Skin burns to the hand is the most likely injury during open beam working but the wearing of heat resistant gloves may be incompatible with the delicacy of the work to be undertaken. The point is that (i) it is difficult to argue against the use of protective eyewear but (ii) if there is a foreseeable risk of bodily exposure to high power laser radiation then some other means of conducting the work should be sought.

EN 12266 Safety of machines using laser radiation to process materials

EN 12266 deals specifically with machines using laser radiation to process materials.

Key features of EN 12266 are:

1. the exposure of people to levels of radiation exceeding Class 1 shall be eliminated during production, or Class 3A during maintenance.
2. a fail-safe laser beam stop (safety shutter) shall be located inside or immediately outside the laser. If the laser beam stop is not lockable in the closed position then an additional beamline beam stop must be provided.
3. the manufacturer shall inform the user of the materials that are intended to be processed by the machine and provide information on the TLV's for these materials and the fume and airborne particulate generated by processing these materials.
4. the manufacturer shall provide a suitable means for removing the fume and airborne particulate, but the user shall be responsible for the safe removal and disposal of the fume and particulate.

Other essential components of a laser machine include a key switch, interlock switches and emission indicators, all of which are needed to satisfy EN 60825, together with a range of other safety features, such as dump circuits which address the control of electrical hazards and condition monitors such as thermal, airflow and coolant level switches.

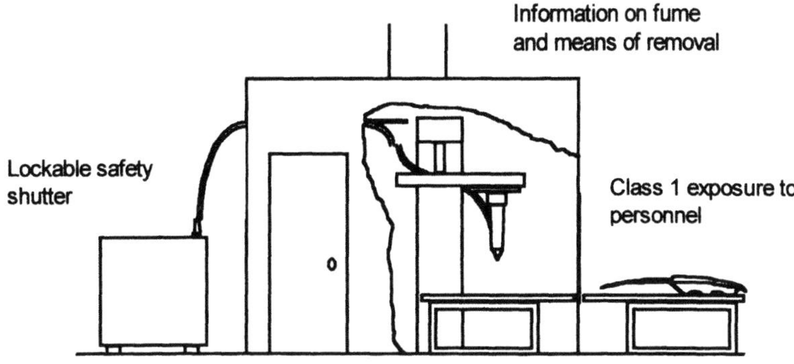

Fig. 7.10 Key features of EN 12266, additional to EN 60825-1.

7.5 CONTROL MEASURES

7.5.1 General approach

The approach to laser safety takes the following steps:

1. Fully enclose hazardous levels of laser radiation if reasonably practicable, so that no administrative control or personal protection is needed. This is often the only safe option for the very high power lasers.
2. If the beam cannot be fully enclosed then at least take a number of recommended steps to control the beam path, and impose administrative controls.
3. If, after imposing engineering and administrative controls there is still a reasonably foreseeable risk of exposure to hazardous levels of laser radiation, eye protection should be worn.

7.5.2 Enclosure of the laser beam

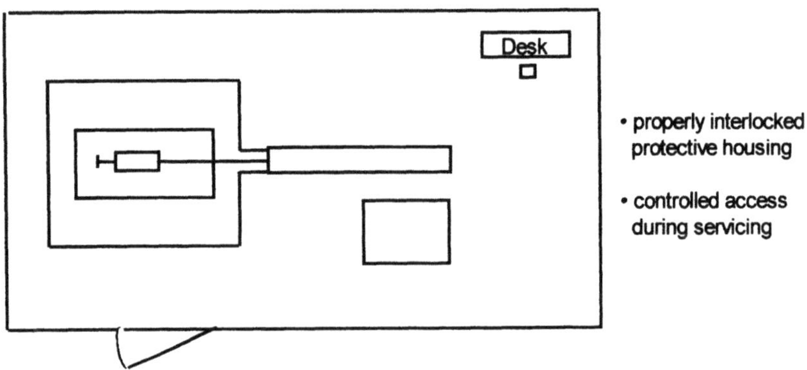

Fig. 7.11 Access to laser area.

A laser product has a protective housing around it to prevent human access to hazardous levels of radiation. This means that it provides a physical barrier to the human body and prevents the escape of hazardous levels of laser radiation, except for radiation emerging from identified laser apertures. Beam enclosure by the user continues this process, and is an important safety measure for Class 3B and Class 4 lasers.

Enclosure of a laser beam can be a simple process, certainly there is a wide variety of opaque materials to choose from: thin metal sheet is universally applicable, but in some wavelength ranges plastics can be used. The key features are robustness, light tightness and access interlocking. The user will generally have to employ administrative controls to overcome some of the weaknesses of enclosure design. Flimsy construction is the most common weakness.

Of course, a room can itself be a protective enclosure, and it is recommended that Class 4 lasers be operated by remote control wherever practicable, thus eliminating the need for personnel to be physically present in the laser environment. This means having the control unit (plus video monitors etc.) in a separate room. For the cost, consumption of time and the general inconvenience of such a form of remote control to be considered practicable, the laser set-up would have to be reasonably permanent and the measurements or observations to be made would have to be amenable to remote monitoring.

Industrial high power lasers used for cutting, welding, surfacing and other forms of materials processing present an example of laser hazard where total enclosure of the beam is usually the only practical solution, especially in the factory floor environment where other forms of control are virtually impossible to enforce. The design of such enclosures is a specialist job.

7.5.3 Beam stops and beam control

It is a cardinal rule that beam paths should be terminated at the end of their useful range, even beams from Class 2 laser products. This applies not only to the primary beam, but also to secondary beams produced, for example, by surface reflections as the beam passes through a transparent window.

Beam dumps for low power beams may not have to dissipate significant power, but as safety critical components they need at least to be robust, and preferably equipped with some means whereby they are securely located in position. In addition, beam dumps for Class 4 laser beams should be blackened and partially shielded to minimise user exposure to diffusely scattered radiation. For high power laser use they may also need to be cooled and fitted with a thermal switch wired into the safety interlock chain.

There are a two basic rules concerning the control of unenclosed laser beams indoors:

1.   beam paths should not be at eye height
2.   beam paths should be clear of surfaces producing hazardous reflections

7.5.4 The laser safety officer

EN60825-1:1994 recommends the appointment of a LSO. The responsibilities and training of a LSO is the cause of much concern and confusion, mainly because the term has such a broad meaning.

EN 60825-1 sub clause 10.1: For installations where lasers of greater than class 3A are operated, a laser safety officer should be appointed. It should be the laser safety officers responsibility to review the following precautions and designate the appropriate controls to be implemented.

<u>BS EN 60825-1 sub clause 10.10</u>: (for class 3A, 3B and 4 laser operators) Because of this hazard potential, only persons who have received training to an appropriate level should be placed in control of such systems. The training which should be given by the manufacturer or supplier of the system, the laser safety officer, or by an approved external organisation, should include, but is not limited to:

a)   familiarisation with system operation procedures;
b)   the proper use of hazard control procedures, warning signs, etc.;
c)   the need for personal protection;
d)   accident reporting procedures;
e)   bioeffects of the laser upon the eye and the skin.

In many industrial situations the main activity of the LSO is to ensure that the operation and maintenance procedures are safe and are being implemented. To comply with EN60825-1 the laser manufacturer or machine supplier should provide all the information and training needed for safe use of the laser equipment. The user should also insist on being provided with a pre-emptive maintenance procedure but this and the operating instructions will need interpreting in the form of written procedures, appointments and regular inspections. Often the equipment supplier is sub-contracted to do the servicing, but the LSO is still required to ensure that the servicing operations are not creating a hazard for fellow employees.

To carry out these duties, the LSO must be aware of the basics of laser safety as they apply to the laser equipment under his or her jurisdiction, the fundamental properties and biological hazards of laser radiation and the basics of hazard control. To cover a full syllabus on laser safety could take several days and even then there would be no guarantee of the competence of the potential LSO to undertake his or her responsibilities. It may be far better, safer and cost effective to train the LSO to appreciate and deal with the hazards of everyday operations and to recognise when to call in a laser safety specialist. A competent consultant can provide the necessary training in-house in a single day, including help in drafting standing orders, conducting a risk assessment and providing health and safety training to operators.

7.5.5 Administrative control

The fundamental issues in administrative control relate to restricting the use of the laser and restricting access to the hazard zone. Training in the use of lasers, their biological hazards and what to do in an emergency is a prerequisite for Class 3B and Class 4 lasers, and to a lesser extent to Class 3A laser operators. Such lasers are fitted with a key switch for use as an aid to restricting use.

Fig. 7.12 Administrative control during servicing.

Indoor applications of lasers are particularly amenable to administrative control since the hazard zone is generally defined by the walls of the room. The general principal, however, is that any person entering the hazard zone should be made aware of the fact, at least by the posting of appropriate warning signs and a physical barrier, perhaps fitted with interlocks, or by someone on duty. In addition, except where the laser is Class 2, it is recommended that warning lights be fitted which are energised whenever the laser is in a state ready to operate. Administrative controls of some kind will need to be enforced within the area and eye protection may well be needed.

The design and content of warning signs to laser areas are not standardised and will clearly depend on the administrative controls in operation within the area and the need for personal protection. At a minimum the signs should include a prominent laser warning triangle. The remote interlock connector of a Class 3B or Class 4 laser can be connected to an emergency stop disconnect or to the room, door or fixture interlock. It is common practice to fit a momentary override to the remote interlock connector to allow access to authorised persons with the laser running.

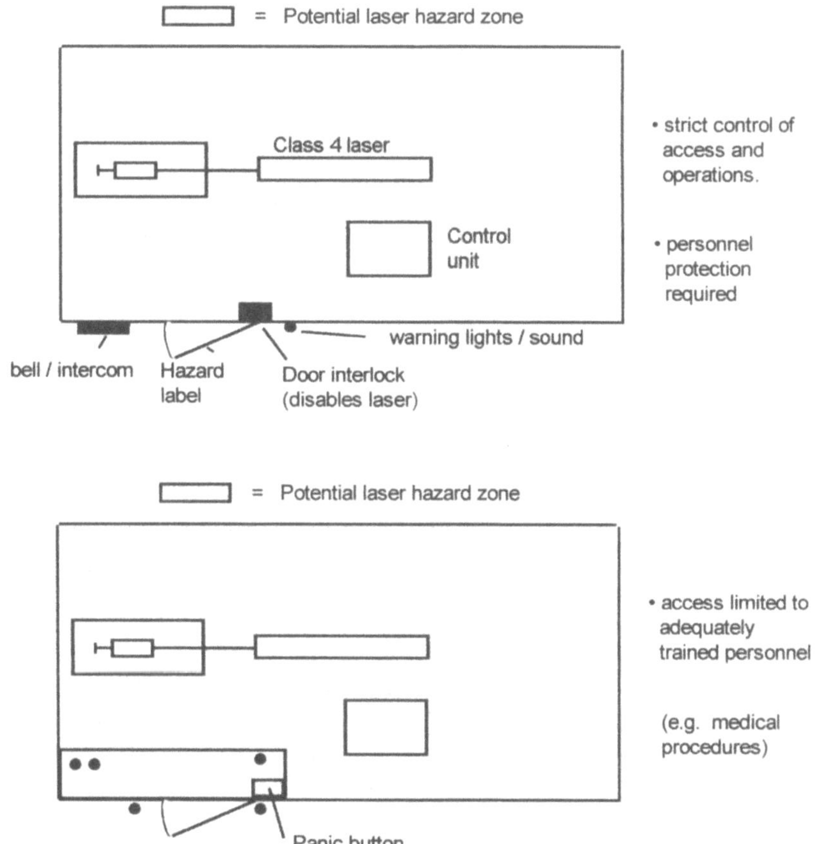

Fig. 7.13 Access to laser areas.

In the life testing of components, or in surgical operations, for example, access to the room is needed without either switching off the laser or requiring someone inside the laser area to activate a door interlock override.

Solutions include engineering control (e.g. key card override) or restricted access through administrative control (training), but then a screen should be fixed inside the room to shield anyone entering. Protective eye wear should be readily available within the room, close to the entryway.

However the access to laser areas is designed, it must clearly allow rapid egress and admittance in an emergency. The review of associated laser hazards can reveal other safety features needed in the laser area, especially in regard to the provision of electrical isolation points, ventilation and fire control. Patrol and emergency services information about the associated hazards present when the laser is in use (e.g. toxic chemicals) should be noted on the main access doors.

Display Standing Orders in a Prominent Position with the Equipment

1. **State the purpose of the work**
2. **State the associated hazards**
3. **State the safeguards that must be attended to**
4. **Be specific in procedure (i.e. you must ... or you must not ...)**
5. **Give details of emergency procedures to follow**
6. **List responsible personnel, their local address and phone number**
7. **Attach all other pertinent documents as an appendix**

7.5.6 Maintenance, servicing and alignment

Administrative control never plays a greater role than during maintenance, servicing and alignment operations. During these operations there can be a change in the hazard zone and for embedded lasers an increase in the laser class. Removal of parts of the protective housing during servicing can, for example, release relatively weak yet hazardous secondary laser beams and can expose other hazards, in particular those of high voltage.

It may well be necessary for the LSO to erect a temporary controlled area, bounded by screens, with a warning sign and lights, to restrict access to the area and to provide additional eye protection. Strict administrative control, embodying the manufacturer's information, will be enforced within the area.

Alignment is a part of servicing and maintenance; it is also a routine procedure wherever a laser beam is used. In all cases it ranks number 1 in the list of hazardous operations.

In addition to the issues connected with radiation hazards during servicing and maintenance, alignment presents a problem with regard to the wearing of eye protection. During alignment the operator's eyes may be at the level of the laser beam, there is maximum uncertainty about the position of the beam and there is every temptation, especially if the beam is visible, not to use protective eye wear. Many laser operators will claim that there is no alternative to using the main laser beam, even though it may be Class 3B or 4, but this is not true.

The infrared beam reduces the fluorescence producing a dark spot on the screen. Alternatively, **liquid crystal** film is available which changes colour (black to red to yellow to blue) as it is warmed up. **Burn marks** can be made on heat sensitive paper inserted in the beam. Alternatively, a visible Class 2 laser can be made collinear with the more powerful beam for alignment or TV cameras or infrared (thermal imaging) cameras are available to **see** the beam.

A preventative maintenance procedure should be instigated to minimise the need for unscheduled replacement of optics, and major realignment.The preferred approach for invisible radiation is to make use of a permanently incorporated low power (Class 2 or 3A) visible CW laser (e.g. a Helium Neon or diode) whose beam is made collinear with the main laser beam.
Occasionally it is necessary to check the collinearity by bypassing the focusing optics and directing the high power beam and alignment beam a considerable distance (say 10 m). This operation can be hazardous and requires the setting-up of a temporary controlled area.

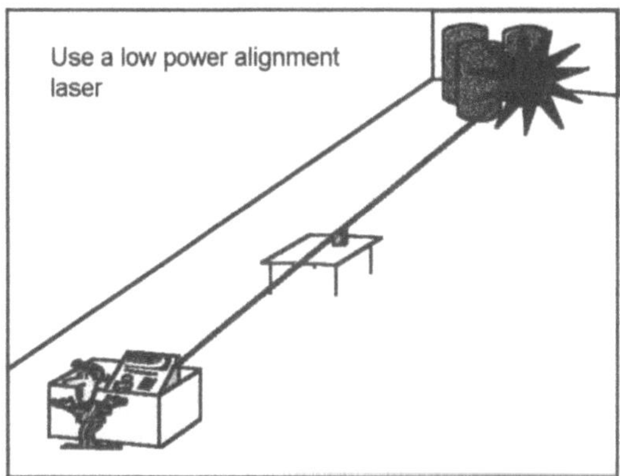

Fig. 7.14 The importance of safe beam alignment.

The most common method of determining beam position during alignment involves taking "burn prints" of the laser beam, but this creates fume which can contaminate optics. The preferred method involves the use of beam position equipment permanently incorporated into the laser beam delivery line. Laser cavity alignment involves analysis of the laser mode as well as the beam power. Mode analysis can be done using a burn print or on line diagnostic equipment.

ALIGNMENT TECHNIQUE
HIGH POWER CO_2 LASER BEAM

1. **Collinear Class 2 laser, beam expanded**
2. **Key system for switching from main beam to alignment**
3. **Safe system of work for throwing high power beam and taking burn patterns**
4. **Align first with Class 2 laser, then high power**

Initial setup may involve repositioning a workpiece, programming a NC machine or robot head (the mechanical hazards of this operation need attention), or making adjustments to a focusing lens after replacement. The key issue here is transfer of control. Part of the set-up can be achieved using the alignment beam, but any operations requiring the main laser power must clearly involve the implementation of several safety measures, which might include:

1. making adjustments then stepping back before firing the laser.
2. the use of a positive pressure laser-on button by the person making the adjustments.
3. adequate personal protection e.g. a full face mask and protective (flameproof) clothing.

Fig. 7.15 Clear transfer of control is important in alignment work.

Alignment using the main laser beam should only be conducted under the following conditions:

1. The Laser Safety Officer or an appointed deputy should conduct or supervise the operation.
2. If possible, beam enclosure should be maintained whenever the laser is emitting. For example, a "burn box" should be incorporated into the beam line for taking "burn prints".
3. If the operation permits access to the laser radiation then the area should be temporarily screened off and warning signs placed around. Access to the hazard area must be restricted, for example by administrative control.
4. An operator should be positioned for rapid manual shut down of the laser.
5. People inside the screened off area should wear appropriate laser safety eye wear. The requirement for skin protection should be balanced by the need for freedom and delicacy of movement.

Following alignment, the Laser Safety Officer or an appointed deputy should check that the equipment is properly restored before operation. This check should include the replacement of all optical components and service panels, and removal of tools and any interlock overrides.

Fig 7.16 Ensure proper restoration of equipment before operating.

5.7. Eye protection

Approved laser safety eye wear comes clearly labelled with information adequate to ensure the proper choice of eye wear with particular lasers. It takes the form of goggles or spectacles fitted with side shields. Such approved eye wear should in general be used or at least available wherever laser radiation greater than Class 3A is accessible. Clear eye wear is available for protection against scattered Nd YAG and CO_2 laser radiation (not the same eye wear for both) and provided it is kept clean and free from scratches, can be worn with little inconvenience. Eye wear damaged mechanically or by radiation should be replaced.

Suppliers of commercial laser safety eye wear will select the most appropriate design for the user. Such eye wear should satisfy the needs of strength and comfort, maximum visible light transmission and good peripheral vision. Regular laser users need the best quality eye wear, and in most cases prescription filter lenses can be made up for users in the form of spectacles with side shields. Despite their limitations for long periods of wear, goggles do provide secure fitting and are generally more robust.

Another consideration in the selection of protective eye wear is protection against collateral radiation, in particular blue light and ultraviolet radiation plasma emissions from above the workpiece during laser processing.

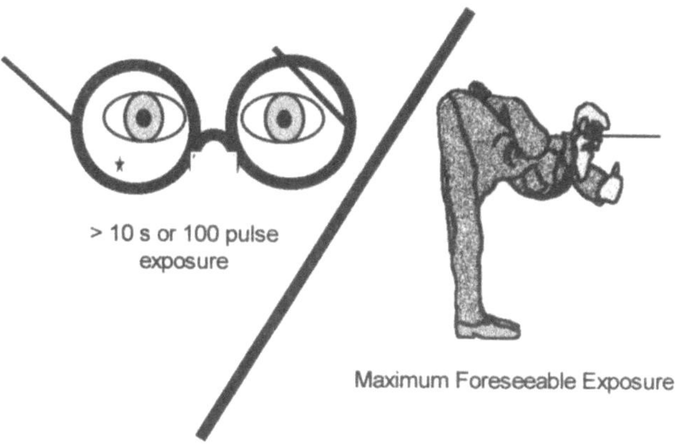

Fig. 7.17 Laser protective eyewear.

The most important property of laser protective eye wear is its Optical Density (or OD) at the laser wavelength. Here again, reputable suppliers will give guidance. In addition to OD requirements, in order for filter material to satisfy the standard it has to provide protection for prolonged exposure (10 seconds or 100 pulses) to the maximum level of laser radiation for which the eye wear is required.

7.5.8 Protective clothing

Protective clothing receives little attention because, except at very high laser powers the injury inflicted by a laser beam on the skin is relatively minor. Only Class 4 lasers pose a skin hazard (and a fire hazard), but unlike eye wear, skin protection can significantly hamper movement and produce significant discomfort, giving rise to the possibility of more mundane non-laser accidents. Workers wearing eye protection and using powerful IR lasers for materials processing (if, exceptionally, unenclosed) occasionally feel the diffusely scattered IR radiation as mild heat on the face. Clearly in these cases where there is much laser beam scatter skin precautions need to be taken on repeat occasions and there are a wide variety of common transparent materials made up in the form of face masks capable of providing adequate protection in the far IR.

7.5.9 Medical surveillance

The value of medical surveillance is a fundamental problem as yet unresolved by the medical profession. A medical examination following an apparent or suspected injurious ocular exposure is recommended, but the main value of pre-, interim- and post-employment ophthalmic examinations of workers using Class 3B and Class 4 lasers is medical-legal. For example, a persons eyes could be diseased or damaged, say, by blue light emission during arc welding before taking up a laser related post.

7.6 RADIATION RISK ASSESSMENT

7.6.1 Introduction to general risk assessment

<table>
<tr><td align="center">Risk

A combination of the probability and the degree
of the possible harm in a hazardous situation

Hazard
A potential source of harm</td></tr>
</table>

The harmonised approach followed in the various EC directives is as follows:

1. **Identify the hazards.** All situations (e.g.. operation, servicing) and events (e.g.. human behaviour, component failure) shall be identified.
2. **Assess the foreseeable probability and severity of harm.** The former depends on frequency of exposure and availability and effectiveness of protection.
3. **Decide if the risk is significant**
4. **If not, implement risk reduction management** until the residual risk is as low as reasonably practicable.
5. **Define safe working practices and PPE** if the residual risk is unacceptable.

EN 60825-1 suggests that the process be broken down into three parts:
1. the laser or laser system's capability to injure personnel
2. the environment in which the laser is used
3. the personnel who may be exposed

In general, the first part is the most important and has most influence in deciding on control measures. It relates to the **laser hazard classification** scheme which gives a good guide in most cases to the laser product's capability to injure by virtue of its emitted laser radiation. **Associated hazards** of lasers, especially high voltage, must not be ignored, but here we are concentrating only on the radiation hazard.

Thanks to the hazard classification scheme, measurements play only a minor role in laser radiation hazard evaluation (unlike many other areas of safety assessment).

The second point, the environment, includes defining (calculating) the area within which a laser radiation hazard exists. The primary calculation is that of the **Nominal Ocular Hazard Distance (NOHD)** of a laser. This is the distance from a laser within which it poses a direct radiation hazard. At the NOHD the beam diameter has expanded so much that the MPE is not exceeded in the beam. In the case of a focused laser beam performing materials processing, there are three NOHD calculations which should be performed.

1. Escape of the **raw** beam. For an unfocused CO_2 laser with a power of 1 kW and a beam divergence of 2 milliradian (0.1 degrees), for 10 second exposure the NOHD is about 1 km. Such an event, which could occur for example during servicing, would create a hazard everywhere in the workshop in which the laser is located.
2. Specular reflection of the focused beam from the workpiece. The NOHD in this case depends on the F-number of the focusing optics. If F-10 optics are being used with a 1 kW beam then after focus it is diverging at an angle of approximately 6 degrees and for 10 second exposure and 100% reflection the NOHD is about 30 m.
3. Diffuse reflection of the focused beam from the workpiece. The NOHD in this case decreases away from the normal to the reflecting surface. For example, assuming 10% diffuse reflection and 10s viewing at an angle of 45°, the NOHD for a 500W CW focused Nd:YAG laser is only about 0.5m and 0.4m for a 5 kW CO_2 laser. Measurements of scattered radiation under normal incidence irradiation have indicated that levels of

radiation escaping interception by the laser focusing head are, as predicted, low.

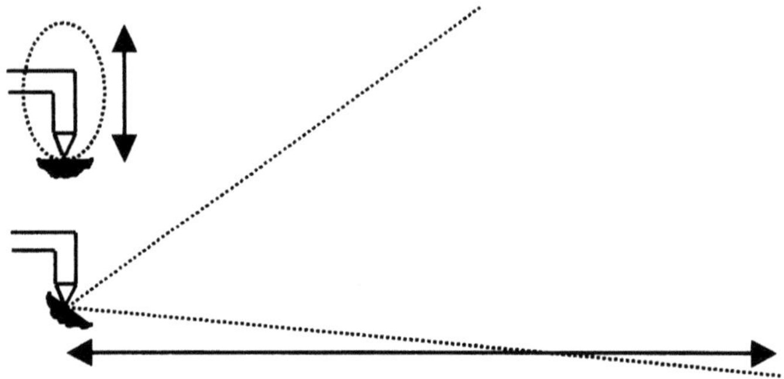

Fig. 7.18 Situations with different values of nominal ocular hazard distance (NOHD).

The region in which a laser radiation hazard can exist is referred to as the **Nominal Hazard Zone (NHZ).** Laser guards limit the NHZ. Determining the NHZ in general requires an assessment not only of the freedom movement of the laser itself, but also of optical components, both in normal and under fault conditions.

The third point relates to the likelihood of personnel being exposed to hazardous levels of laser radiation i.e. of being within the NHZ. The personnel involved will include the laser operator, other workers within the NHZ and those inside the NHZ not involved in operations. For such categories of personnel issues including training and the provision of protective eye wear must be considered.

A basic, if obvious principle, is that unprotected personnel must be excluded from within the NHZ. This will be achieved by a combination of barriers and administrative control. A key issue is the degree and nature of administrative control which is required to prevent human access to laser radiation.

7.6.2 Examples of risk assessment

There are a large number of potential factors to consider in any risk assessment. Here we list some examples of some frequently encountered situations to be considered.

Workshop environment

SAFETY CONSIDERATIONS FOR INDUSTRIAL LASER APPLICATIONS

- Life threatening injuries
- Hazardous secondary reflections
- Large hazardous zone around laser installation
- Sources of errant beam
- Problems of enclosure
- Problems of alignment

Use of heavy mobile equipment around the laser can threaten the integrity of the laser enclosure, especially overhead beam tubes. The presence of overhead walkways and crane operation threatens the integrity of the laser enclosure. In this case, beam enclosure must prevent access to laser radiation from above. Contamination of optical surfaces from a dusty workshop environment can give rise to component failure when surfaces are exposed to high power laser radiation. Floor transmitted vibration can cause misalignment of optical components.

Secondary hazards

1. **Electricity**. High voltages are used in all the industrial lasers and may be exposed during servicing operations.
2. **Fume**. Emissions during materials processing can be highly toxic and can be produced in copious quantities. Extraction can be particularly difficult in dynamic situations, particularly robot controlled processing. Particle sizes in laser produced fume are characteristically small, with a relatively high concentration of sub-micron diameter. The fume needs to

be effectively extracted and treated before release into the atmosphere. The atmosphere inside **walk-in** enclosures is a particular concern: excess oxygen or shield gas (nitrogen, argon, helium) can cause dizziness or asphyxiation.

3. **Fire.** There is an obvious fire risk from any flammable material exposed to high power laser radiation. Emissions during materials processing sometimes ignite. The escape of a high power laser beam can pose serious risk of fire or explosion.

4. **Secondary (Collateral) radiation.** UV and blue light produced during welding can present more of a hazard than scattered laser radiation. Some lasers are excited by high power RF and leakage of this radiation needs to be addressed.

5. **Mechanical.** The edges of X-Y tables can present mechanical traps. Robots pose lethal mechanical hazards.

Modes of operation

1. Operation, servicing and maintenance require specialist training. Several shifts of staff may be involved. Administrative control of personnel in the laser area can present problems

2. Pressure of work schedules means alignment is generally a rushed job. Class 1 embedded lasers products can become Class 4 under some conditions during servicing. Dubious and even some downright dangerous alignment procedures are sometimes used.

3. **Walk-in** Workstations pose additional problems of laser operation with personnel inside as well as the usual hazards of working in confined spaces.

Control and monitoring of the laser beam

Most basic laser systems include an on-line power meter at the laser output plus a burn box downstream. Errant and non-errant beam detectors can play an invaluable role in ensuring beam control, while on-line beam quality monitors combine safety and performance functions. Beam delivery lines are generally equipped with a safety shutter and a process shutter. Panels which provide access to high power laser radiation as well as access doors to **walk-in** workstations should be power interlocked to the safety shutter.

Enclosures for materials processing applications
- containment of radiation

Several standards addressing such issues as the design of enclosures (a problem in principle for beams able to **burn** their way through opaque barriers) and associated hazards are currently in preparation. The Standard BS EN 60825-1 includes an informative annex E **High power laser considerations particularly appropriate to materials processing laser products** which is particularly relevant in this context. Enclosure is the best way of dealing with laser radiation hazards. Metal sheeting and (for UV and far-infrared wavelengths only) some clear transparent materials can be used. Thin walled enclosures will **keep fingers out** and block any weakly scattered laser radiation, but a risk assessment will be required and additional measures taken as necessary to ensure that fault conditions do not arise which would expose the enclosure to the full power of the laser beam. Regular maintenance of optics and the provision of auxiliary cooling, for example, become important safety measures at high laser powers.

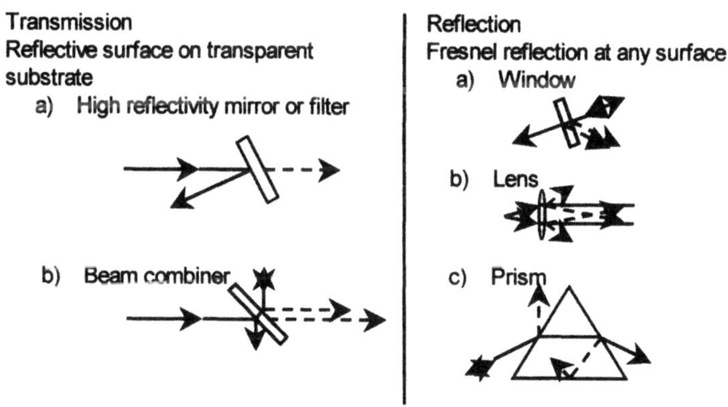

Fig. 7.19 Common sources of errant beams from beam forming components.

The usual arrangement for cutting and welding operations is for the incident laser radiation to be normal to the workpiece surface, so most specularly reflected radiation will be directed back towards the laser although some radiation will always be diffusely scattered at larger angles. Flat sheet CO_2 laser cutting machines often provide no enclosure for such radiation, but while this may in future raise problems of conformity with the Class 1 requirements of the standard for laser processing machines, simple calculation shows that an operator is unlikely to receive more than the MPE from diffuse scatter.

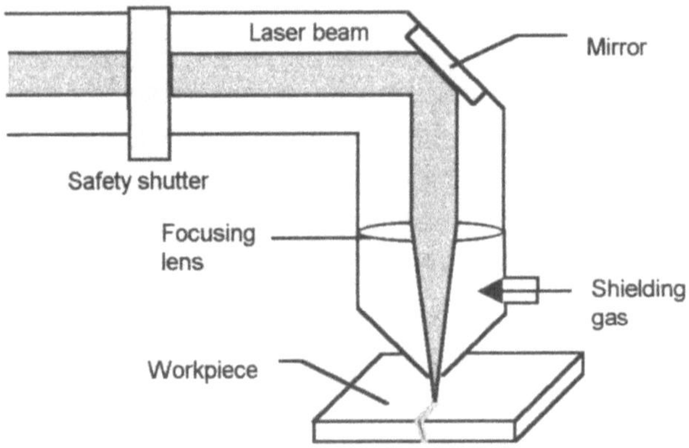

Fig. 7.20 Laser beam path.

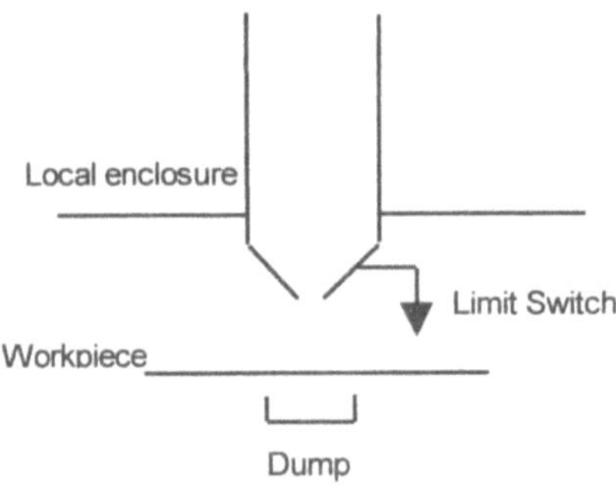

Fig. 7.22 Local enclosure.

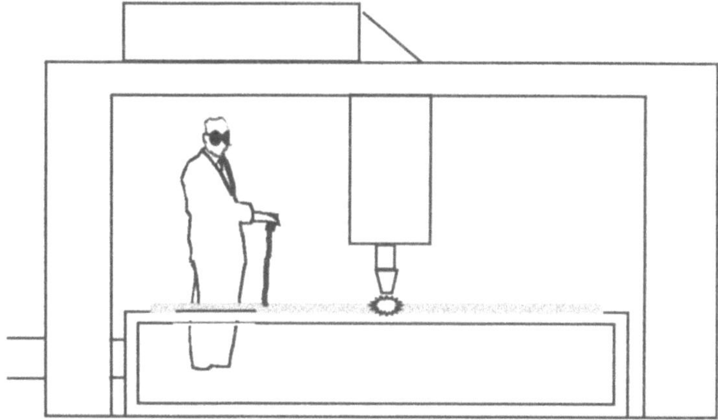

Fig. 7.22 Open XY cutting table.

An alternative or supplement to complete enclosure is local enclosure, a substantial guard located close to the point where the laser radiation interacts with the workpiece. Together with the workpiece, this guard almost completely surrounds the process zone, generally leaving only a small gap to allow unimpeded relative movement of the two components. In some application, where total enclosure is not reasonably practicable by virtue of the size of material to be handled, local enclosure alone **might** be acceptable if all the following conditions are satisfied:

1. the local enclosure is robust, allows only a minimum gap (for safety) with the workpiece, and is regularly inspected for internal damage
2. eye protection is worn
3. the operator keeps well away from the area of operation, others are excluded from the area and no other work takes place in the close vicinity.
4. process conditions are closely controlled and result in strong absorption and/or diffuse scattering of the laser radiation.

In the context of safety screening it is convenient to define a Foreseeable Exposure Level (FEL). The FEL is the level and duration of laser radiation exposure of a screen, assessed under normal and reasonably foreseeable fault conditions, which most threatens the integrity of the screen.

Fig. 7.23 Screening requirements.

One general approach to screening in general can be summed up by three questions:

1. **What is the foreseeable exposure level of the screen to laser radiation?**
   A preliminary assessment of a given laser situation will identify the conditions which maximise the exposure at some intended position of a screen. Assuming this corresponds to a fault condition, an iterative process generally follows in which steps are taken to eliminate fault conditions or minimise the level and/or duration of the resulting exposure at the screen. The intention of the following two questions is to identify how such conditions are sensed and responded to.

2. **How are the conditions sensed which give rise to the FEL?**
   If some form of intervention is required to limit the exposure of the screen, then some form of sensing will be needed. If the sensing is automatic then, depending on the risk assessment, a degree of fail-safe operation may be required. If the sensing involves direct human monitoring then the visibility of the onset of laser damage becomes an important issue

3. **What is the response to exposure at the FEL?**
   The answer is inevitably to strive to limit the duration of laser exposure when the screen is threatened. The reliability and speed of terminating the laser emission is the key issue. Generally, a Class 4 laser is provided with a beam stop. In the case of a laser processing machine it is normal practice for a beam stop/safety shutter to be directly linked to

an external interlock chain for manual operation or for linking to any sensing devices with safety functions.

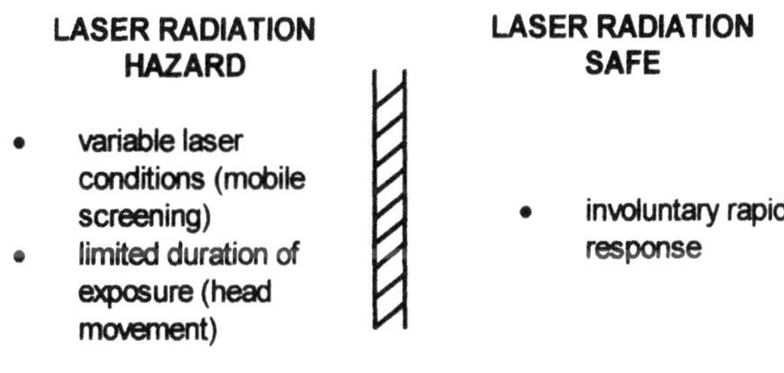

Fig. 7.24 Protective housing as defined in EN 60825-1.

Fig. 7.25 Protective eyewear.

Enclosures for materials processing applications - laser guards

For a Laser Processing Machine the assessment of FEL may include cumulative exposure during each part of the processing cycle of the machine as well as the most demanding combinations of irradiation and exposure duration. For materials with significant thermal conduction not only the irradiance (or radiant exposure) but also the area of exposure can be important, while for grazing incidence the direction of polarisation of the laser radiation can have a dramatic impact.

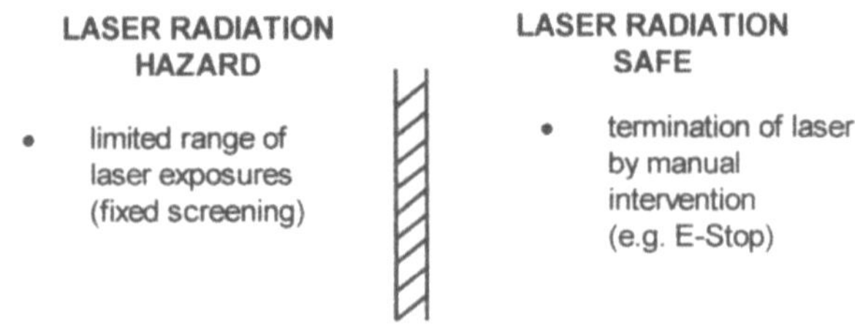

Fig. 7.26 Laser machine guarding.

Fig. 7.27 General purpose safety screens.

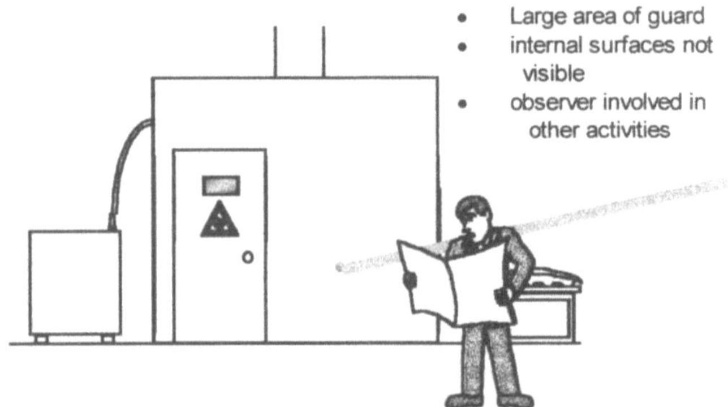

Fig. 7.28 Laser guards – problems with administrative control.

The basic requirement for a laser guard is that, with one surface exposed to laser radiation at the Foreseeable Exposure Level, the accessible laser radiation on the opposite surface shall not exceed the 100 s Class 1 Accessible Emission Limit. The preferred approach to enclosing high power laser radiation, as outlined in Annex E of EN60825-1, is to find engineering solutions to potential fault conditions that could give rise to powerful errant laser beams, otherwise to limit the duration of such faults by some means of automatic detection and termination of the laser radiation. To cover both the latter situations and others (including normal operation of the machine) where laser exposure is not prevented by automatic detection and shut-down, the basic performance requirement for a passive laser guard must be met over the time between successive safety maintenance inspections of the laser guard, and not less than eight hours; otherwise an active laser guard must be used.

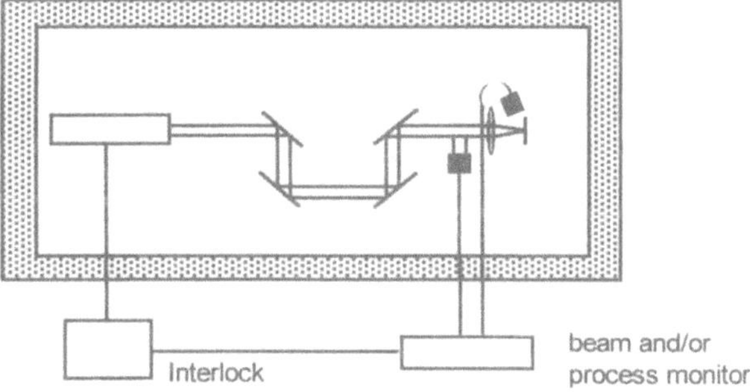

Fig. 7.29 Passive enclosure.

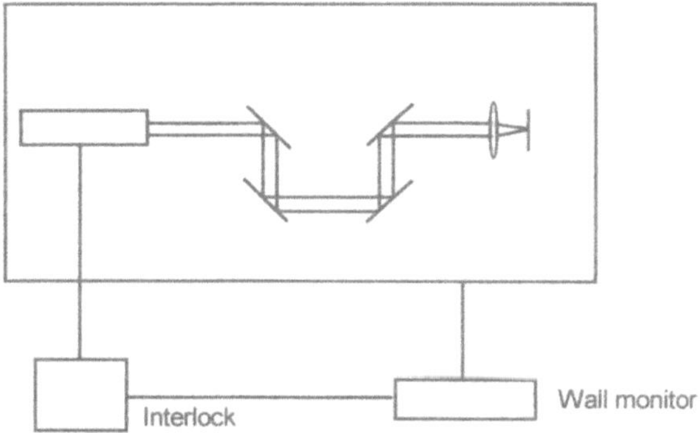

Fig. 7.30 Active enclosure.

The Provision and Use of Work Equipment Regulations effectively places the same risk assessment requirements on the laser user designing safety screening as the Machinery (Safety) Directive places on the manufacturer designing laser guarding on a laser processing machine. The approach of assigning a FEL and undertaking an assessment based on the three key considerations as outlined above, is consistent with the risk assessment approach and has been shown to lead to practical solutions for all situations considered.

8

Beam Manipulation

R.C. Crafer

8.1 SETTING THE SCENE

8.1.1 The requirement for automation

Compared to conventional production technologies, the industrial laser is relatively new. Although the first lasers were developed in the opening years of the 1960's, it was not until towards the end of that decade that viable industrial systems and processes were first developed. The commercial development of the industrial laser therefore covers at most a quarter of a century, in comparison to almost a century for electric arcs and many centuries for gas flames and furnaces. Nevertheless, the industrial laser has been accepted as the chosen technology for a large number of applications across a wide range of industries world-wide. The detailed reasons for the rapid uptake of such a comparatively new technology are explored in other chapters in this Handbook of the Eurolaser Academy.

The burgeoning numbers of successful applications have lead to total annual sales figures for industrial laser systems that are currently of the order of 1.3 billion ECU. In the 1980's, sales reached double figure percentage annual growth rates, and are expected to continue to grow in the 1990's, although at a somewhat reduced rate. These sales figures should properly be viewed in the context of static or decreasing system prices, resulting from rising production, more cost effective designs, increased competition between manufacturers, and the current recession in certain geographical market areas, particularly Europe. The overall picture is therefore one of a technology that is still in a growth phase, and has yet to reach the static sales levels of a mature industry.

About 55% of the world's industrial lasers are currently supplied by the USA, with the remaining 45% being divided fairly equally between the Pacific region and Europe (Fig. 8.1).

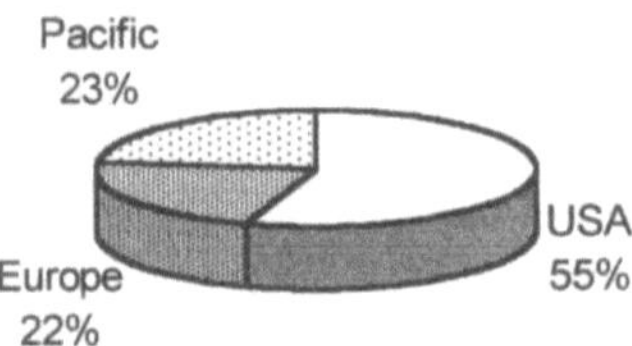

Fig. 8.1 Analysis of worldwide laser sales value by region (source Laser Focus World, Jan 1997).

Currently, the lasers supplied from the rest of the world are not in numbers significant enough to affect these figures although this could change over the next few years. This represents a complete reversal of the situation in the early 1990's, when Japanese laser manufacturers dominated the market.

Figure 8.2 shows how the world production of industrial lasers is currently divided between the three principal non-diode types, namely: Carbon Dioxide, Nd:YAG and Excimer, by total sales revenue.

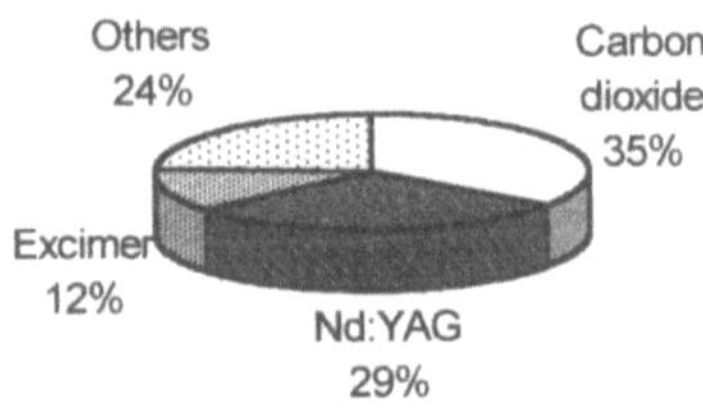

Fig. 8.2 Analysis of worldwide laser sales value by type (source Laser Focus World, Jan 1997).

The majority of all expenditure on industrial laser systems is still on carbon dioxide, although not as much as say five years ago. The distinction is not so extreme when production is analysed in terms of units rather than revenue, but carbon dioxide still has over 50% of total sales.

Lasers however are very expensive devices when compared to conventional energy sources. Whereas an industrial 5kW arc welding power source can be purchased for a cost of say 10,000 ECU, a laser of equivalent power will cost at least 10 times that amount, depending on the source and the quality.

So given its high capital cost, why is the laser such a popular technology? The ultimate answer must be economic, in that it performs certain tasks for less cost per unit product than the competitive technologies. Undoubtedly the most important underlying reason for this is that laser processes are ideally suited to automation.

8.1.2 Conventional automation

Automation in its broadest sense, from the early production lines of the Ford Company in the 1920's to the flexible manufacturing factories of today has revolutionised the manufacturing of all kinds of goods. The standard of living which we enjoy today stems directly from automation. One has only to consider the comparative costs of production line and hand built automobiles to gain a graphic insight into the consumer benefits of automation. Conventional (non laser) automation however is hampered by three particular limitations, and it is indeed a tribute to the ingenuity of production and automation engineers that conventional automation has been so successful. The three limitations are:

1. Tools are subject to wear and mechanical failure. Cutting tools such as mills and drills become progressively blunt with use. In extreme cases they can fail mechanically if overstressed. Provision must be made for their frequent replacement and maintenance.
2. Conventional machining places reaction stresses on head stocks, tail stocks, carriages and bearings. These need testing and adjusting on a regular basis to avoid positional errors.
3. Processes are workpiece dependent. Take the case of GMA (Gas Metal Arc, MIG) welding of stainless steel using filler wire. The arc on which

the entire process depends varies considerably from batch to batch of nominally the same material. Other dependencies are also known.

8.1.3 Where lasers differ in their requirements

1. Tool is not subject to wear and mechanical failure. In the case of the laser, the tool is simply the region where the beam is brought to a focus, or in the case of certain forms of heat treatment, impinges on the workpiece. Incorrect interaction between the tool and the workpiece does not cause the tool to become blunt or to break. When the correct interaction is restored, the process continues as before. The tool is always correctly set and sharp.
2. Mechanical forces. Since there is (almost) zero reaction force on either the workpiece or the tool caused by the process itself, it is not necessary to build large robust structures to handle the laser beam. Robustness is only necessary to offset distortion of the system caused by acceleration of the various moving masses involved in manipulating the beam and workpiece. In a similar manner, the almost complete absence of reaction forces on the workpiece can greatly simplify fixturing and jigging.
3. Greater workpiece independence. The laser beam is unique in appearing to the workpiece as a heat source of essentially infinite temperature and small dimensions. (In practice nothing is ever infinite, but this is about as near as you can get). It is similar to an industrial electron beam, but without the need for a vacuum environment, X-Ray shielding and a large vacuum chamber. Due to its apparent high temperature which is a property of the laser alone and not, for example, the property of an arc of vaporised workpiece material, the transfer of heat, and therefore the conduct of the process, is very much less dependent upon subtle differences in the workpiece material. Thus, although the process will indeed vary from metal to metal, it will not change appreciably between different batches of nominally the same metal.

8.1.4 Advantages of automating the laser.

The principal advantages of laser automation stem directly from those mentioned above, but there are others. The power at the tool tip can be varied very rapidly, in some cases up to hundreds of kilohertz, thus the automated system can be highly responsive to the demands of the process. A particular application of this very rapid response is in the reprographic industry where lasergravure is gradually replacing conventional photogravure in the manufacture of printing rolls for a variety of applications. Also, laser processing can be very fast. For example continuous laser welding can be accomplished at speeds orders of magnitude faster than those available with conventional arc processes. At these speeds, not only is automation desirable, but also necessary, because control of the process is beyond the capabilities of manual control. Finally (but not exhaustively), automation is necessary in order to comply with the strict international safety requirements of industrial practice.

8.1.5 Disadvantages of laser automation

There are very few disadvantages to the automated laser process. Those that can be enumerated are more in the nature of technical challenges rather than disadvantages.

Parameter ranges, processability diagrams

Conventional machining is very tolerant to specific parameters. In the particular case of mechanical milling for example, it is quite acceptable for the x, y or z feed rate of the milling cutter to become zero for an extended length of time. Although this is not an economical way to operate, it will not have any adverse effects on the workpiece. In laser processing, however this is not the case. With any form of laser processing there is a definite range of interrelated values that any parameter is allowed to assume. Beyond that range and the process fails.

Figure 8.3 shows a typical processability diagram with respect to power and speed for CO_2 laser welding of specific types and thicknesses of steel. Similar diagrams with differing degrees of tolerance may be expected to hold for other laser processes. For optimum quality one should stay well within the appropriate shaded areas, but on no account should the parameters be allowed to fall outside the shaded areas.

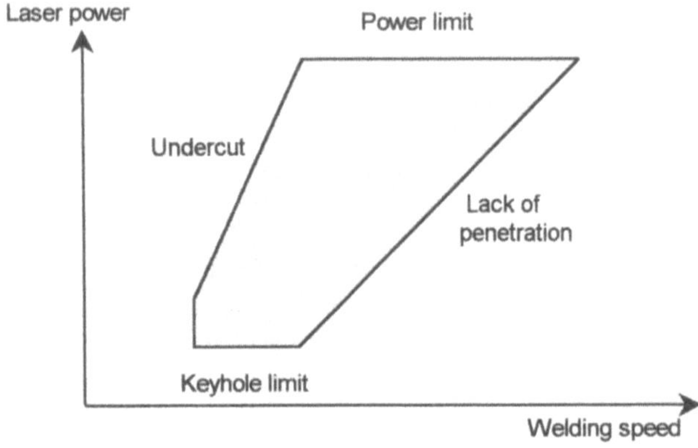

Fig. 8.3 Typical processing diagram for steel.

Tolerances

Tolerances will also have subtle differences when compared with conventional processing. Particularly when comparing laser cutting with oxy-fuel cutting or laser welding with arc welding, it will be found that the positional tolerances for the tool point are much tighter with the laser than with conventional processing. Taking welding as an example, in arc welding it is probably not significant if the welding gun deviates by up to ~1mm from the joint line, whereas with the laser, if the deviation exceeds say 0.25mm in thin material, the laser keyhole will completely miss the joint, with ensuing lack of fusion and a failed weld. Likewise the narrowness of the kerf of a laser cut demands much higher positional accuracy than with oxy-fuel cutting.

8.1.6 Summary

Laser processing is a technology that not only benefits from automation, but in most cases demands it in order to fulfil the full potential of its capabilities.

8.2. SMALL SCALE APPLICATIONS

8.2.1 Overview of manipulation techniques

Masking and imaging

The simplest technique is optical imaging of a laser illuminated mask onto a workpiece. The process is similar to photographic projection, only in reverse (Fig. 8.3).

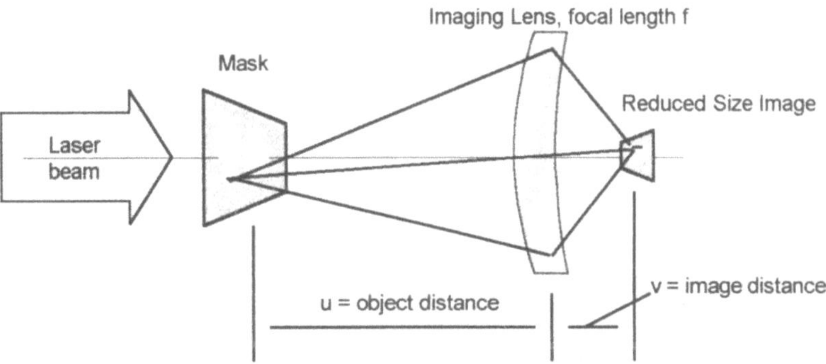

Fig. 8.3 Imaging of a Laser Illuminated Mask.

The mask is significantly larger than the mark to be imaged onto the material of the workpiece. The normal laws of classical optics apply in this case. The controlling equations are:

$$Magnification = \frac{v}{u}; \quad \frac{1}{u} + \frac{1}{v} = \frac{1}{f} \tag{8.1}$$

The laser here is used purely as a source of intense radiation. None of the coherence properties are utilised in this method, indeed they can be a severe disadvantage as they lead to fringing and degradation of the image.

Moving workpiece

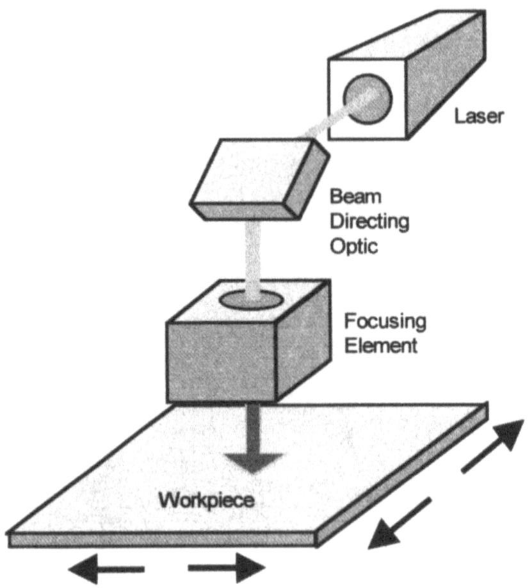

Fig. 8.4 Moving workpiece.

This is the simplest of the coherent focusing techniques, in which the optics remain stationary and the workpiece moves relative to the focus (Fig. 8.4). Since all optical parameters are constant, the focal spot size, intensity and depth are also constant and predictable. This technique is best applied under the following conditions:

1. Small, low mass workpiece, or;
2. Any workpiece moving with quasi-constant motion.
3. Large, high mass workpiece with constant processing speed in one axis only.
4. Since most industrial lasers are large, massive, and are used in the horizontal position, it is customary, though not essential to insert a beam directing optic into the beam path as shown in Fig. 8.4.

Moving mirrors

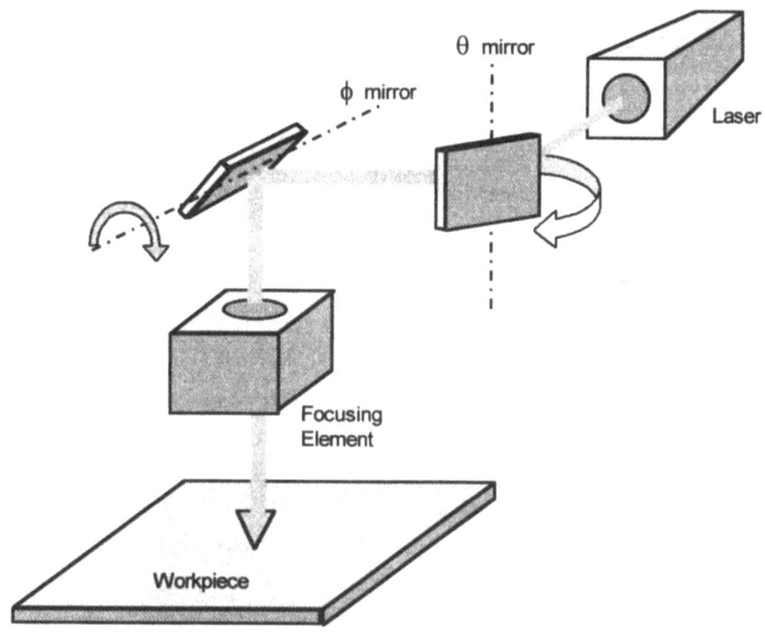

Fig. 8.5 Moving mirrors with pre-objective scanning.

The most commonly used arrangement in which both the laser and the workpiece remain stationary while the optics route the beam from laser to the appropriate location on the workpiece. Figure 8.5 shows the most usual arrangement for two dimensional beam manipulation over a flat workpiece employing pre-objective scanning to achieve a planar focal surface. As with all pre-objective scanning techniques, the focusing optic is constrained to work at large aperture, or small f/number. One of the skills in designing such systems is to minimise the aberrations caused by working at small f/numbers. Other arrangements with fewer or more axes of movement are also possible.

Moving lens

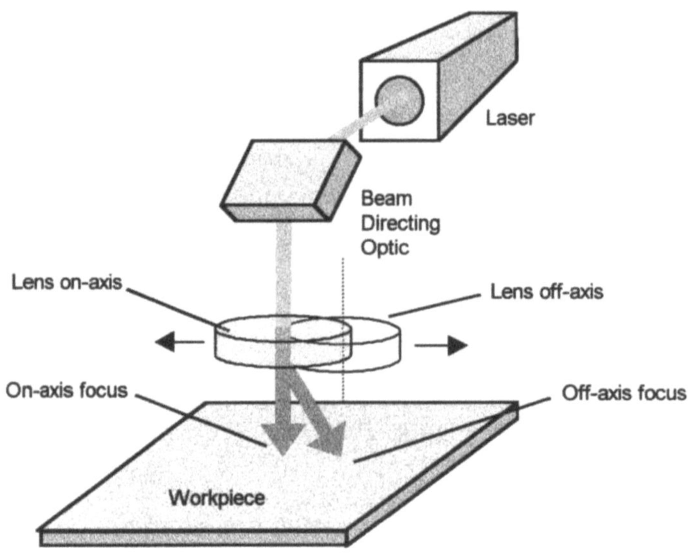

Fig. 8.6 Moving lens.

This is a specialist technique included for completeness rather than normal application. It transcends all the conventional rules of optical manipulation and should only be used where a specific application demands it (Fig. 8.6). It is based on the principle that any light ray incident upon a lens surface and travelling parallel to the optical axis, will be brought to a focus on the optical axis at the focal plane. The disadvantages are that for large axial displacements, the focal plane is not indeed a plane but a rather complex surface, and that the size and shape of the focal spot, and hence the power density available for processing is a very strong function of the axial displacement. This latter effect is known as spherical aberration, and the particular aberration involved here is known as coma, from its likeness to the shape of a comet.

Post-objective scanning

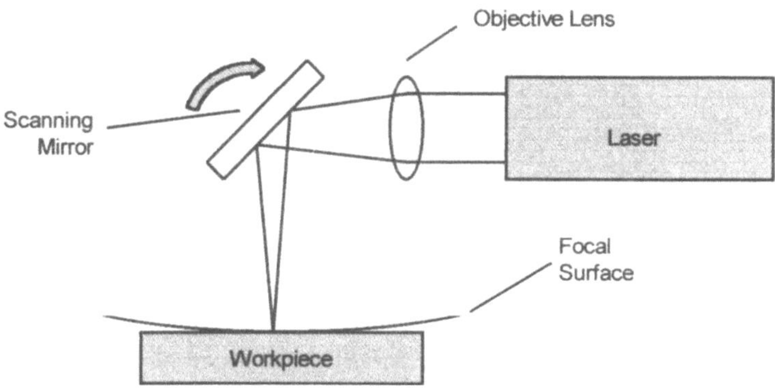

Fig. 8.7 Post-objective scanning.

A superficially attractive method of small scale scanning in which the laser beam first encounters the focusing lens, and is then subsequently deflected by the mirror system (Fig. 8.7). The main advantage is that the lens diameter needs to be no larger than the diameter of the laser beam itself, and hence cost is minimised.

The main disadvantage is that since the total distance between lens and focus is fixed (focal distance = focal length), and the beam is deflected in an angular fashion, the focal surface is not a plane, but approximates to a sphere. If the workpiece has a spherical geometry this is ideal. However most real workpieces for small scale laser processing are planar (printed circuit boards, SMDs etc., and therefore this technique is seldom used.

A secondary disadvantage is that the power density on the scanning mirror(s) is increased, since the partially focused beam is always smaller in diameter than the raw beam. However with the relatively low average powers involved, this is not a major problem.

Pre-objective scanning

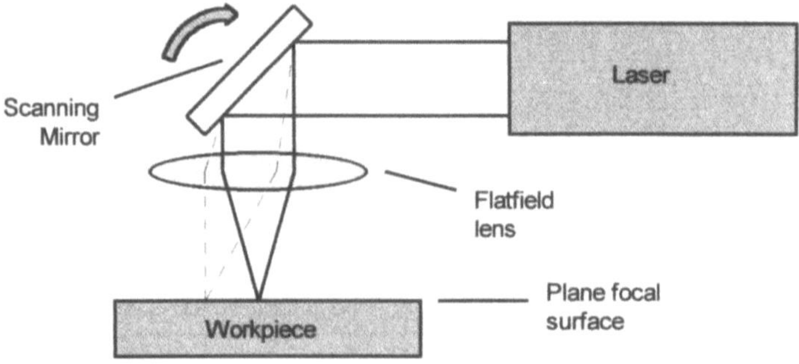

Fig. 8.8 Pre-objective scanning.

One solution to the problems of post-objective scanning is to locate the scanning mirrors prior to the focusing optic. This is called pre-objective scanning. The scanning mirrors simply deflect the beam rather than bring it to a focus, so that the beam incident upon the focusing optic is always similar. To allow large workpieces to be used, a special kind of lens called a flat-field lens is employed.

Fibres

The ideal solution to all problems of beam manipulation would be to dispense with deflecting mirrors, flatfield lenses and rigid gantries/robots, and use a flexible light-conducting cable connected to the laser at one end and a simple robot/manipulation system at the other. Additionally, this cable would also preserve the laser mode and coherence properties of the laser beam.

Such cables, known as optical fibres or fibre optics are already well known in the optical communications field where their transmitting properties are now so good that signals can be sent tens or hundreds of kilometres without the necessity for amplification. Typical absorption figures would be in the region of 0.1 dB/km. The power levels involved however are minute, and communications fibre developers are not concerned with problems of damage arising form absorbed power. It is also possible in the communications field to use solid state (i.e. diode) lasers tuned to the optimum wavelength for fibre transmission. This is typically around 1.3μm to 1.4μm in the infrared region of the spectrum.

Developers of fibres for industrial processing do not presently enjoy all the benefits accorded to workers in the communications field. In particular, the wavelengths of interest are well defined at 1.06μm for Nd:YAG and at 10.6 μm for CO_2, and the power levels involved are orders of magnitude higher at kilowatts or tens of kilowatts cw, and higher still for pulsed operation. Nevertheless, suitable fibres have been developed for Nd:YAG lasers with average power capabilities of several kW. These fibres are used successfully in applications such as welding where large lateral and angular movements of the beam are required at high power levels. Present fibres are not able to retain any of the mode structure of the transmitted beam, but this is not a serious problem since high power Nd:YAG lasers operate in a highly multimode configuration where detailed mode structure is not the dominant parameter.

At longer wavelengths, power transmitting fibres at still at the early stage of development. Some progress has been made in the 5 μm wavelength band where the CO laser operates, but considerable progress is needed before useful fibres are available for use at 10.6μm. Figure 8.9 shows the method of using fibres with Nd:YAG lasers. At the laser end, the output mirror of the laser is imaged onto one end of the fibre using a focus lens, and at the process end, the end of the fibre is imaged onto the workpiece.

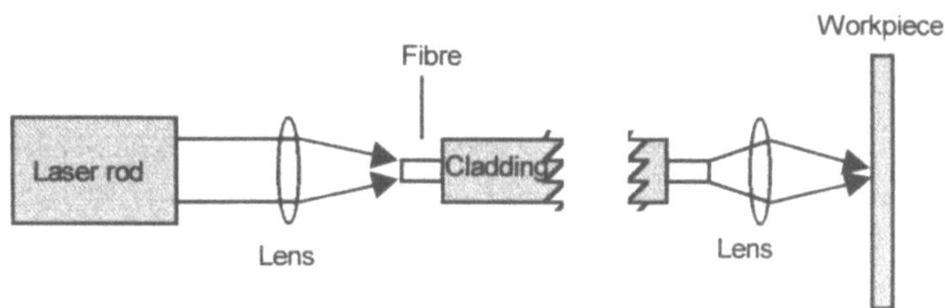

Fig. 8.9 Fibre optic coupling scheme.

Rotating mirrors

The archetypal small scale application is in the electronics industry where lasers are used, inter alia, for resistor trimming, cable and component marking, and soldering of surface mount devices onto printed boards. These applications require very fast response, and the best solution so far is to use rotating mirrors.

Operating Principles

Rotating mirrors operate on the galvanometer principle shown in Fig. 8.10. A multiturn coil in a radial magnetic field experiences a turning moment proportional to the current passing through the coil, the number of turns, and the magnetic field.

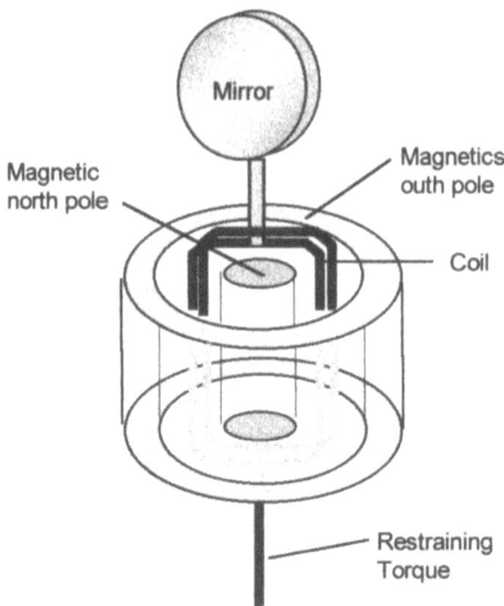

Fig. 8.10 Schematic of galvo mirror.

If the rotation of the coil is restrained by a device such as a hairspring or a torsional suspension, then an equilibrium position will eventually be achieved such that the angular displacement is proportional to the applied current. To obtain optimum response, the coil should also be connected to a damping device so that the whole assembly is critically damped.

If the current is I, the magnetic field H, and the number of turns N, then the electromagnetic turning moment M will be given by:

$$M \propto NIH \tag{8.2}$$

But this will equal the restraining torque applied by the spring or the torsional suspension, so that the resulting angular displacement θ is given by:

$$\theta = \frac{NIH}{k} \tag{8.3}$$

where k is the spring constant.

Speed of response

The response speed of galvanometer mirrors can be very fast indeed. Scanned marking systems will typically write up to ~60 type sized characters per second, and have linear writing speeds of 1 m/s.

Tool tip accuracy

Resolution and repeatability of scanning systems are of the order of 5μm and 25μm respectively for a 100x100mm field. This is a common size of field for current marking systems, although systems with larger fields, but lower resolution and repeatability are being introduced.

Normal systems - paraxial rays

In so called normal systems, light rays are paraxial, i.e. nearly parallel to the optical axis of the lenses and mirrors being used, and deviating from that axis by only small angles. Figure 8.11 shows the scheme. Under these conditions one can make the geometrical approximation that:

$$\sin \theta \cong \tan \theta \cong \theta \approx 0 \tag{8.4}$$

where θ is the angle of incidence/refraction of any ray upon any optical surface.

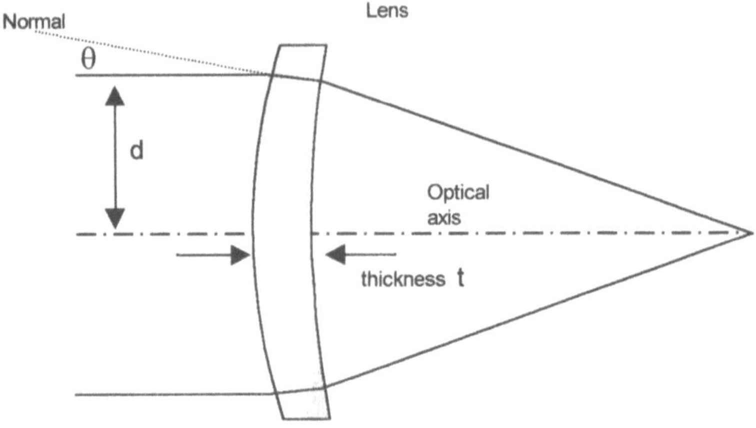

Fig. 8.11 Geometry of the paraxial ray approximation. Parameters θ, d and t are all small.

It follows in this approximation that all rays are focused to the smallest possible focal spots, and that all such spots fall on a single plane called the focal plane. A further requirement for these conditions is that all lenses are thin, that is their maximum thickness is negligible compared to the diameter. This is sometimes known as the thin lens approximation.

Non - paraxial systems

To achieve the thin lens approximation in practice is not always possible. For example the requirement for compact equipment and small focal spots requires the use of low f/number optics. Similarly, a lens may be required to withstand a large differential pressure, for example in a coaxial gas jet cutting system, and must of necessity be thick, i.e. not thin. In the case of rotating mirror scanning systems in particular, using the pre-objective technique described earlier, it is impossible to achieve the paraxial approximation. In order to achieve the required small focal spot sizes, the spherical surface optics which are usable according to the paraxial approximation can no longer be used. A special type of aspheric (= not spherical) lens has to be used. This is called a flat-field lens.

Flat-field lenses

Flat-field lenses, sometimes known as f/0 lenses, are a particular solution to working at small f/numbers when the principal requirement is a constant focal spot size and a plane focal surface. Modern flat-field lenses can operate at an aperture up to 200 x 400mm.

Working "on the fly"

Processing can take place on the fly when the relative motion between focal region and workpiece is less than the required process resolution in the time frame of a single processing pulse. Conceptually this is very similar to photographing a moving object using a short shutter speed. Mathematically, we can express this as follows:

$$\frac{\text{pulse duration x relative velocity}}{\text{process resolution}} << 1 \tag{8.5}$$

The highly significant advantage of this mode of operation is that it is not necessary to stop the motion of the workpiece while processing takes place.

8.2.2 Examples of applications

Scribing

The scribing process involves making a line of holes, which may partially or fully penetrate the material, in order to produce a plane of weakness. This allows the material to be mechanically broken to precise dimensions. It is analogous to cutting glass by scoring with a tungsten carbide tool and then snapping.

Scribing was developed to process alumina substrates for the electronics industry and has been in production use since the early 1970's. Sub-kilowatt CO_2 lasers are used for the process in a pulsed mode. The pulsed output means that the material removal mechanism is by vaporisation, thus ensuring no debris is left on the substrate. A good beam quality is important in order to give narrow scribed lines. Although the process uses relatively low power lasers, processing speeds are high (in the order of m/min) as it is a relatively low energy process. Its advantages include this speed, cleanliness, and a minimal risk of substrate cracking.

The process is well established with little development work being carried out, and forms a significant proportion of laser usage (see section 1). It is rivalled by more conventional cutting using CO_2 or Nd:YAG lasers when complex substrate shapes are required. However, this is a small part of the current industrial demand.

Scribing using Nd:YAG lasers has been developed more recently for the processing of silicon and germanium electronic devices, as they readily absorb the $1.06\mu m$ wavelength. These are more sensitive to contamination than alumina substrates, and there can be a problem with re-deposition of evaporated material on the device. However, several manufacturers in North America are reported to be using the technique.

Marking

Laser marking methods can be classified according to how the mark profile is generated by the beam, the type of laser used, and the response of the material that gives the mark. Two methods are used to generate the mark profile. In one, the unfocused beam is pulsed through a mask, shaped as the negative of the required mark. The beam is then partially focused and imaged onto the workpiece. The complete mark is usually made by a single pulse of the laser, taking fractions of a second. The method is therefore very suitable for marking "on the fly", as part of a production line. TEA CO_2 and Excimer lasers are used for this marking method. The second method involves moving the focused laser beam over the workpiece surface, so that its path describes the required mark. Pulsed CO_2, Q-switched Nd:YAG, and Excimer lasers are all used for this process. The beam movement is usually achieved by reflecting it off two computer controlled, galvo-driven mirrors, which move it in orthogonal directions. It is then focused onto the workpiece surface by a flat-field lens.

CO_2, Nd:YAG, and Excimer lasers are all used for marking. The laser type and power will depend on the material, marking mechanism, and required speed of a particular application.

Pulsed CO_2 marking lasers are usually between 20 and 200 W, and have pulse frequencies of up to 5 kHz. A typical Nd:YAG marking laser has a maximum cw rating of 60 W, and can be Q-switched at frequencies up to 25 kHz. However, for the deep engraving of metals, 150-250 W Nd:YAG lasers with conventional pulsing are being introduced. Excimer lasers used for marking are general less than 50 W average power, with a pulse repetition rate of up to 300 Hz.

The marking rate of single shot mask marking systems will depend on the pulse repetition rate available from the laser, at the pulse energies required for the mark. This can be in tens or even hundreds of marks per second for small marks on easily treated materials. For example, an Excimer laser based cable marking system can mark the cable insulation with 8 or 16 characters, produced in a single laser pulse, at 60 m/min with the mark every ~50 mm.

Scanned marking systems will typically write up to ~60 type sized characters per second, and have linear writing speeds of 1 m/s. Resolution and repeatability of the scanning system are of the order of 5 μm and 25 μm respectively for a 100x100 mm marking field. This is a common size of marking field for current systems, although systems with larger fields, but lower resolution and repeatability are being introduced.

Laser marking competes against techniques such as ink-jet, stamping, etching, engraving, and adhesive labelling. Compared with these, the laser marking process is versatile, clean, fast, and inexpensive, giving a mark that can be tailored to the application, and be virtually unremovable if necessary.

Specific advantages are:

1. No tool wear;
2. Minimal deformation or microcracking;
3. No fixtures or jigs necessary;
4. Able to mark inaccessible places, and curved and fragile surfaces;
5. Suitable for production line use.

The laser marking process is used for an enormous variety of applications, in a wide range of industries. Examples of products marked include vernier calipers, aircraft parts, car radio bezels and dashboard switches, surgical implants, and electricity meters. One application that affects every household is the date marking of food packaging.

Technical developments to further increase marking speed can be expected, as companies use higher speed galvos in their systems. This will reduce the cost per mark, but there is a danger that writing speed will become the main criteria by which systems are judged, and legibility of the mark will be overlooked.

Soldering

The laser soldering process was specifically developed for the electronics industry, to join the connectors on components to printed circuit boards. It was not intended to replace the existing methods such as wave and bath soldering, which make many joints simultaneously, but to be used when a sequential technique is required.

The process is relatively simple, with solder paste that has been screen-printed or extruded onto the joint region being remelted by a pass of the laser beam. The molten solder flows to the metallic joins, and therefore the paste can be run across a line of connections without risking bridging of the insulation once laser remelted. The process is often carried out by two beams in parallel, on the opposite sides of a component, to avoid lifting of the component as the solder cools and contracts. Both CO_2 and Nd:YAG lasers are used for the process, operating in cw mode to avoid vaporisation of the solder. Recently there has been a trend to favouring the use of Nd:YAG lasers, as the beam can be transmitted by fibre optics thus simplifying machine design, and the 1.06 μm wavelength is better absorbed by the solder but less absorbed by the printed circuit board material.

Laser powers are typically up to 20W, and reflowing the solder for one joint takes 20-40ms. The process is used for prototype and specialist production of circuit boards, and laser soldering systems usually have good CNC and memory in order to easily cope with the variety of product. It is the fastest serial soldering method, operating at up to 30 joints/s. No applications of the process are known outside of the electronics industry, and after its initial development work, it appears to be a routine application of laser technology, with little or no further development work needed.

8.3 MEDIUM AND LARGE SCALE APPLICATIONS

8.3.1 Overview of Manipulation Techniques

Power and Energy

The major factors differentiating medium and large scale applications from those at the small end of the scale are:

1.   The power and energy present in the beam.
2.   The varying distance between laser and workpiece.

The significance of power and energy resides in the effect that absorbed power may exert on the optical properties of the various elements within the overall beam transmission system. At power levels of several kW, an absorbed fraction of only a few percent of the transmitted power can produce considerable thermal distortion in components that must remain accurate to a fraction of a micron. Considerable effort has therefore been expended on reducing the absorbed fraction as much as possible, and on removing whatever is absorbed in a way that distorts the optic as little as possible.

The trend for Nd:YAG applications at power levels up to and including several kilowatts is now to use fibres as the beam transmission system of choice to permit flexibility in beam delivery, whilst reserving mirrors and lenses for coupling, focusing and beam splitting/recombining. The basic fibre optic coupling scheme for Nd:YAG lasers is shown in Fig. 8.9. These fibres can be sheathed in protective coverings to allow them to coexist with all the usual hazards of the workplace, and can also have built in safety features to protect personnel from the optical hazards of laser light escaping from a damaged fibre.

With CO_2 lasers at powers up to tens of kilowatts, suitable fibre materials do not yet exist. Beam manipulation is therefore carried out using mirrors carried on gantries or robots, or incorporated into semi-flexible beam tubes. Particularly with gantry systems, the total optical path changes with the position of the tool point, often by considerable multiplicative factors such as 2 or 3. Since real laser beams do not propagate as parallel light, but obey Maxwell's Equations (see Chapter 2) and change in size and local divergence, precautions must be taken to prevent these changing distances having an adverse effect on the intended process.

Moving laser

A simple solution to the problems of changing optical distances is to move the laser and optics as a single unit relative to the workpiece. For mechanical reasons this is restricted to lasers of relatively low mass and size. This technique is seldom used in its pure form, but is sometimes used in conjunction with either moving optics or moving workpiece. A typical application might involve the laser being mounted on the X motion of a gantry, with moving optics occupying the Y motion and the workpiece remaining stationary.

Moving workpiece

The principles of this method have already been discussed in Section 8.2 and illustrated in Fig. 8.5.

Moving mirrors

By far the majority of CO_2 applications employ the moving mirrors method. This allows both laser and workpiece, which are usually fairly large and heavy to remain stationary, while the optics direct the beam from laser to workpiece. Unlike the small scale method of rotating the mirrors to move the tool point, in this case the mirrors are translated over quite large distances. Figure 8.12 illustrates the method.

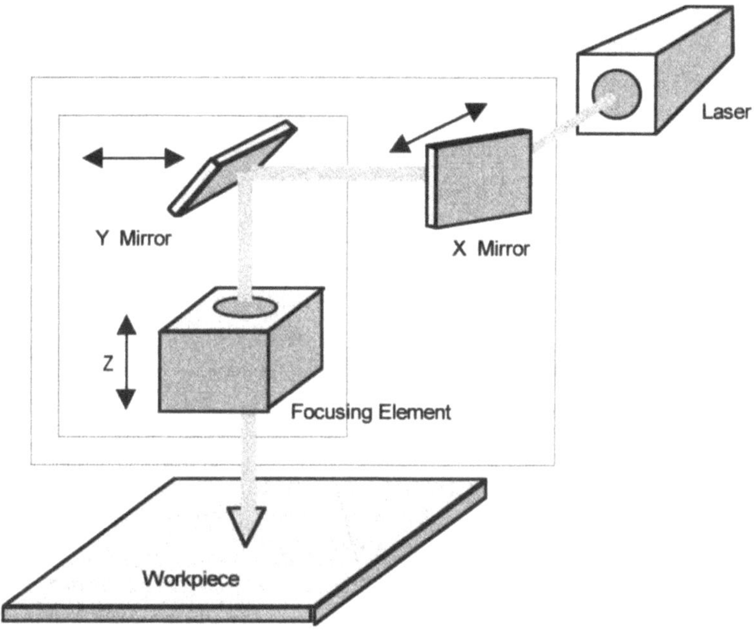

Fig. 8.12 Moving mirrors.

Both X and Y mirrors travel together on a carriage which moves parallel to the X axis. The X mirror is fixed rigidly to this carriage in such a position that it intercepts the beam from the laser which itself is directed parallel to the X axis, and deflects it accurately through 90°. The Y mirror travels on its own carriage along the Y direction in such a way that it intercepts the deflected beam and directs it downwards towards the workpiece. Both the X and Y carriages are driven by DC feedback motors or stepper motors under direct computer control. Control of the Z (vertical) position of the tool point (focus) is achieved by moving the focusing element in either the up or down direction, again under computer control.

If control of the tool point direction (θ and ϕ) is required, this is achieved through the use of an additional set of rotating mirrors, typically in the pre-objective position. Examples of this will be seen in the applications laboratory. This final assembly is often known as the wrist.

Translating mirrors

These mirrors are driven along the appropriate axis by DC feedback motors or by stepper motors. In some cases a lead screw is employed, in other cases a directly driven carriage is used with a linear optical encoder to determine position and velocity. Translating mirrors are mainly used with CO_2 lasers, and are therefore manufactured either from copper or silicon substrates. Copper mirrors are usually diamond turned, as this imparts a durable and highly reflecting surface, but sometimes a polished surface is used with an enhanced reflectivity coating. Silicon mirrors are almost without exception provided with an enhanced reflectivity coating since the natural reflectivity of this material is very low.

Whether manufactured from copper or silicon, the mirrors will require cooling, usually with water. Copper mirrors may have cooling channels built - in. Silicon mirrors are often bonded to cooled copper backing plates. In any case, the mirror assembly, with its adjusters and cooling apparatus is large and heavy, and requires substantial power to move and accelerate it. Likewise, the gantry assembly must be very stiff to resist any deformation due to attempted acceleration of the mirrors. To gain an idea of the masses involved, a large gantry system can have carriage masses of several hundred kilograms.

Power Loading

In a well designed beam transmission system, the mirror diameters should be 2 or 3 times as large as the nominal size of the laser beam. This is because the laser beam itself extends well beyond its nominal size, and any attempt to use optical apertures less than 2 or 3 beam diameters in size will seriously degrade the quality of the focus. A typical beam transmission system for powers up to 10 kW will have an aperture of 50 mm, giving a peak power loading of up to 4 kW/cm^2.

Absorption

In a typical industrial environment, a mirror in a beam transmission system can remove up to 2 or 3% of the incident power. Of this figure, around 1% could be due to the mirror materials and their condition, and the rest due to contamination from vapours and particulate material. Although 2 or 3% per mirror may not seem a large figure, it has to be realised that in a large CO_2 system, there could be around ten mirror surfaces, so that about one quarter of the total power can easily be lost by absorption. If that system utilises a 10 kW laser, then the losses are equivalent to a 2.5 kW laser costing around 100,000 ECU!

Cooling

Cooling is of paramount importance in the design of industrial lasers and laser systems. In a 10kW system typically between 100 and 300W of laser power is absorbed in each beam transmission mirror. Furthermore, the intensity profile of the laser beam (the so-called Mode Profile) is highly non-uniform. It follows therefore that not only is there a non-uniform rate of heat input, but also that its derivative with respect to time is also non-uniform. Unless effective cooling is applied to the mirror, distortion will seriously affect the beam quality. There are four main types of mirror cooling.

Integral Labyrinth Cooling. This is predominantly used in high power CO_2 mirrors which are manufactured from copper. In a typical case, the mirror is manufactured from two parts, one of which is machined to contain a labyrinth water cooling channel as shown in Fig. 8.13a. The parts are joined by a suitable technique such as brazing, soldering, electron beam welding, or diffusion bonding to form a thick disc shaped blank. The blank is then turned and polished in the usual manner.

Bonding to a Cooled Base. This is predominantly used in the case of Silicon mirrors which by virtue of their mechanical properties are not suitable for conventional machining. A water cooled base, Fig. 8.13b is manufactured similar to a labyrinth cooled copper mirror but without the optical mirror finish. The Silicon mirror is then attached to this cooled base using a thin layer of high conductivity adhesive.

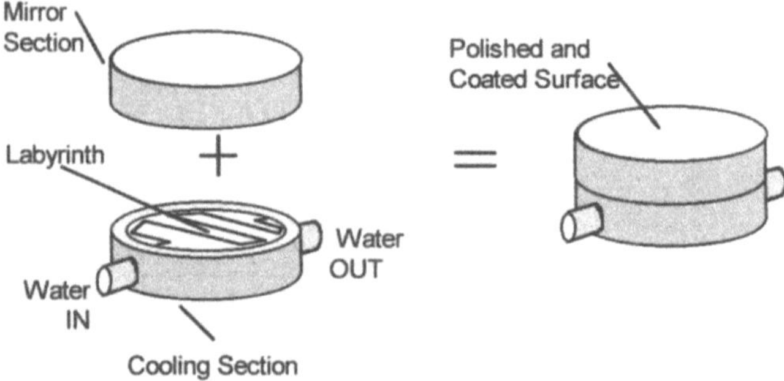

(a) Construction of Labyrinth Cooled Mirror

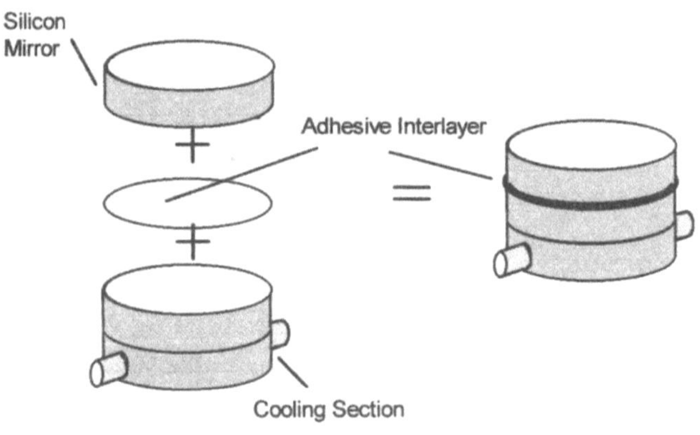

(b) Construction of Bonded Cooled Mirror

Fig. 8.13 Details of high power laser mirrors.

Internal Jet Cooling.   A method applicable to the cooling of unusual geometries under high thermal stress.   Instead of a labyrinth channel with slowly flowing water coolant, a jet of coolant (probably water) impinges on the rear surface of the optical element with improved heat removal due to the scouring action of the turbulent water jet.   This method is not in common use.

Gas Jet Cooling. Commonly used for highly stressed transmissive elements (windows, lenses). Gas jets are aimed at the faces to remove heat and prevent contamination. Mainly used in coaxial gas jet laser cutting and coaxial shroud welding where the cutting or shielding gas is first passed into the region around the lens before being directed onto the workpiece.

Distortion

Distortion in laser mirror systems is due to two main causes:

1. Uneven clamping forces between the mirror and the carriage;
2. Uneven temperature distributions throughout the mirror material.

Clamping force distortion can be designed out of a system by careful attention to detail. Modern mirror mounts use metal to metal contact between the mirror and a precision ground seat. Thus not only is the alignment of the mirror in the mount reliable and repeatable, but forces tending to bend and distort the mirror itself are minimised. Older mounts, derived from research machines featured the cooling channels as part of the mount, with the mirror sealed to it by an elastic "O" ring. In an endeavour to assure water-tight mounting, it was easy to over-stress the mirror and cause distortion. Thermal distortion is a fact of life and can only be minimised. Static distortion is brought about by uneven heating of the optic by a laser beam which of necessity is concentrated on a relatively small fraction of its surface (the reader will recall from previous discussions that the optic aperture should ideally be between 2 and 3 times the size of the laser beam). The most critical case is that of the transmissive object clamped (lightly of course) in a water cooled edge mount as shown in Fig. 8.15. There are three main heat sources:

1. absorption at the incident surface due to the coating and contamination
2. absorption throughout the bulk of the substrate material due to lattice imperfections of various kinds
3. absorption at the exit surface due to the coating and contamination
4. Under normal circumstances, all of these will be linear absorbing terms, i.e. the heat gained per unit area or per unit volume will simply depend on the intensity of the laser beam at that point. Figure 8.15 shows the form of the temperature distribution.

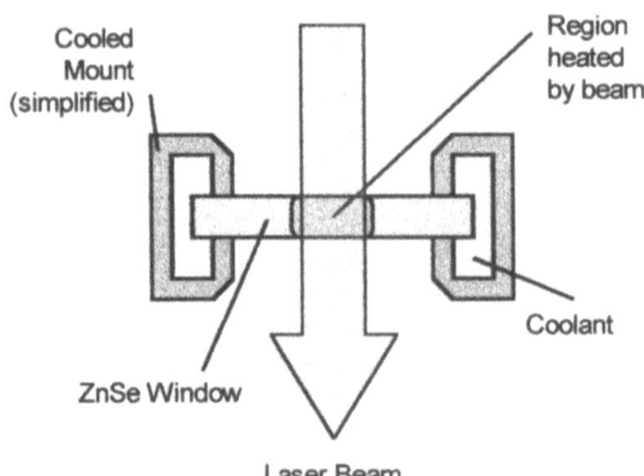

Fig. 8.14 Simplified schematic of edge cooled ZnSe window.

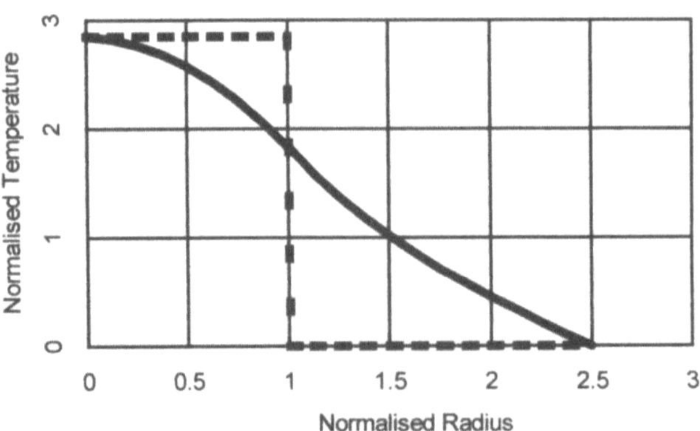

Fig. 8.15 Computed (solid) temperature rise as a function of radius for an edge cooled window and a 'top hat' beam (dashed). Radius normalised to beam radius; Temperature normalised to $\alpha P/4\pi K$ where α = absorptivity; P = power; K = thermal conductivity.

The form of the temperature profile is as follows:

$$T = \frac{\alpha P}{4\pi K}\left\{1 + 2\ln\left(\frac{A}{w}\right) - \frac{r^2}{w^2}\right\} \qquad r < w$$

$$T = \frac{\alpha P}{4\pi K}\left\{2\ln\left(\frac{A}{r}\right)\right\} \qquad r \geq w$$

where T = temperature rise above coolant
α, A, K = window absorptivity, radius and thermal conductivity
   P, w = beam power, radius

The astute reader will immediately notice that the maximum temperature, which occurs on axis, depends only on the ratio of beam size to window size and not on the absolute sizes. This contrary is to the intuitive idea that the more concentrated the beam power (= high intensity), the higher the temperature. The reason of course is simple, and is that the heat has a smaller distance to travel, and therefore the thermal gradient varies inversely with the size.

More significant however is the value of the thermal gradient, and this clearly does depend on the size, becoming increasingly larger as size diminishes. This determines (to a first approximation) the deviation of the window surface from its intended shape caused by thermal expansion. Likewise it also determines the angular distortion of the transmitted beam caused by such distortion.

A different condition holds for cooled copper and silicon mirrors. Here the major heat flow is not radial but from front to back, over an essentially constant distance, and therefore the temperature at the front surface depends linearly on the intensity, in direct contrast to the case of the edge cooled transmitting optic. There is thus a double contribution to the angular distortion caused to the reflected beam, one from the intensity increasing as mirror size decreases, and an additional distortion due to the scale effect on the thermal gradient. Fortunately the thermal conductivities of both copper and silicon are greater than those of transmitting materials such as ZnSe or KCl, so that if mirrors are kept in good condition, distortion is acceptable. However if the condition is allowed to deteriorate, then beam distortion can occur rapidly with disastrous effects.

Working at varying distances

In medium to large CO_2 systems, the z (axial) position of the tool point (beam focus) is adjusted by moving the focusing element in an axial direction. The distance moved can be typically from around one metre to upwards of ten metres, (see Fig. 8.12). In such a distance the beam as seen by the focusing element can change its parameters in a very significant manner.

Beam Propagation

The propagation of laser beams through space in the absence of distortion caused by heated mirrors and other effects, is given by the well known derivatives of Maxwell's equations. The beam from a typical CO_2 laser operating in a combination of low order modes has a beam diameter vs distance behaviour as shown in Fig. 8.16.

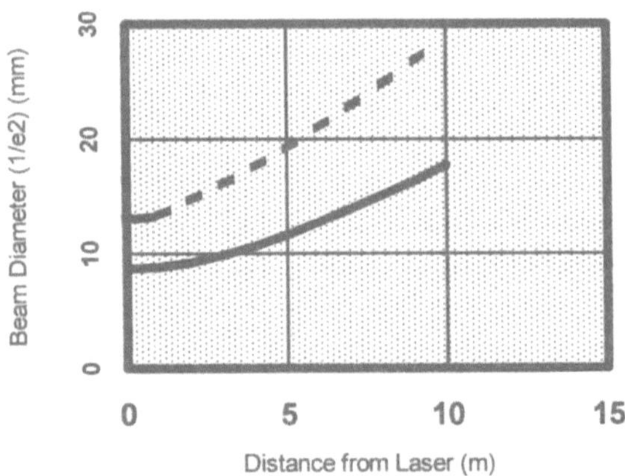

Fig. 8.16 Sample computed beam diameter as a function of distance from the output window (TEM$_{00}$ solid, real beam dashed lines).

Note how the beam appears to a focusing lens placed at different distances from the laser output window. Near to the window, at up to 1 m, the beam appears parallel, and geometrically emanating from a point at -∞. The focus of the beam will therefore be a distance f to the right of the lens, where f is the geometrical focal length. Now move the lens to 10 m from the window. The beam now appears to be emanating from a point approximately 5 m to the left of the window, i.e. 15 m to the left of the lens. If the lens has a focal length of 250 mm, the focal position will have moved 5 mm to the right. Furthermore the spot size at the geometrical focus position and the intensity there will have changed considerably with significant changes to the process itself.

Beam expanders and telescopes

One solution to the problem of changing focus is to modify the beam with an inverted telescope known as a beam expander. This operates on the principle that an invariant property of a laser beam is the product of beam waist size and far field divergence. Thus by expanding the beam and then 'refocusing' to produce a large beam waist centrally disposed about the limits of the working distance, the defocusing effects of the non-parallel beam can be minimised. A typical beam expander configuration is shown in Fig. 8.17

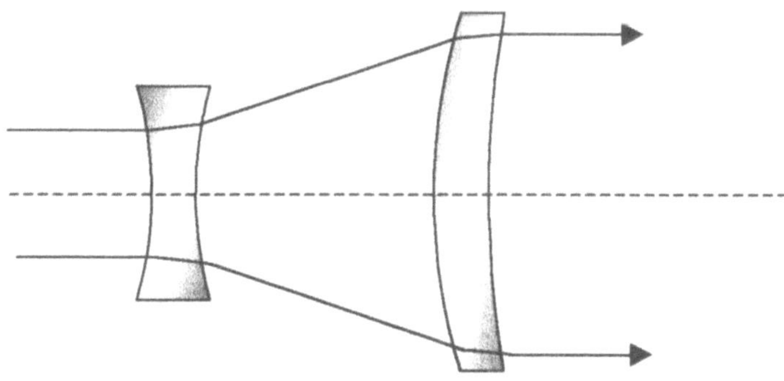

Fig. 8.17 Typical beam expander configuration (Galilean type).

Example

A worked example from Laser Processing in Manufacturing (1993) concerns a laser cutting system in which the gantry has an operating envelope of 5 m x 3 m, with the nearest cutting position at 3 m from the laser and the furthest position at 11 m. Using the 'raw' beam and a 100 mm focal length ZnSe meniscus lens, the power density varied by a factor of ~2, and the depth of focus by a factor of ~4 over the range of the operating envelope. Using a 1.5x magnification beam expander suitably adjusted, and the same lens as before, the power density remained sensibly constant at twice the best value of the original system, whilst the depth of focus remained roughly at the mid range value of the original system.

Scanners and Integrators

In certain applications such as surface treatment, it is necessary to redistribute the beam power over a large area. A number of methods of achieving this are described in the following three sections. All three methods have been successfully employed in the fields of transformation hardening, alloying and cladding.

Rotating/Vibrating Mirrors. The focal spot can be scanned in a raster pattern over an area of side several tens of millimetres by the use of rotating or vibrating mirrors. In the simplest technique, one such mirror would be employed to scan the focus in the x direction, whilst using the motion of the workpiece as a travelling y scan. A more flexible method employs two mirrors scanning at right angles to produce a rapidly repeated raster scan.

Multifaceted Integrators. An alternative technique employs a single fixed mirror divided into facets. Each facet is usually plane, and individually oriented to build up the desired intensity profile at the workpiece. Such a mirror would be used in conjunction with a focusing element. This technique is typically employed with very high power CO_2 lasers.

Axicons. As an alternative to the multifaceted integrator mirror, variants of the 'axicon' lens can be used. This is a lens of which one surface is spherical in order to focus the beam, and the other surface is conical in order to redistribute it. Depending on the position of the workpiece with respect to the lens, a number of different intensity distributions are available.

Fibres and "Flexible" beam guides

Fibres for Nd:YAG. Low loss optical fibres are now the transmission system of choice for most high power Nd:YAG applications. These have already been described in considerable detail in chapter 4, and principles of coupling into and out of the fibre are identical to those described in section 8.2.

Jointed Arm beam guides for CO_2. Materials with suitable optical and elastic properties for fibre production are not yet available for CO_2, although progress is being made in this direction. An alternative technique which possesses some of the advantages of fibres is the jointed arm guide. It is designed for use either with a robot for industrial purposes or manually for medical purposes, and comprises a series of tubes connected by ball joints. Each ball joint contains a plane mirror pivoting around an axis orthogonal to the plane defined by the two tubes, and set at an angle symmetrical to the two tubes. By having a number of such links, it is possible to direct the beam in any direction at any position, although the CNC control algorithms may be quite challenging.

"Flexible" beam guides for CO_2. A recent development which partially bridges the gap between jointed beam guides and fibres is the *Laser Power Plus + Dynaflex cable* from US based Laser Power Optics. The makers claim the cable is effective over several metres, allows 360 degree manipulation, and preserves mode quality and performance. It is claimed to enable precise delivery of 10 kW beams from CW industrial lasers.

8.3.2 Examples of Applications

Drilling

Drilling, like cutting, was an early use of the laser. Ruby lasers, the first type made to work, were used to drill diamond dies in the 1960's. Their use for drilling has been superseded by other pulsed laser types, but it is believed that there are some of the original lasers still in use. Drilling forms a few percent of industrial laser applications, but this belies its importance to specific industries. There are three main approaches to laser drilling:

1. Single shot;
2. Percussion;
3. Trepanning.

In single shot drilling, an individual laser pulse is used to drill a hole whose diameter is determined by the laser beam diameter and the pulse parameters used. It can obviously be used with a stationary workpiece but, because of the speed of the process with the single pulse, is often used to drill multiple holes in a continuously moving component. This drilling "on the fly" will cause some elongation of the drilled holes, dependant on the laser pulse duration and the component velocity. Its use is therefore limited to applications where this is acceptable.

Percussion drilling uses multiple pulses, but is otherwise similar to single shot drilling. Collectively, the two methods are known as direct drilling. A coaxial assist gas is generally used with both methods, which protects the lens from debris, and reduces the recast layer in the hole bore by shear force on the liquid film. This mechanism will, however, only apply to fully penetrating holes. Oxygen is often used as the assist gas when drilling metals, as it is effective at reducing the recast layer, probably because of the fluxing properties of the material oxide. It is not thought, however, to add significant energy to the process, in contrast to its role in laser cutting.

For trepanning, a central hole is directly drilled, and the required hole is then made by moving the focused beam from this central hole along a path that describes the required hole, while continuing to pulse the laser. It is suitable for hole diameters from 0.25 mm upwards, and can produce non-circular holes. The results are generally more precise than those obtained by direct drilling, but the minimum size hole is greater, and only fully penetrating holes can be made. Trepanning always uses an assist gas to ensure that liquid debris does not block the drilled hole.

Laser drilling is carried out using all three common types of industrial laser, i.e. CO_2, Nd:YAG, and Excimer as described elsewhere in this book. CO_2 lasers are usually used for single shot drilling of polymers and paper, often "on the fly". Material thicknesses are less than 1 mm, and the drill rate can be as high as hundreds of holes per second in thin material applications, such as the perforation of cigarette paper. Although it is possible to drill thicker materials with a CO_2 laser, it is not widely applied nor a precision process. The drilling of metals is generally left to the more effective Nd:YAG laser. These are used to drill holes from 0.01mm upwards, although the minimum size hole will depend on workpiece thickness for direct drilling and, as discussed previously, the minimum hole size for trepanning is ~0.25 mm. Maximum hole size is unlimited for trepanning, but will again depend on material thickness for direct drilling. A typical range of hole sizes for Nd:YAG metal drilling is 0.1-1.0 mm. Metal thicknesses of up to 25 mm are regularly laser drilled, and holes of over 100 mm depth have been achieved experimentally. However, industrial applications are more usually less than 10mm. Metal drilling rates may be in tens per second, as in the drilling of de-icing panel holes in thin titanium for aircraft wings at more than 50 holes/s, or be several seconds per hole for thicker metals.

Applications for Nd:YAG laser drilling are predominantly in aero engine manufacture. Here the application is to drill cooling holes in combustion chambers, nozzle guide vanes, and rotor blades. A modern gas-turbine engine may have in excess of 50,000 such holes, and although not all are laser drilled, a significant proportion are. Rolls-Royce, for example, has five Nd:YAG laser drilling systems in-house, and there are a further 15-20 installed at their subcontractors. Applications also exist in the electronic, airframe, and automotive industries, but not in such a widespread manner.

Cutting

Laser cutting was one of the first industrial laser processes to be developed, and has been used by industry since the late 1960's. It is suggested that the early adoption of the process, and its ensuing growth, is because it can often be a direct substitute for an existing process, offering speed, flexibility, and a higher quality product, with minimal disruption to the rest of the manufacturing sequence.

Laser cutting is a focused beam process, for which the development breakthrough came when the use of a co-axial gas-jet was proposed in the late 1960's. The focused beam passes through a small (~1mm) nozzle immediately prior to the minimum waist of the beam. This minimum waist is positioned at or near the workpiece surface, with the nozzle being typically 0.5-1mm above the surface. The gas-jet from the nozzle assists in the cutting process and also protects the focusing optics from fume and spatter. Where a lens is used as the focusing optic, it is also the top of the chamber through which the pressurised cutting gas is fed to the nozzle. However, when a mirror focusing optic is used, this is obviously not possible, and the cutting process is therefore less effective. The cutting process can be divided into three variants:

1. Vaporisation;
2. Fusion;
3. Reactive fusion.

Vaporisation cutting occurs when the laser beam heats and vaporises material in the cut zone, which is ejected with the aid of the gas-jet. This is the cutting method used for materials that do not melt, such as wood and some polymers, and also for cutting with a high peak power pulsed beam. The cutting gas is usually air or nitrogen, although argon is also used for some materials.

Fusion cutting involves the expulsion of molten material from the cut by the gas-jet, thus needing a lower energy input than would be required for vaporisation cutting. Again, the cutting gases are typically air, nitrogen, and in some cases argon. Cutting gas pressures have traditionally been of the order of a few bar, but the last few years has seen the development of high pressure cutting, using pressures of up to 30 bar to achieve high quality cut surface finishes on metals. Both metals, and non-metals that melt, are cut using this method and, because of the lower energy input, cut widths and heat affected zones on the parent material are usually very narrow.

Reaction fusion cutting uses the additional heat source of an exothermic reaction between the cutting gas and the workpiece as part of the process. It is typically used to cut steels and stainless steels, and is analogous to an extremely precise oxy-fuel cutting process. Oxygen or an oxygen-rich gas mixture is used as the cutting gas. The exothermic reaction adds energy to the cutting process and speeds can be more than double those achieved using normal fusion cutting. The cut edge is oxidised by the process, but for mild steel the oxide layer is often easily removed. For stainless steel, a more viscous slag can be formed which adheres to the bottom surface of the cut material as a less easily removable dross. Titanium and its alloys cut very well using an oxygen cutting gas. However, this results in a more crack-sensitive cut surface because of oxygen entrainment, which may well be unacceptable for the high quality applications to which this material is usually put.

CO_2, Nd:YAG, and Excimer lasers are all used for laser cutting applications, however, the CO_2 laser is the predominant type. Although CO_2 lasers up to 25kW have been used for cutting, a typical cutting laser is 0.5-1.5 kW. Pulsed Nd:YAG lasers of a few hundred watts average power are used for precision cutting applications, and the use of higher power pulsed and cw Nd:YAG lasers (up to 2 kW) with fibre-optic beam delivery is becoming of interest to industry. Excimer lasers cannot compete with the CO_2 and Nd:YAG lasers for conventional thermal cutting, because of their lower power levels and higher cost. However, they are finding a niche in ultra-precision cutting and in ablative cutting of organic materials.

Welding

Although predicted as being impossible by some commentators, laser welding was demonstrated in the late 1960's, and found its first application in the early 1970's. This was for the unlikely task of welding lead battery grids, which is still thought to be in current production. Rapid adoption of laser welding was then forecast, but the reality has been a little slower, with the gradual introduction of the process for more conventional applications, which now form just less than 20% of industrial laser use.

The focused laser beam, incident on the surfaces of the workpieces to be joined, heats, melts, and vaporises the material, forming a vapour filled "keyhole". This "keyhole" is surrounded by molten material, and may fully or partially penetrate the thickness of the workpieces. As the laser beam is moved along the joint line, molten material flows from front to back of the "keyhole" forming the weld. In order to ensure the "keyhole" is formed with the focused beam, good fit-up between the parts to be joined must be ensured. A wide range of joint configurations are possible, provided the fit-up requirement can be met.

A gas shield is normally used to protect the workpiece surface from oxidation, to stop spatter and fumes from reaching the focusing optics, and to reduce the formation of beam absorbing plasma above the workpiece surface when welding slowly.

Both CO_2 and Nd:YAG lasers are used for welding. CO_2 lasers are used in the cw mode, and Nd:YAG lasers are usually used in the pulsed mode. The minimum laser power for welding with a CO_2 laser is generally considered to be ~500 W, whereas Nd:YAG lasers are used to make fine spot welds at average power levels as low as tens of watts. This is achievable as the peak power in a pulsed Nd:YAG laser beam is many times the average power.

Laser welding is virtually always used on metals, although thin polymers are welded industrially, and ceramics have been welded experimentally. For welding metals, the normal metallurgical considerations of fusion welding apply, and allowances must be made for the comparatively rapid thermal cycle. Some of the more reflective metals do, however, pose problems for the process, particularly when using the CO_2 laser. Restricted welding performance is possible with aluminium alloys, but copper and gold are not realistic materials for the CO_2 process. The Nd:YAG laser can overcome these problems with its shorter wavelength and high peak power pulsing.

8.4 SENSORS

8.4.1 Introduction

In addition to all the usual sensors asscociated with complex CNC systems, laser systems and laser processes require an additional degree of monitoring and control. This is brought about to some extent by the nature of the laser source itself in which a number of parameters combine to determine the strength and quality of the output, and also by the nature of the interaction in which variables outside the control of the operator may influence the quality of the process itself. These problems are not unique to laser processing, but many of the techniques employed to overcome them will not be encountered in other processing technologies. The detailed methods by which the laser beam is sensed, and feedback employed will depend on the particular application, and such details will not be discussed here. It is however essential to have a knowledge of how beams can be measured, and this information forms the basis of this part of the lecture. Laser beams can be characterised by four principal parameters:

1. Beam Power
2. Intensity distribution
3. Quality factor
4. Direction

8.4.2 Beam Power

Beam power is measured routinely in lasers used for materials processing. It is a principal variable for processes such as welding, cutting, surfacing etc. Almost all commercial lasers feature a built in power monitor.

In Nd:YAG lasers a known fraction of the beam is sampled, either by a partially reflecting beam splitter or by a lossy resonator mirror, and the sampled fraction measured using one of two methods. In the thermal method, the beam is absorbed on a cooled block of known thermal conductivity, and the power derived from the steady state temperature difference across the block. This method can handle powers from tens of milliwatts to a few kilowatts. In the photoelectric method, a small fraction of the beam falls onto a photodiode, producing either a current or a voltage proportional to the beam power. In either case the measured power is divided by the sampling fraction to provide an estimate of the total beam power.

In CO_2 lasers, the photon energy is too small to have a photoelectric effect, so the thermal method is almost invariably employed. At low powers it is possible to sample a portion of the beam as in the Nd:YAG laser, only using different optical materials appropriate to a wavelength of 10.6 μm. (materials such as ZnSe, GaAs, Ge, and KCl can be used). The sampled beam could be directed to a thermal detector as described above. At high powers however, it is unusual to use transmissive beam splitters due to the distortion they would introduce into the beam itself. The simplest option is to sample the entire beam power directly using a water cooled calorimeter. This is probably the most accurate method of beam power measurement since no assumptions are made concerning the sampled fraction, but it does have one limitation - since it absorbs the entire beam, it cannot be used on-line or in-process. In spite of its limitations, it is frequently used as an accurate means of setting the laser power, calibrating in line monitors, and it also acts as a beam dump while the process is not taking place.

8.4.3 Intensity Distribution by Beam Splitting

A number of methods are available for determining the intensity distribution of the laser beam. They all rely on sampling a small proportion of the wavefront of the beam. At low powers, a transmissive beam splitter directs the sample onto a detector array of typically up to 64 x 64 elements.

For Nd:YAG lasers the array elements could be photodiodes or pyroelectric devices, for CO_2 lasers, the devices will be pyroelectric or utilise some other thermally based mechanism. The elements are addressed electronically and the resulting output displayed (after computer processing) on a video monitor. Photoelectric elements produce a signal directly proportional to the photon flux, or power. Pyroelectric detectors, owing to the transient nature of the detection process, require some form of mechanical beam chopping before the sampled beam reaches the detector. Unlike photoelectric devices, pyroelectric elements and arrays are broadband and will detect from the visible to the far infrared. At present, array techniques are the only method of obtaining true real time imaging of the entire beam profile. Very few of these are in use owing to the prohibitively high prices, and the low damage thresholds of the pyroelectric material.

At higher powers it is necessary to use reflective beam splitters to overcome the problems of thermal distortion of transmissive elements. They can also of course be employed at low powers. Two of these are particularly worthy of note.

1. The grating mirror replaces one of the 45° mirrors in the transmission system. This is a solid, usually copper or silicon, water cooled mirror and performs the function of a normal mirror as far as the process is concerned. Its surface is a lightly ruled diffraction grating of a suitable pitch and blase to diffract a small portion of the beam at an angle that takes it out of the beam path. The diffracted beam is then monitored in the usual way.

2. The hole matrix mirror replaces one of the 45° mirrors in the transmission system, but this time allows a portion of the beam to be transmitted through a matrix of holes drilled into the mirror itself. The analysis of the transmitted beam is somewhat more complicated than for the case of transmissive or grating splitters, but nonetheless it is possible to produce an intensity profile for the beam.

8.4.4 Intensity Distribution by Scanning

At up to and including the highest powers (~10 kW) two types of in line monitor are available commercially. These have the advantage that no special mirrors or beam splitters are required, and that the sampling method introduces no significant perturbation into the laser beam or into the outcome of the process.

1. The first to appear on the commercial market was the rotating needle laser beam analyser developed by Prof. W.M.Steen at the Imperial College of Science, Technology and Medicine in London. This employs a polished rotating needle to reflect a portion of the beam into two pyroelectric detectors. By suitable signal processing, the X and Y profiles of the intensity distribution can be determined in real time and displayed on a video monitor or an oscilloscope. It is claimed that a real time full 2 axis intensity distribution can also be derived, however it is not clear what mathematical assumptions are necessary in order to achieve this. Both focused and unfocused beams can be measured using this technique, but there is an upper limit on the intensity due to absorption and damage to the rotating needle, particularly when measuring near to focus.

2. The second monitor to appear was developed at the Fraunhofer Institut für Lasertechnik in Aachen and is manufactured and marketed by Prometec GmbH in Germany. This device also employs a rotating needle, but this time the needle is hollow, and has a small hole at one end to sample the beam. The hole is resistant to high intensity laser beams, and the small fraction of the beam which penetrates the hole, is reflected along the hollow interior of the needle until it is incident upon a pyroelectric detector. As the needle rotates, its centre of rotation is translated so that the hole scans out a quasi-linear raster across the beam. A single beam scan takes several seconds. One advantage of this device is the directness of the sampling method. No mathematical assumptions are required to derive the full 2 axis intensity distribution. Its one disadvantage is the length of time taken to scan the full size unfocused beam, during which time the distribution itself may have changed, it can not therefore be described as operating in real time.

8.4.5 Incident beam monitors

Any of the above methods can be used to monitor the beam as it is incident upon the workpiece. Ideally the monitor should be placed as near as possible to the interaction zone to reduce errors due to downstream optics.

8.4.6 Reflected beam monitors

When measuring reflected power as either a process feedback variable or as a diagnostic, two methods may be used. With Nd:YAG lasers a beam splitter placed prior to the workpiece will split off any reflected power. This can be the same optical element as used to sample the incident beam. The disadvantage is that only reflected power travelling paraxially will be captured. With any laser, including Nd:YAG, a better solution is the hole mirror. The hole allows the incident beam to pass through, and any reflected power outside the solid angle subtended by the beam can be directed to a detector. Since only total reflected power is normally measured, various designs of mirror are possible which will capture the reflected power and direct it to the detector. Any suitable beam power detector can be used, but care should be taken to filter out any optical radiation from the process at wavelengths other than that of the laser itself.

8.4.7 Transmitted beam monitors

In certain processes such as cutting and welding, measurement of the transmitted power can be a valuable feedback parameter for process control. In keyhole welding of metals under conditions of complete penetration for example, a transmitted fraction of about 10-20% is often sought to give correct weld profile. This is a more challenging measurement than reflected power as the measurement region contains high velocity molten metal droplets and other process debris. Any suitable beam power detector can be used, but care should be taken to filter out any optical radiation from the process at wavelengths other than that of the laser itself, and of course any physical debris that could damage the detector or invalidate the measurement.

8.4.8 Plasma luminosity monitors

A considerable amount of research has been carried out into the correlation of plasma luminosity and process results. No definitive results have yet been achieved. It is known, however that under given process conditions, the plasma luminosity can be a useful indicator of deviation from the correct conditions. The radiation from hot plasmas at 10000–50000 K is almost entirely in the visible and ultraviolet regions of the spectrum where photodetectors are used. Care should be taken, particularly with Nd:YAG lasers to filter out any laser radiation that might damage the detectors or invalidate the measurement.

8.5 ACKNOWLEDGEMENTS

The author wishes to acknowledge the following sources of information used in the preparation of this chapter.

Baasel Lasertechnik
Eurolaser UETP State of the Art Review, Comett Project 6735/A, 1993
General Scanning Ltd
Laser Lines Ltd
Laser Processing in Manufacturing, R.C.Crafer and P.J.Oakley eds, Chapman and Hall 1993, ISBN 0 412 41520 8
Peter. J. Oakley, Consultant
Laser Focus World, January 1997
TWI Laser Centre

Index

MIX
Papier aus verantwortungsvollen Quellen
Paper from responsible sources
FSC® C105338

If you have any concerns about our products,
you can contact us on
ProductSafety@springernature.com

In case Publisher is established outside the EU,
the EU authorized representative is:
Springer Nature Customer Service Center GmbH
Europaplatz 3, 69115 Heidelberg, Germany

Printed by Libri Plureos GmbH
in Hamburg, Germany